Using and Conquering the Watery World in Greco-Roman Antiquity

Also available from Bloomsbury

Conceptions of the Watery World in Greco-Roman Antiquity by Georgia L. Irby
Mountain Dialogues from Antiquity to Modernity edited by Dawn Hollis
and Jason König
The Ancient Mediterranean Sea in Modern Visual and Performing Arts
edited by Rosario Rovira Guardiola
The Horse in the Ancient World: From Bucephalus to the Hippodrome
by Carolyn Willekes

Using and Conquering the Watery World in Greco-Roman Antiquity

Georgia L. Irby

BLOOMSBURY ACADEMIC
LONDON • NEW YORK • OXFORD • NEW DELHI • SYDNEY

BLOOMSBURY ACADEMIC
Bloomsbury Publishing Plc
50 Bedford Square, London, WC1B 3DP, UK
1385 Broadway, New York, NY 10018, USA
29 Earlsfort Terrace, Dublin 2, Ireland

BLOOMSBURY, BLOOMSBURY ACADEMIC, and the Diana logo are trademarks of
Bloomsbury Publishing Plc

First published in Great Britain 2021
This paperback edition published 2023

Copyright © Georgia L. Irby, 2021

Georgia L. Irby has asserted her right under the Copyright, Designs, and
Patents Act, 1988, to be identified as Author of this work.

For legal purposes the Acknowledgments on p. xiv–xv constitute an extension of
this copyright page.

Cover design: Terry Woodley
Cover image: Neptune, Roman god of the sea carrying his trident and riding in chariot
pulled by horses with dolphin tails. 2nd century ad mosaic. Bardo Museum, Tunis. World
History Archive / Alamy Stock Photo

All rights reserved. No part of this publication may be reproduced or transmitted in any
form or by any means, electronic or mechanical, including photocopying, recording,
or any information storage or retrieval system, without prior permission in writing
from the publishers.

Bloomsbury Publishing Plc does not have any control over, or responsibility for, any
third-party websites referred to or in this book. All internet addresses given in this book
were correct at the time of going to press. The author and publisher regret any
inconvenience caused if addresses have changed or sites have ceased to exist,
but can accept no responsibility for any such changes.

A catalogue record for this book is available from the British Library.

Library of Congress Cataloging-in-Publication Data

Names: Irby, Georgia L. (Georgia Lynette), 1965– author.
Title: Using and conquering the watery world in Greco-Roman antiquity / Georgia L. Irby.
Description: London, UK ; New York, NY, USA : Bloomsbury Academic, 2021. |
Includes bibliographical references and indexes.
Identifiers: LCCN 2021004154 (print) | LCCN 2021004155 (ebook) |
ISBN 9781350155848 (HB) | ISBN 9781350155862 (ePDF) | ISBN 9781350155855 (eBook)
Subjects: LCSH: Hydraulic engineering–Greece–History–To 1500. | Hydraulic engineering–
Rome–History–To 1500. | Naval art and science–Greece–History–To 1500. | Naval art and
science–Rome–History–To 1500. | Greece–History, Naval. | Rome–History, Naval.
Classification: LCC TC16 .I73 2021 (print) | LCC TC16 (ebook) | DDC 627.0938—dc23
LC record available at https://lccn.loc.gov/2021004154
LC ebook record available at https://lccn.loc.gov/2021004155

ISBN:	HB:	978-1-3501-5584-8
	PB:	978-1-3502-5078-9
	ePDF:	978-1-3501-5586-2
	eBook:	978-1-3501-5585-5

Typeset by RefineCatch Limited, Bungay, Suffolk

To find out more about our authors and books visit www.bloomsbury.com
and sign up for our newsletters.

Contents

List of Illustrations	x
List of Abbreviations	xii
Acknowledgments	xiv
Introduction: Using and Conquering the Watery World	1
Part One Controlling and Harnessing Water	5
1 Water Rights	7
Introduction	7
Water Regulation in the *Poleis* (City States) of Classical Greece	8
Access	8
Water and Property Rights	9
Willful Contamination	11
Water Regulation in Rome	12
Water Rights and Roman Law	12
Private Property	13
Drainage	15
Municipal Oversight	15
Provincial Water Administration	16
Fees	17
Standardization	17
Abuse and Sabotage	17
Public Maintenance	19
Polluting the Waters	19
Common Access	20
Conclusion	20
2 Water Quality and Urban Planning	23
Introduction	23
Water Quality	23
Water Freshness	24
Drinking Water	25
Water Flavors	26

	Purifying Water for Human Use	28
	Masking Unpleasant Odors	28
	Straining Out Particulates	29
	Boiling	30
	Water and Urban Planning	31
	War and Water	32
	Lead Poisoning	34
	Conclusion	35
3	Urban Hydraulic Engineering	37
	Introduction	37
	Finding, Conveying, and Storing Water	37
	Wells	39
	Cisterns	40
	Aqueducts	42
	Early Initiatives	42
	Greek Initiatives (Sixth to Second Centuries BCE)	42
	Roman Initiatives	45
	Roman Provincial Initiatives	46
	Operation	47
	Private Aqueducts	48
	Fountains	49
	Mégara	50
	Athens	51
	Hellenistic Innovations	51
	Free-standing Waterspouts	52
	Monumental Fountains	52
	Baths and Showers	53
	Aquae Sulis	56
	Latrines	57
	Drainage and Sewers	58
	Wetland Reclamation	59
	Industrial Water Uses	61
	Water-lifting Devices and Irrigation	61
	Hydraulic Machines	62
	Waterwheels and Water Mills	62
	Hydraulic Mining	63
	Conclusion	65

4	Marine Hydraulic Engineering	67
	Introduction	67
	Anchorages and Harbors	67
	Piraeus	71
	Portus	73
	Dredging	74
	Canals	76
	Lighthouses	78
	Pharos	78
	A Coruña	81
	Conclusion	82

Part Two Engaging with the Watery World   83

5	Sailing and Navigating	85
	Introduction	85
	Officers of the Deck	87
	Kubernetes/Gubernator	88
	Keleustes/Pausarius	89
	Prorates/Proreta	90
	Launching the Vessel	91
	Anchoring	92
	Divers	92
	Propulsion	93
	The Military Sailing Season	96
	Winds	97
	Navigation	101
	Latitude and Longitude	102
	Bearings	103
	Shoals	103
	Pilots	104
	Navigability	105
	Celestial Navigation	106
	Maritime Highways and "Charts"	107
	Conclusion	109

6	Maritime Trade and Travel	111
	Introduction	111
	Trends in Ancient Mercantilism	112

	Greek Trade	113
	Bronze Age Wrecks	113
	Athens	114
	The Fourni Wrecks	115
	Trade in the Roman Era	115
	The *Periplus* ("Coasting Guide") *of the Erythraean Sea*	115
	Trade Routes	119
	The Merchant Sailing Season	122
	Piracy	126
	Ships and Cargo	131
	River Barges	133
	"Passenger" Ships	135
	Conclusion	137
7	Harvesting the "Barren" Sea	139
	Introduction	139
	Fishing	139
	Fish and Foodies	142
	Tuna	143
	Fish Farms	143
	Mollusks: Oysters and Mussels	146
	Guilds, Dues, and Access to the Waters	148
	Salt, Fish Salting, and Garum	148
	Murex Purple Dye	151
	Conclusion	153

Part Three The Sea and "National" Identity: The Political Manipulation of the Watery World 155

8	Minoan Thalassocracy, Archaic Expansion, and Maritime Iconography	157
	Introduction	157
	Minos and Minoan Thalassocracy	157
	Homer's Catalogue of Ships and "National" Identity	162
	Archaic Settlement: Migration and the Development of Naval	
	Power and Coinage	163
	Return of the Heroes	163
	Exploration and Expansion	164
	Aquatic Themes on Archaic Greek Coins	165
	Dolphins	166
	The Pinnipeds (Seals) of Phocaea	169

	The Tuna of Cyzicus	169
	The Octopods of Eretria	170
	The Turtles of Aegina	170
	The Crabs of Acragas	171
	The Ships of Phaselis	171
	Conclusion	172
9	Hellenic and Hellenistic Thalassocracies	175
	Introduction	175
	Fifth-century Naval Strength	175
	The Battle of Salamis	175
	The "Delian League" and the Peloponnesian War	177
	The Ship of State	179
	The Panathenaic Festival	180
	Hellenistic Thalassocracy	181
	Demetrios, the Son of Poseidon	183
	The Nike of Samothrace	184
	Conclusion	186
10	Rome: *Oceanus Domitus*	187
	Introduction	187
	The Punic Wars	188
	Numismatic Iconography in the Second Century BCE	189
	Pompeius Magnus (Pompey the Great)	190
	Sextus Pompeius	191
	The Battle of Actium	194
	Gaius Caligula	197
	The "Conquest" of Britain	198
	Naumachiae	199
	Oceanus Domitus	201
	Conclusion	203
11	Conclusion	205
Appendix		208
Notes		214
Bibliography		249
Index of Places Cited		274
Index of Authors and Sources		282
General Index		287

Illustrations

Maps

1	Poleis and Regions of the (Greek) Eastern Mediterranean	xvi
2	The Mediterranean	xvii
3	Provinces of the Roman Empire	xviii

Figures

3.1	Scheme of an ideal Roman aqueduct, after Connolly and Dodge 1998: 131	43
3.2	Inverted Siphon, after Hodge 1992: 148 #102	44
3.3	Athletes at a public washbasin: Hamilton and Tischbein 1791: 1.plate 58	54
3.4	Women showering at a fountain house. Panofka 1843: 9, 18	55
3.5	Reconstructed water mill at Barbegal (with overshot wheels). After Hodge 1990: 109	64
4.1	Piraeus with long walls	72
4.2	The octogonal harbor at Portus	74
4.3	Dolphins lead ships into harbor past Ostia lighthouse: Ostia Antica Statio 23	79
5.1	Attic black-figure hydria showing the three deck officers. Paris, Louvre E735. Photo credit SCALA/ Art Resource, NY (20535)	91
6.1	Emporia of the Erythraean Sea	116
6.2	Torlonia relief, Museo Torlonia #430	134
6.3	Romano-Gallic bas-relief showing a river barge, second/third-century. Musée Lapidaire d'Avignon, Fondation Calvet	135
8.1	Minoan Seals: Casson 1995 #37, 40	160
8.2	Stylized aristocratic horse: Evans 1935: 827 #805 = Casson 1995: #52	161
8.3	Scylla intaglio: Evans 1935: 952, #921	161
8.4	Dolphin-rider coin, Taras: Kraay 1976: #673–7	168

8.5	Oceanus with crab-claw headdress. Marriage of Peleus and Thetis, drinking cup by Sophilos: BM 1971,1101.1	172
9.1	Silver *drachma* of Demetrios Poliorcetes, from Tarsus, ca. 298–295 BCE Newell 1937: #41	184
9.2	Nike of Samothrace	185
10.1	*Denarius* of Sextus: Naval Trophy: Crawford 1974: #511/2b	192
10.2	*Denarius* of Sextus: Scylla smashing a ship's rudder: Crawford 1974: #511/4	193
10.3	*Denarius* of Octavian with globe and aplustre: Zanker 1990: 41 and figure 31a	194
10.4	*Denarius* of Octavian with rostral column: RIC 271; BMC 633; C 124	196
10.5	"Gorgon" Pediment at Aquae Sulis	202

Tables

5.1	The Winds and Their Cardinal Points	100
6.1	Lengths of Some Historical Sails	121

Abbreviations

AE	*L'Année épigraphique*
AJA	*American Journal of Archaeology*
AJP	*American Journal of Philology*
AWP	Hippocratic *Airs, Waters, Places*
BM	British Museum
CIL	*Corpus Inscriptionum Latinarum* (Berlin 1863–)
CJ	*Classical Journal*
CMG	*Corpus Medicorum Graecorum*
CPh	*Classical Philology*
CQ	*Classical Quarterly*
DK	Hermann Diels and Walther Kranz (eds.), *Die Fragmente der Vorsokratiker, griechisch und deutsch* (Berlin, 1951–52)
FGrHist	F. Jacoby, *Fragmente der griechischen Historiker* (Leiden, 1923–): https://referenceworks.brillonline.com/browse/die-fragmente-der-griechischen-historiker-i-iii
HH	*Homeric Hymn*
I.Ephesus	*Die Inschriften von Ephesos* (Bonn, 1979–1984)
IC	M. Guarducci, *Inscriptiones Creticae* (Rome, 1935–1950)
IG	*Inscriptiones Graecae* (Berlin, 1873–1939)
ILS	H. Dessau (ed.), *Inscriptiones Latinae Selectae* (Berlin, 1892–1916)
IosPE	*Inscriptiones antiquae Orae Septentrionalis Ponti Euxini*. Third edition online: http://iospe.kcl.ac.uk/index.html
JHS	*Journal of Hellenic Studies*
JRA	*Journal of Roman Archaeology*
JRS	*Journal of Roman Studies*
LIMC	*Lexicon Iconographicum Mythologiae Classicae* (Zurich, 1981–)
LSCG	F. Sokolowski, *Lois sacrées des cités grecques* (École française d'Athènes, Travaux et mémoires des anciens membres étrangers de l'École et de divers savants 18) (Paris, 1969)
OGIS	Wilhelm Dittenberger *(ed.), Orientis Graeci Inscriptiones Selectae* (Hildesheim, 1903–1905)

PECS	William L. MacDonald and Marian Holland McAllister (eds.), *Princeton Encyclopedia of Classical Sites (Princeton, 1976)*
RIB	R.G. Collingwood, R.P. Wright, R. S. O. Tomlin, et al. (eds.), *The Roman Inscriptions of Britain* (Oxford, 1965–): https://romaninscriptionsofbritain.org/
RIC	H. Mattingly, E. A. Sydenham, and others (eds.), *Roman Imperial Coinage* (London, 1923–67); revised edition of vol. 1: C. H. V. Sutherland and R. A. G. Carson (1984)
RPC	Burnett, Andrew, Michel Amandry, Chris Howgego, Jerome Mairat, *Roman provincial Coinage* (London, 1992–): https://rpc.ashmus.ox.ac.uk/
SEG	*Supplementum Epigraphicum Graecum* (Leiden 1923–2002)
SVF	*Stoicorum Veterum Fragmenta* (Leipzig, 1905–1924; reprinted Stuttgart, 1968)
TEGP	Daniel W. Graham 2010. *The Texts of Early Greek Philosophy: The Complete Fragments and Selected Testimonies of the Major Presocratics*. 2 vols. (Cambridge, 2010)

Acknowledgments

No work of scholarship is produced in isolation, and this project has its genesis in a course that I have the privilege of teaching to first year students at William & Mary, "Why Water Matters," where students learn about the fluidity of critical inquiry and the scholarly process as an interplay between primary sources, academic training, and the scholar's own cultural and political biases. Together we investigate the critical question about human engagement with the natural world, our sources of information, the advantages and disadvantages of those sources (often fragmentary, poorly preserved, inadequately contextualized, and insufficiently curated), our understanding, perception, and interpretation of those sources, how social, cultural, and political factors influence our perception of the environment, and how our own attitudes to the natural world shape our culture, art, literature, and politics. This book is very much the product of that conversation. In particular, I thank those students in my Fall 2018 class. Enduring an uncivilized 8:00 a.m. curtain call, this talented and enthusiastic group read drafts of the chapters in various stages of polish, bringing slips to my attention, and making countless salient suggestions for improving transitions, adding material, and clarifying arguments. They have earned their mention here: Greg Arrigo, Elizabeth Ashley, Theo Biddle, Jonathan Broady, Marcus Crowell, Zack Daniel, Daniel Gittings, Caileigh Gulotta, Christian Gulotta, Zack Johnson, Cole Kim, Rebecca Klinger, Abhishek Mullapudi, Freddie Nunnelley, Charlie Perry, Clay Shafer, Ben Sharrer, Percy Skalski, Ann Grace Towler, and Tammy Yin. I should also like to recognize Emma Grenfell, Abby Maher, Jake Morrin, Lindsey Smith, and Georgia Thoms, students from my Fall 2019 class who vetted the penultimate version of this manuscript, and brought to my attention a number of lingering, pesky infelicities.

My gratitude extends also to Alex Wright (now at Cambridge University Press), whose gentle prodding guided this project from scattered lecture notes to book manuscript, and to Alice Wright and Georgina Leighton, my editors at Bloomsbury, who took up the gauntlet midstream, and their assistant Lily Mac Mahon. The manuscript has benefited immensely from the helpful observations and suggestions of the anonymous reviewers. Thanks are also owed to Senior Production Editor, Rachel Walker at Bloomsbury, Merv Honeywood, the project manager at RefineCatch, copyeditor Susan Dobson, and the page setters, whose combined efforts behind the scenes facilitated the publishing process and added sparkle to the text. I am grateful to the Inter-Library Loan Staff at Swem Library at William & Mary, who sometimes even seem to anticipate the needs of their patrons, as well as the cheerful, knowledgeable, and efficient staff in Information Technology who keep my digital resources in good working order. Also meriting recognition is my research assistant, Keegan Sudkamp-Tostevin, who painstakingly double checked many of the primary references, caught slips both great and small, and helped bring my vision into greater focus. His deep curiosity and enthusiasm were inspirational as my energy levels waned. Individual chapters have benefited from the scrutiny of Robert Nichols and Jessica Stephens, my colleagues at William & Mary. Andrew Ward's comments on Samothrace have been invaluable. The adroitly graceful comments of Molly Ayn Jones-Lewis, especially on the medical material, have resulted in many improvements, a more nuanced treatment, and a more coherent structure of the entire manuscript. Duane W. Roller read the entire manuscript in draft form. As always, his eagle-eye caught many infelicities, and his encyclopedic comments, leading to new avenues of inquiry, have improved the substance and spirit of the text. Any errors that remain are my own.

Many colleagues, students, and friends have cheered the project on from its inception. I would like to single out Tejas Aralere, Joyce Holmes with her endless supply of smiles and hugs, and John Oakley who always has time for a chat about Greek vases. I should also like to thank Jennifer Andrews-Weckerly, Cary Bagdassarian, Charlie Bauer, Michael Bryant, Mike Crookshank, John Donahue, Lu Ann Homza, Bill Hutton, Michele and Les Hoffmann, Martha Jones, Jasmane Ormand, Steve Otto, Jessica Paga, Huntley Polanshek, Linda Reilly, Jessamyn Rising, Rebecca and Marshall Scheetz, Wayne Shaia, Molly Swetnam-Burland, Joshua Timmons, Gene Tracy, Ben Zhang, the crew of the Godspeed at Jamestown Settlement, Va., and my mentors James C. Anderson and Christoph F. Konrad. I am, as ever, indebted also to John L. Robinson, my nautical mentor and best friend, and my mother, Patricia A. Irby, for her daily support and encouragement. My father, from whom I inherited my love of boats and capacious curiosity, is, as ever, woven deeply into the fabric of these pages.

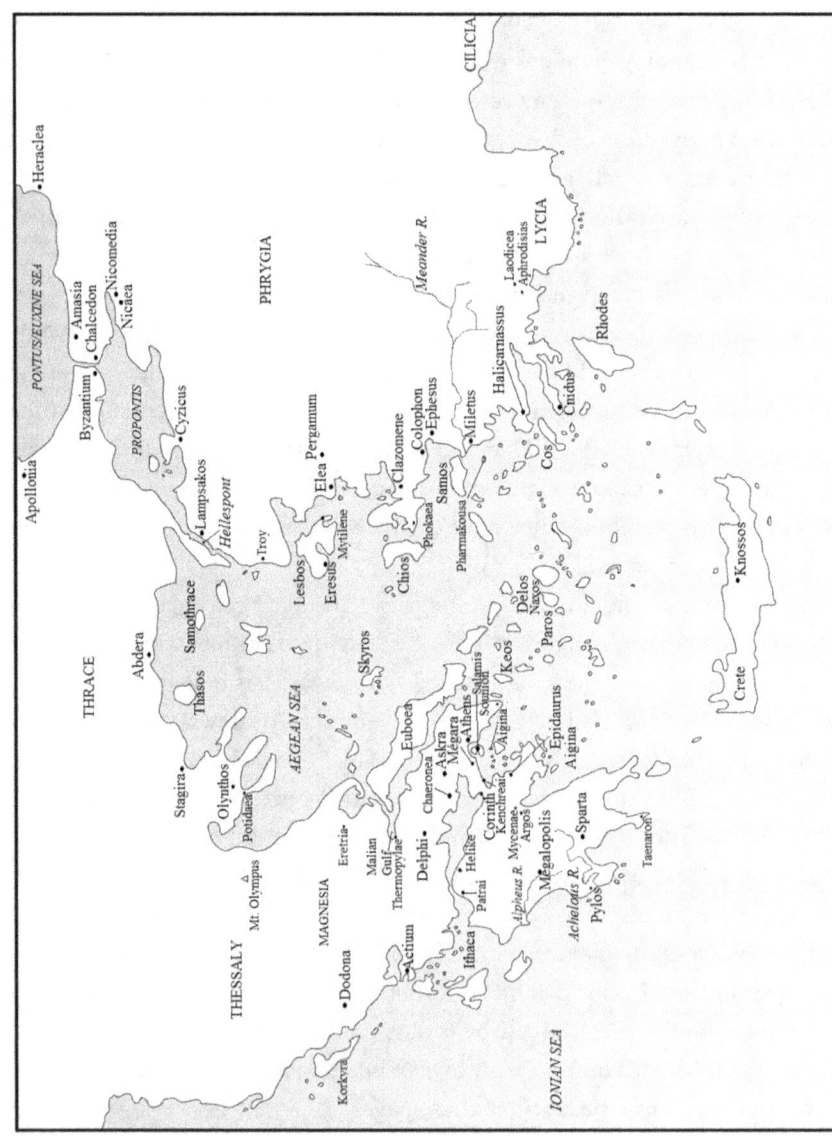

Map 1 Poleis and Regions of the (Greek) Eastern Mediterranean

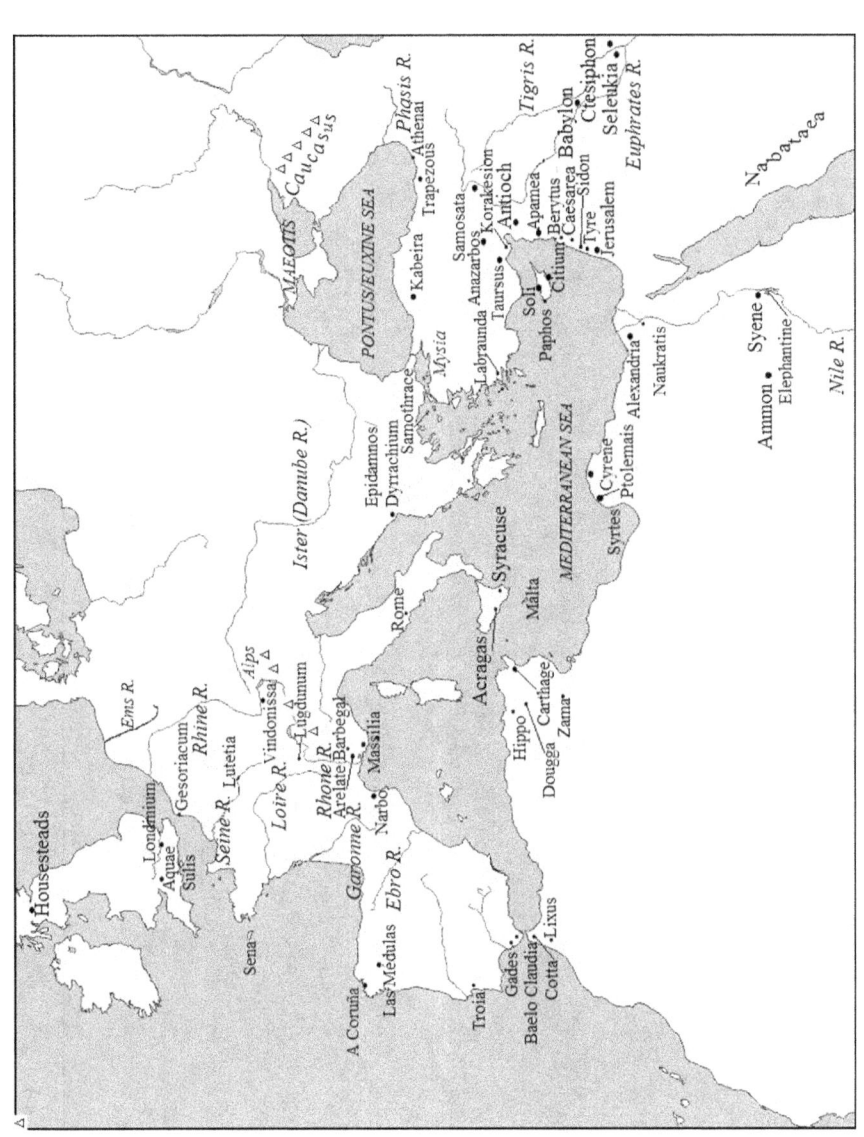

Map 2 The Mediterranean

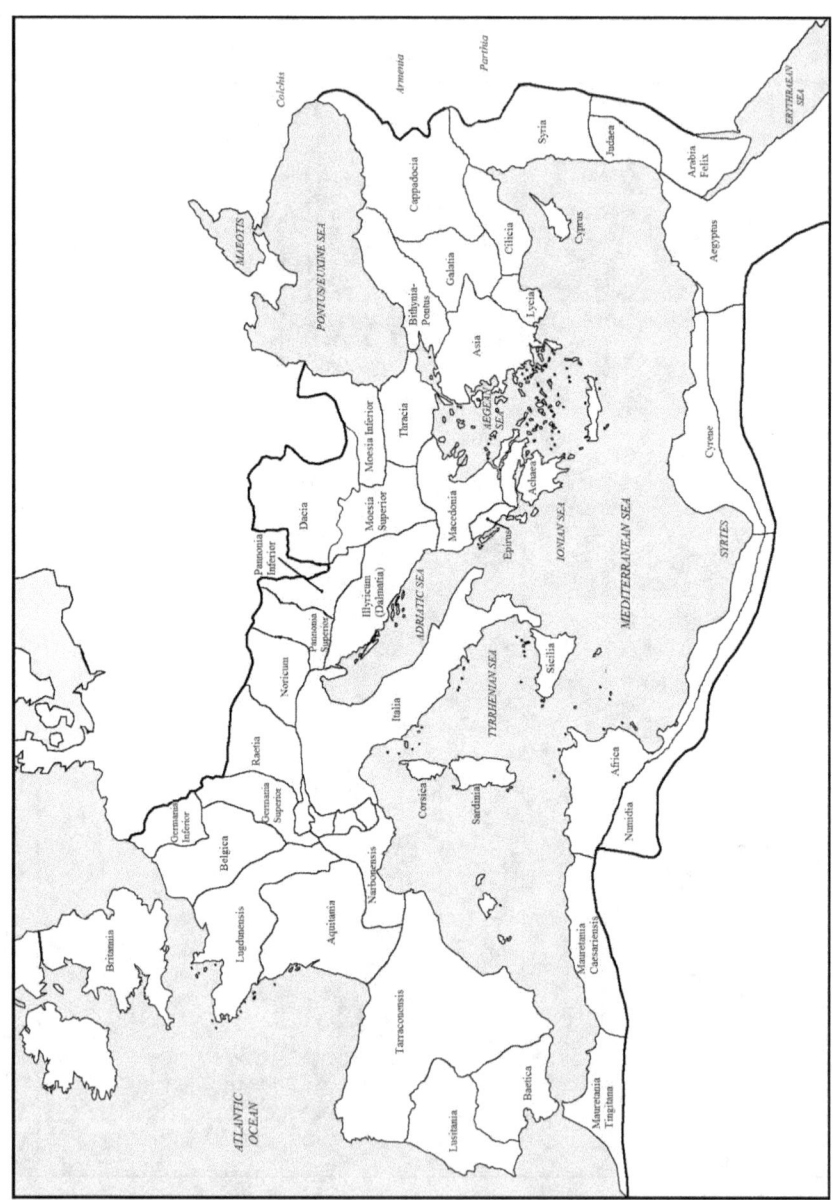

Map 3 Provinces of the Roman Empire

Introduction: Using and Conquering the Watery World

In *Conceptions of the Watery World in Greco-Roman Antiquity* (*CWW*), we considered water in the imagination of Greek and Roman thinkers: how they interpreted the nature of water, its place in the framework of the cosmos, their explanations of watery weather and water-borne disease, their understanding of real and imagined creatures of the aqueous sphere, and water's spiritual and divine aspects as a link between the human and spiritual worlds. On the philosophy of Thales (sixth century BCE), water was viewed as a fundamental component of the world's fabric or as an agent of change. Water was one of the four elements of the popular theory that explained the physical world and change within it from the fourth century BCE onwards. Some early thinkers even envisioned that human beings were originally creatures of the sea (see *CWW*: Chapter 1). Water is the backdrop of Greek mythology (Herakles' far-ranging adventures, Jason's maritime quest for the golden fleece on the *Argo*, and the Trojan War). The Mediterranean Basin, by its very geography, is moreover dominated by the sea: "The Sea" to the Greeks, "Our Sea" to the Romans (see *CWW*: Introduction).

Water is essential to biological life, as we have seen, but it is also a powerful tool of human endeavor, used for industry, trade, hunting, and fishing, as an element in luxurious aesthetic installations. Just as the Mediterranean Sea shapes the coastlines and is in turn affected by the inflow of terrestrial waters and earthy sediment, so too were the peoples who dwelt on those shores forged by their local waters. Understanding and explaining the nature of water enabled control over it. Prometheus' gift to humanity was technology, the ability to control the natural world. The Titan taught humanity the importance of numbers, and the rising and settings of the stars, together with the art of sailing, a knowledge that liberated early humanity and enabled people to interact profoundly not just with coastal waters but also with distant lands, even beyond the Pillars of Herakles.[1] These *technai* ("skills") facilitate human control over the natural world, a virtue according to Plutarch (46–120 CE), who remarked that the efficacy of water is enhanced when manipulated by fire ("Whether Fire or Water is more Useful:" *Moralia* 955d–958e). The natural world, in fact, becomes all the more valuable when "improved" by human intervention, according to Plutarch.

The relationship between humanity and the watery environment was complex and pervasive, touching on every single aspect of life in the ancient world, from mundane acts of collecting water for the household, to private and public issues of comfort and health, to the identity of the state writ large, or the spiritual well-being of the individual and the community. Here we turn to the banausic and the political: how the ancients controlled and harnessed water, how they engaged directly with water by means of sailing and maritime trade, and how they politicized waterways and naval engagements in global initiatives to spread hegemony (authority).

Armed with both theory and technology, the ancients aimed to control this essential resource (water) through law, feats of engineering, and shows of force. Hydraulic installations (aqueducts, dams, canals, harbors, fountains, baths) proved indispensable, to be sure, but their function as symbolic of authority and munificence was just as important. In the late-first century CE, Frontinus (*Aqueducts* 1.16), for example, famously compared the pragmatic advantages and impressive achievements of the Roman aqueduct system with the famous but idle Greek and Egyptian monuments.

Waterways, especially the Mediterranean Sea and navigable rivers, were integral to the fabric of Greek and Roman society. The sea was the lifeline of Greek communities, essential to *polis* (city state) identity, subsistence (through fishing), commerce (the people of the island of Rhodes were prosperous owing to the sea: Strabo 14.2.10), and defense (Sparta alone of the prominent Greek *poleis* was landlocked). Seas and rivers fed consumers with tuna and other delicacies. Trade was conducted along the roads of the Mediterranean. Empires were built on the waves. Explorers ventured beyond the safe barrier of the Pillars of Herakles into the wave-tossed Atlantic. Settlement ("colonization") was largely a maritime enterprise, as were great voyages of exploration under Hanno, Pytheas, Seleukos, Alexander, and others.[2] Athens relied on its fleet against Sparta during the Peloponnesian War, and Pompey's success against the Cilician pirates in the 60s BCE helped to secure for him a loyal client base who would come to his aid against Caesar during the Roman Civil War (49–45 BCE), to cite only two of many examples.

Our topic is vast, and we do not claim comprehensivity. Hydraulic infrastructure and "water culture" have been particularly fertile nodes of scrutiny. Other studies have focused on aspects of the ancient world as they relate to the sea, including marine resources (Marzano 2013), piracy (de Souza 1999), naval warfare,[3] navigation,[4] and ships.[5] We shall temper breadth with selectivity, highlighting resonant case studies in order to explore the pervasive importance of water to the Greeks and Romans, largely from their own points of view. Our evidence comes from the literary record, supported (or disputed) by artistic and material remains. Although Athens and Rome remain our cultural and political centers, Roman provincial evidence is especially rich (wherever the Roman army went, so followed Roman engineering). We follow the relevant and interesting evidence surviving from a vast chronological period, from the Minoan era (ca. 3000–1100 BCE) to late antiquity (e.g., Justinian, d. 565 CE). All too often, centuries separate events from the best accounts of them.

The material falls into three thematic categories. In four chapters we consider how the Greeks and Romans controlled or harnessed water: the state-legal codes that restrained or granted access to water (Chapter 1: Water rights); efforts to "improve" water quality for civic or private use (Chapter 2: Water quality and urban planning); the installations (both state sponsored and privately built) that conveyed water into settlements that then fed public fountains, water-intensive luxuries, and water-powered heavy industry (Chapter 3: Urban hydraulic engineering); and the initiatives to improve access between land and water through marine installations, including artificial harbors (Chapter 4: Marine hydraulic engineering).

In three chapters, we explore how the Greeks and Romans engaged directly with large bodies of water and the watery world: the protocols and mechanics of traveling on water (Chapter 5: Sailing and navigating);[6] and the reliance on waterways as both a conduit for trade (Chapter 6: Maritime trade and travel) and as a source of foodstuffs and luxury items to be traded (Chapter 7: Harvesting the "barren" sea).

In three chapters, furthermore, we probe the political manipulation of the watery world, where the sea and maritime expeditions shape "national" identity, justifying thalassocratic ambitions and cascading hegemony. We investigate the epic cycle as an analog to Greek expansion during the Archaic Era (Chapter 8: Minoan thalassocracy, Archaic expansion, and maritime iconography). We then consider Greek thalassocratic initiatives at Athens (from the Persian to Peloponnesian Wars) and in the eastern Mediterranean, as Alexander's successors vied for maritime hegemony (Chapter 9: Hellenic and Hellenistic thalassocracies). Finally, we close with a survey of Roman efforts to establish and maintain their authority by naval conquest (Chapter 10: Rome: *Oceanus Domitus*).

The transliteration of Greek names is a subject that has become contentious: whether to retain the traditional anglicized spellings or to Hellenize with the view to fidelity to the original language. In the interest of accessibility, we employ "traditional" anglicized orthography of most Greek names. Although Sokrates (Socrates), and Empedokles (Empedocles) or Theophrastos (Theophrastus) should cause no problems, Aristoteles (Aristotle), Epikouros (Epicurus), and Dioskourides (Dioscorides) may raise eyebrows, and Herakleitos (Heraclitus) may seem like a mystifying string of unpronounceable letters. Despite existing conventions and every effort at consistency, T. E. Lawrence's thoughts on the transliteration of Arabic names is as revealing as it is entertaining and remains applicable to any effort at transliteration:

> Arabic names won't go into English, exactly, for their consonants are not the same as ours, and their vowels, like ours, vary from district to district. There are some 'scientific systems' of transliteration, helpful to people who know enough Arabic not to need helping, but a wash-out for the world.
> *Seven Pillars of Wisdom*. New York: Doubleday reprint 1991, 21

All translations are my own, except for Seneca the Younger's *Natural Questions*, for which I rely on the superb translation of Harry M. Hine (2010), Strabo, expertly translated by Dwayne W. Roller (2014), and where a previously published translation simply could not be bettered (as cited). Cross-references between chapters are indicated by chapter number (e.g., Chapter 1). Cross-references to the first volume are indicated by acronym and chapter number (e.g., *CWW*: Chapter 5).

Part One

Controlling and Harnessing Water

1

Water Rights

Introduction

The United Nations estimates that in the twentieth century, water usage increased more than twice the rate of population growth,[1] and we are already seeing the stresses caused by competition for clean water on a planet with a rising population. The Indian subcontinent is facing a water crisis because of rapid urban growth, privatization of water resources, agricultural demands (India is a major grain producer), and industrial and human waste (among other issues). There is, additionally, concern that India's aquifers may not be replenishable in the long term.[2] Similar stresses are faced by Africa's rural poor. As human population increases globally, the supply of safe water will be further burdened and will no doubt become a political flashpoint.

Another anxiety regards access to and ownership of water sources. The Pima of the Gila Valley in south-central Arizona had long been successful agriculturalists because of their knowledge of crop irrigation, until white immigrants began to settle above the Pima villages in the 1890s, creating stresses on the water supply, threatening Pima prosperity, and compelling them to rely on the US government for subsistence aid.[3] Pima frustration was expressed in Clay Southworth's 1931 series of interviews conducted in order to "support Pima rights to the waters of the Gila River."[4] Here follows one representative expression of the general Pima grievance:

> We Indians on both sides of the river were getting all the water we wanted for irrigation and were self-supporting people. But when white people began to take the water above, we were reduced to poverty and sought aid from the government.
>
> Havelena of the Blackwater District[5]

Before the federal government imposed its settlement policies on the Pima, the land was sufficiently watered and quite capable of sustaining the agricultural needs of the population. Once white settlers arrived, the waterways were diverted away from Pima land, parching their territory, undermining their agriculture, and destroying their livelihoods.

The experience of the Pima raises important questions regarding who has the right to access water. Does water belong to the land or to the people who control that land? Greek and Roman law addressed these concerns, safeguarding landowners against such diversions. The anxieties regarding water access resonated with Greco-Roman thinkers in many genres. In Ovid, for example, Latona (Leto in Greek), who had recently given birth to the divine twins, Apollo and Artemis/Diana, expressed irritation similar to the Pima's when she sought a drink of water from an inviting pond on a hot day in Lycia (Turkey). When the locals tried to drive her away from their precious water, she delivered a vitriolic rant on water rights:

> Why do you prohibit us from the water? The use of the waters is shared. Nature has made neither the ground, nor the air, nor the delicate waves its own: I came to public works. I who ask, however, as a suppliant, that you give (me a drink of water. I was not preparing to wash my joints and tired limbs here, but to relieve my thirst. My mouth lacks the moisture for speaking, and my throat is dry, and scarcely is there a path for my voice. A drink of water will be nectar for me, and I will confess to receiving life at the same time: you will give life from the water.
> *Metamorphoses* 6.349–365

Latona, then, fittingly turned the sullen Lycians into amphibious frogs. The goddess may have been better served visiting the Red Sea where the Trogodytes collected rainwater explicitly for use by travelers in the mid-first century CE (Pliny 6.189).

Here we shall survey the regulations that guided water use and oversight in the ancient Mediterranean. The topic is vast and complex, and evidence is marshaled from sources that cover a very wide chronological period. Our treatment is necessarily selective.[6]

Water Regulation in the *Poleis* (City States) of Classical Greece

Access

Solon would have agreed with Latona that water should be made freely and widely available. Elected as Archon of Athens in 594 BCE, the legendary statesman was responsible for reforming the Athenian political system, enacting legislation against economic and political corruption, and laying the groundwork for the Athenian democracy. Some of his laws were transmitted by Plutarch in the second century CE, including one on access to water:

> Since nearby water is sufficient for the place neither in free-flowing rivers or lakes or generous springs, but most folk use constructed wells, Solon wrote a law that where there was a public well within a *hippicon* (ca. 2300 feet = ca. 710 meters), the people should use it. But where a well was further away, they should seek their own water. If they did not find water, after digging to a depth of ten *orguia* (ca. 18 meters [ca. 60 feet]), then they could take from their neighbor twenty hydrias (ca. 20 liters [ca. 5 gallons]) twice each day. For he thought it necessary to furnish resources against lack but not against idleness.
>
> Solon 23.5

Public wells were communal, and those who had to travel an inconvenient distance to collect their water and whose lands lacked viable sources had the right to obtain modest amounts of water from their neighbors. The law likely remained intact at least into the Classical period.[7] River water was also regulated by municipal decree. At Gortyn in Crete, in the fifth century BCE, residents were allowed to draw water from the Mitropolianos (Litheos) River, so long as the water level did not drop, but the documentary evidence does not specify if river water was intended to benefit only the farms abutting the river or also those further away.[8]

Plato (428/427–348/347 BCE) surveyed "excellent old laws" in Athens that regulated water rights (*Laws* 844a–c). In sum, farmers could bring water onto their properties from public resources, but they could not damage someone else's land by doing so. Individuals could procure from a neighbor's supply only as much drinking water as required by the household. Farmers could not restrict the flow of water to downstream plots nor conversely damage the fields of the downstream plot with excessive flooding. Plato also referred to laws that secured property owners against water theft and poisoning in Athens (*Laws* 845e).

Water and Property Rights

The Greek legal code was not as well curated as its Roman counterpart, but scattered evidence suggests oversight of the water rights for property owners. Landholders were liable for accommodating the natural flow of water towards downslope estates, as we see in one case preserved in Demosthenes' convoluted, *Against Callicles* (fourth century BCE). Callicles' neighbor had built a wall around his own property some fifteen years earlier, which—Callicles alleged—blocked a gully and had thus caused flooding and damage to his (Callicles') fields during a recent storm. Callicles brought suit against the neighbor's son, Teisias, seeking damages in the amount of 1,000 drachmas.[9] Teisias denied that a natural watercourse existed across his land, countering that Callicles had built his own wall that subsequently raised the level of the road separating the two estates and increased the threat of flooding. Callicles' suit was inappropriate—so argued Teisias—motivated by years of strife between the neighbors. The case, moreover, so Teisias maintained, was not supported by Callicles' "facts." The jury's decision has not come down to us.

Regulations restricted personal use of private property in order to preserve access to water. Laws similar to Solon's decrees would later be codified under Roman legislation. For example, water-hoarding fig and olive trees could not be planted within 9 feet (2.7 meters) of property boundaries, and other (less thirsty) trees were to be planted at least 5 feet (1.5 meters) from the property line. Ditches were required to be as far from the property line as they were deep.[10] Evidence from *horoi* ("mortgage"-stones) also indicates that the right to use water was not necessarily linked with property leases, but separate legal agreements might be required to guarantee access to the lessee.[11] Plato added that owners of upslope properties were responsible for taking measures that would minimize any damage caused by flooding (*Laws* 844c).

Water supply and drainage were largely the responsibility of the *polis* or community at large, but storage was the obligation of the individual.[12] Distinctions were made between public and private installations. Landowners were required to maintain wells and cisterns in good condition, in order to protect the overall water supply, especially during times of war.[13] Plato, nonetheless, mentioned city stewards whose obligations included the surveillance of rainwater drainage (*Laws* 6.779c). Access, however, was generally regulated. An elected *hydaton epistates* ("supervisor of the waters") oversaw the use and distribution of the city water supply (Aristotle, *Athenian Constitution* 43.1), and under him a number of guards kept watch over public fountains and wells. Serving as *hydaton epistates* before his first archonship (493 BCE), the populist statesman, Themistocles, imposed fines on individuals who had diverted water from the public supply in order to irrigate their private crops, according to his second-century-CE biographer (Plutarch, *Themistocles* 31.1). The office endured, and Pytheas, the *hydaton epistates* of 333 BCE, restored several much-needed aquifers, for which he received a gold wreath.[14]

Willful Contamination

Regulations were intended to secure the water supply against willful contamination. In the early-third century BCE at Ceos (one of the Cyclades islands in the Aegean), the *boule* (governing body) decreed that the waters supplying the sanctuary of Demeter were to remain clean:

> If anyone cleans himself or washes anything in the fountain, the *epimeletes* (water superintendent) shall have the authority to impose a penalty of 10 drachmas to a free man and to the children of free men and to whip the slaves.[15]

Demeter's water was reserved for ritual purposes, and fines were imposed according to the means and status of the offender (corporal punishment was common for slaves).[16] In southern Italy at Heraklea (a Greek *polis*), farmers who rented lands from the sacred precinct of Apollo were required to clean the watercourses as necessary to preserve the supply for farms beyond.[17]

Surviving at Pergamon in Asia Minor, a lengthy second-century-CE decree on municipal maintenance prohibited, among other things, the installation of uncovered pipes in the city streets (where the water supply might become contaminated), as well as watering animals and washing utensils or clothing in public springs.[18] The dumping of dirty water was a municipal concern with far-reaching health concerns (see *CWW*: Chapter 5), and laws, as at Pergamon, shielded some watercourses from gray and black water (respectively dirty water from household use and water contaminated with fecal matter).

In Athens, the tanning industry, notorious for its stench,[19] was singled out in an early-fifth-century-BCE decree:

> it is not permitted to allow hides to rot in the Ilissos River above the temple of Herakles; nobody is permitted to tan hides or to throw litter into the river.[20]

Elsewhere, public streets were safeguarded from water pollution. For example, in the early-fifth century BCE, residents of Thasos (in the north Aegean Sea) could be fined for failing to keep the streets in front of their homes free of detritus.[21] In Paros (in the Cyclades, in the southern Aegean), a worshiper could be sued 51 drachmas for tossing dirty sacrificial water into the street.[22] Laws also aimed to insulate property owners from contamination and water theft. Plato proposed the following legislation in order to defend private water in his ideal state:

> Let this, then, be the law concerning [the protection of private water]: if anyone willingly destroys the water of another person, if in a spring or collected in a well, by means of poisons or digging or theft, let the injured party bring legal

action in front of the city magistrates, registering the value of the damage; and if anyone is condemned with using poisons, in addition to a fine, let him clean the springs or water-basins according to how the laws of the advisors deem it necessary that the cleaning occur each time with respect to each plaintiff.[23]

Water sources, thus, were damaged by toxic dumping, excavation, and over use, which could result in debased property values and utility. It is not clear who would have received the funds from any incurred fines or to what use those fines might have been applied.

Finally, Plato distinguished between those who fish inland and those who fish in the seas, and he would restrict fishermen only from harbors and sacred waterways (rivers, marshes, and lakes: *Laws* 7.824b–c). Lytle observes that there is no evidence to support the long-held "common assertion" that *poleis* could enforce special taxes and regulation on fishermen and their catches.[24] He further remarks that "for fishermen the sea was freely accessible, a fact that reflects the limited regulatory reach of ancient *poleis*." Fishermen comprise a community that largely exists "outside the legal and social structures of the *polis*."

Water Regulation in Rome

Water Rights and Roman Law

Complex, organic, evolving, and occasionally contradictory, Roman law is perhaps that society's most enduring legacy, borne from the class conflict in 450s BCE that resulted in the passage of the *Twelve Tables*, which in turn formalized the right of appeal for all citizens regardless of class or income. The legal code and the court of law helped define the Roman sense of self: rhetoric—the art of speaking persuasively—was the cornerstone of Roman upper-class education, and most Roman elite, especially those with political ambitions, had some practical experience in court (e.g., Marcus Tullius Cicero, 80–43 BCE). Over the years, the legal code grew, developing into a genre of technical legal literature. In the fifty books of his *Digest of Roman Law*, Justinian (ruled 527–565 CE) systematically collated and published the vast body of Roman law organized according to topic, including material drawn from the ad hoc *Twelve Tables* (449 BCE), down to the legal code in use in his own day. In turn, Justinian's codification, together with Frontinus' *On the Aqueducts* (written under Trajan some four centuries earlier: Chapter 3), helped to shape water policy in the United States.[25]

Water legislation was a concern in Rome from the city's earliest days. In the *Twelve Tables,* legislative measures were established for mitigating the damage caused by flooding winter rainwater. Roman water legislation also aimed to preserve access to water (*Digest* 8.3) through orderly and regulated distribution and maintenance (*Digest* 3.5.30.7), as well as to prevent deliberate property damage because of poor water management (*Digest* 8.2.18, 39.3). Early Roman legislators recognized that the following resources were common by natural law: "air, flowing [drinkable] water, and the sea, and through it the shores of the sea" (*Digest* 1.8.2.1). According to Frontinus (*Aqueducts* 2.94–95), water legislation, at least during the Republic, was intended more "for the general good than private luxury." Frontinus here emphasized the legendary frugality and self-reliance of Rome's forefathers, a popular trope in imperial literature of the first century CE. There were fees, nonetheless, for access to the water supply in Rome (*Digest* 30.39.5), and water was sold elsewhere, especially in the water-poor provinces of Africa and Egypt (*Digest* 34.1.14.3), and at Palmyra where visitors (private individuals or entire merchant caravans) were charged 800 denarii annually for the use of two water supplies.[26]

Private Property

In agricultural areas, water is essential, and water sources have a profound effect on the productivity of the land.[27] Cato the Elder's (234–149 BCE) advice on selecting a site for a farm includes ensuring dependable sources of water (*Agriculture* 1.2–4). Cicero endorsed his brother Quintus' decision to purchase the Fufidian farm near Arpinum, because the estate boasted a copious water supply with great practical and aesthetic potential. Cicero suggested installing fountains.[28]

But not every property possesses all the necessary resources, and some properties might lack a reliable water supply (Columella 1.2.4). The code concerning private access to private waters was thus based on servitudes (the legal right to use waters owned by someone else): here, the right to use or channel water from or across a neighbor's property, ensuring access to the resources necessary for agriculture, with the aim of achieving self-sufficiency.[29] Usufruct rights (pertaining to using and taking the profits of properties that belong to a neighbor, as guaranteed by servitudes) covered the right of water access and ownership across estates, the discharge of waters, the manipulation of watercourses, and the use of common walls. But all parties involved were legally prevented from actions that might damage or significantly change the water supply, including building, digging, and planting (*Digest* 43.20.1.27). Property owners at least had an ethical obligation to share in the expenses and labors of mutually beneficial repairs.[30]

Servitude legislation, however, was inconsistent. In some cases, water rights might not transfer to heirs (*Digest* 8.3.37) or with the sale of a property. But over time, laws were updated. By Frontinus' day, "the benefit of every water grant is renewed with the owner" (*Aqueducts* 2.107), who would retain the rights and obligations of servitudes when land was purchased, as contractually specified (e.g., *Digest* 8.4.12, 10.3.19.4, 18.1.66). A new owner of a servient property had the right of access, and a new owner of the dominant property was legally compelled to grant it. When properties were inherited, the transfer of servitudes was tacit. Servitudes, however, could be lost because of non-use (*Digest* 8.5.10, 8.6.7). The period of non-use was pre-determined, cumulative, and, furthermore, transferred to new property owners:

> the transfer of the period of non-use made it all the more important for a buyer to know of servitudes, because he could lose his right rather quickly if the previous owner had not been using it.[31]

Similar legislation was in place for access to public waters. For any particular body of water, rights might be granted to several individuals who could draw water only at specified times or simultaneously, provided that the supply was sufficient (*Digest* 8.3.2.1–2). Where rights for access to public water did not transfer, the pipes and canals did (*Digest* 18.1.49). Any items necessary for the operation of a fountain—gutters, conduits, basins—remained on the property (*Digest* 33.7.12). If a spring or aquifer that had previously dried up became active again, property owners would need to petition for the restoration of water rights (*Digest* 8.3.35).

Waterways were also variable: springs might flow and stop according to some regular schedule (Pliny 2.228–29, 232), because of earthquakes (Pliny 31.53–54), or owing to the effects of farming. Such events raise interesting legal questions, especially regarding the potential loss of servitudes by non-use. For example, a stream ran dry at Sutrium (in Etruria north of Rome) on property belonging to Statilius Taurus (consul in 26 BCE).[32] When the water re-appeared, the neighbors resumed their use of it. But Taurus invoked the non-use codicil, and his neighbors petitioned the emperor, Augustus (formerly Octavian: ruled 27 BCE-14 CE), for restoration of their access.[33] According to contemporary documents preserved in Justinian's *Digest*, the emperor ruled in favor of the plaintiffs:

They petitioned me that—because they had lost their right not from negligence or fault, but because they were not able to draw water—their water-rights be restored to them. Since their request did not seem unjust to me, I thought that they should be helped. And so because they had the right at that time when the water was first not able to come to them, it seems pleasing that the water-rights be restored to them.

Digest 8.3.35

Drainage

Drainage is another perennial concern, especially in regions with a wet season. Rainwater drainage is cited in two tantalizing fragments of the *Twelve Tables* (7.8a–b), establishing that the state is liable for damage caused to private property by waterways that lead through public places. *Digest* 39.3 is devoted to the same issue. In short, property owners are liable for damages to neighboring fields if caused by their efforts to control run-off: either by artificial installations that would change the flow of water (flooding the neighbor's fields or withholding water from them), or by channels that were installed solely for water drainage and were otherwise unnecessary for cultivation. Owners were also liable for keeping their drains clean and well-maintained. Recorded by the late-Republican jurist, Labeo, one famous case regards an ancient ditch that the owner of a downstream estate had not kept free of debris, thus potentially harming the upstream estate with back-flowing water. It was the legal responsibility of the owner of the downstream estate either to clean out the ditch himself, or to allow the owner of the upstream estate to do so (*Digest* 39.3.2.1). A property owner could thus undertake maintenance or build installations to prevent flooding so long as the work did not damage someone else's property.[34]

Municipal Oversight

Every aspect of the water system in Imperial Rome was managed, and water oversight in Rome fell generally to the praetors (judicial magistrates). In 33 BCE, Octavian's lieutenant and friend, Agrippa, assumed oversight of hydraulic infrastructure, among his other duties, serving informally as the city's first water commissioner. Agrippa took steps to increase the overall water supply and to distribute water as equitably as possible between public works, private homes, and businesses. As the population increased, together with demands for water-intensive luxuries (baths, latrines, fountains), water remained a precious resource, and, despite hydraulic initiatives in the late-first century BCE, such as the Baths of Agrippa, there is no indication of Augustan-era surpluses or deficits in Frontinus' *On the Aqueducts of the City of Rome*, our best source for Roman aqueducts and water policy. Price scales are also uncertain for public baths and latrines. There was a modest fee for Agrippa's baths until his death in 12 BCE, when he left them to the Roman people for use free of charge (Dio 54.29.4).

After Agrippa's death, Marcus Valerius Messalla Corvinus was appointed *curator aquarum* ("supervisor of the water-supply"), but Augustus seems to have assumed legal oversight of the waterworks, including granting licenses, which remained an imperial prerogative. Water policy was quickly formalized (Frontinus, *Aqueducts* 2.104–106): public fountains, the pre-eminent concern of imperial water policy, were to flow night and day; the number of fountains within the city was to remain static; private citizens could not divert public waters to private buildings without imperial permission (and then only in compliance with certain restrictions), but private citizens could connect their own reservoirs to the public supply in alignment with supervisorial regulations (e.g., positioning of the reservoir, pipes, and taps).

Provincial Water Administration

Water administration in the Roman provinces seems to have followed the ad hoc Republican approach.[35] In the Alpine territory of the Salassians, for example, farmers and gold miners were constantly in contention over water rights. Gold was washed with waters that farmers could then no longer employ for irrigation. When the Romans gained authority in the area, they expelled both local miners and farmers, but the residents maintained control of the mountain heights and, consequently, access to the water that they sold to Roman mining contractors (Strabo 4.6.7).

Water supervisors are also known in the provinces.[36] For example, colonists, at least in Iberia in 44 BCE, had to petition the municipal administration for the right to install private water taps. We do not know, however, if such taps were for private homes or businesses, nor if the inhabitants were required to pay for access.[37] As in Rome, the provincial Iberian aqueduct was government controlled, and the magistrates had the authority to distribute surpluses.[38] Documentary evidence also details the authority of water administrators in Hispania Baetica, as at Irni where local administrators could change the routes of watercourses provided that private property remained unaffected.[39] At Genetiva Iulia (Urso, also in Hispania Baetica), the town council arbitrated both on the course that would bring public waters into the town, as well as the procedures for distributing overflow waters from municipal fountains.[40] Documentary sources preserve the quantity of water to be allotted to certain public buildings, along with their distances from the municipal cistern at Lucus Feroniae (north of Rome).[41] In the second century CE, Pausanias (10.4.1) was appalled that Panopeus, 20 stades (3.7 km/2.3 miles) from Chaeronea, altogether lacked widespread public amenities, including government offices, gymnasia, theaters, market places, and "water descending to a fountain," implicitly suggesting that proper Roman provincial cities should provide water from public installations.

Fees

Water from the public fountains was free in Rome[42] and Thasos in the north Aegean.[43] But fees were in place for the convenience of piping water into some homes and businesses, but perhaps not all.[44] In addition to overseeing distribution and determining fees for water usage, local administrators in Venafrum (near Rome) had the authority to license private taps at no charge.[45]

Standardization

The standardization of pipes under Augustus facilitated oversight and use of the network. Frontinus described many regulations, including the maximum diameter of the adjutages (discharge tubes or nozzles) that would determine the water flow, the level at which adjutages could be set, and the diameters and lengths of pipes (50 feet [15 meters]) by which private citizens could tap into the public water supply (*Aqueducts* 2.106, 112–113). These factors (adjutages and pipe length) affected how much water would flow from the public pipes into private homes or businesses. In the first century CE, licenses were granted and taxed for private access to public waters. Requiring the emperor's approval, moreover, these licenses prescribed how much water could be drawn.[46]

Abuse and Sabotage

Adjutages and pipes that exceeded the legal limits would naturally draw off more than the allowed quantity, and abuse of the waterworks was an ancient problem as demand and population grew against a diminishing supply. As censor in 184 BCE, Cato the Elder cut off the supply to private buildings and fields from public watercourses, giving notice of thirty days.[47] The sources are silent as to why. In

Plutarch, Cato's austerity measures were intensified as a response to his jealous senatorial detractors.[48] We do not know if Rome was suffering from a legitimate shortage, but the water may have been diverted to private houses and gardens, a use that the disapproving Cato would have considered extravagant.[49] In 144 BCE, abuse is clear. After the removal of conduits that had illegally tapped into two of the city's aqueducts, the senate ordered repairs on the network (Frontinus, *Aqueducts* 2.76.1). In 50 BCE, a case of collusion was prosecuted by Caelius Rufus whose address to the senate is preserved in Frontinus:

> We have discovered that irrigated fields, shops, even attic apartments, and finally all brothel houses have been fitted with constant running water.
>
> *Aqueducts* 2.76.2

Rufus noted that the misdemeanors involved forgery and "double-dipping:" legitimately contracted waters were going unused in preference for alternative sources that had been illegally accessed (*Aqueducts* 2.76.4). By 9 BCE, the *Lex Quintia* mandated fines of 100,000 sesterces for the deliberate sabotage of hydraulic infrastructure or illegal appropriation of public water. The guilty party was also responsible for repairs (*Aqueducts* 2.129).[50] Crawford assumes that malicious intent, rather than ignorance, had to be established before the fine could be imposed.[51] Moreover, the right to a private conduit was neither inheritable nor could it be sold (*Aqueducts* 2.107–108).

Hacking was inevitable, much to Frontinus' annoyance:

> There are many lengthy and diverse spaces through which hidden pipes meander under the paving stones in the city. I have discovered that the pipes, damaged by punctures, were conveying water to all the businessmen in their path by means of private branch pipes, resulting in scant measures of water for public use. How much water has been stolen in this way, I estimate by means of the fact that a certain amount of lead has been retrieved (presumably from scavengers) owing to the removal of this kind of branch-pipe.
>
> *Aqueducts* 2.115

The problem was not restricted to the Empire's capital. In the early-second century CE, the aqueduct at the Roman imperial city of Ephesus apparently suffered damage from sabotage. The Ephesian proconsuls issued two separate edicts that reiterated a safety corridor of 10 *pedes* on each side of the aqueduct (9.7 feet/2.9 meters).[52] Violation of the safety zone incurred a steep fine of 50,000 sesterces payable to the city, with the same amount payable to the emperor.[53]

Property owners were furthermore responsible for maintaining any public infrastructure that happened to cross their land, keeping the areas free of tree roots and other hazards. Private owners, it seemed, would deliberately build roads, place tombs, or plant trees with their damaging roots near the installations in order to camouflage them (Frontinus, *Aqueducts* 2.126–127; *Codex Justinianus* 11.43.1). The problem was perennial. In the fourth century CE, Decimus Secundinus as proconsul wrote to the people of Locris in southern Italy with a reminder of his order that they clean the aqueducts.[54]

Public Maintenance

Maintenance of public hydraulic infrastructure also fell under the public purview, and legislation guaranteed ease of access for working crews. As at Ephesus, aqueducts, in particular, were further cushioned by a corridor of 15 *pedes* (14.5 feet/4.4 meters) outside the city on either side of above-ground installations and 5 *pedes* (4.8 feet/1.5 meters) for subterranean installations within the city to ensure that maintenance crews had adequate access for cleaning and repairing public fountains, aqueducts, and drains: "for drains choked with filth threaten pestilence of the atmosphere and ruin" (*Digest* 43.23; cf., Frontinus, *Aqueducts* 2.127).

Polluting the Waters

Frontinus recorded steep fines (10,000 sesterces) for the willful pollution or sabotage of the public water supply (*Aqueducts* 2.97.5–6; cf., *Digest* 47.11.10). Action could be taken against those who corrupted private water sources, including wells or cisterns, the contamination of which was equated with theft:

> Regarding someone who has poured anything into a neighbor's well so that he might contaminate the water by this action, Labeo says that he is liable by the interdict against force or stealth: for living (fresh) water is seen as part of the property.
>
> *Digest* 43.24.11preface[55]

The deliberate pollution of the water supply ("injuries against good practices") could even result in capital punishment (*Digest* 47.11.1.1).[56]

In the first century BCE, an injunction was issued against a neighbor who established a tannery on his property, dumping the waste into the spring on his own property. Polluted water then overflowed onto the neighbor's downstream land. The case is included in the actions regarding warding off rainwater, that is, water that falls naturally from the sky, regardless of whether it later flows into other waters. The tannery operator, however, was not judged liable in actions to ward off rainwater, according to Trebatius who wrote the decision, but other (unnamed) legal authorities believed that the tannery owner could be restrained if he channeled water into his stream or compromised the waterway (*Digest* 39.3.3 preface).[57] Trebatius seems to have dismissed it as such, but the neighbor was interested not in the drainage of rainwater but rather in the toxic waste that was being dumped onto his property (cf. the tanning industry at Athens, above). Wacke sees an analog with a case brought by Cerellius Vitalis against his neighbor on the downstream estate whose cheese factory was polluting the air with fumes and smoke on the coast at Minturnae, between Rome and Naples. Without a servitude and in parallel with water law, air pollution was, likewise, illegal, just as "it is not permissible to discharge water or any other substance from the upper onto the down-stream property" (*Digest* 8.5.8.5).[58] In addition, legislation also forbad the forceful prevention of access to lakes, wells, and fishponds (*Digest* 43.22).

Common Access

Although some waters were public (seas, lakes, and most rivers) and others were private (some rivers and most springs),[59] residents expected reasonable access to water and waterways. The farmers along the Hiberus River (Ebro, on the Iberian Peninsula), for example, had organized themselves to seek (and receive) imperial sanction for irrigating their fields with water from the river.[60] Under Marcus Aurelius and Lucius Verus (co-emperors 161–169 CE), access to public rivers for irrigating fields was also guaranteed and the amount of river water allocated to each property was "in proportion to the size of the fields" (*Digest* 8.3.17). Restrictions were in place, including prohibitions against taking water from navigable rivers or those debouching into navigable rivers.[61] Roman fishermen in particular were granted legal access to public waters through private property, provided that they kept clear of buildings and monuments (*Digest* 1.8.4. preface 1).

Conclusion

The political resonance of this basic right to water remains robust, as it was in antiquity. For the Persians of Herodotus' day, "earth and water" were powerful symbols of their dominion and superiority. Those who granted demands for "earth and water" to Darius the Great (ruled 522–486 BCE) had thus forfeited their rights over their own lands and their very lives.[62] Their surrender to the Persian invaders was unconditional. The tyrannical Darius had even posited the Hellenic refusal of "earth and water" as his excuse to invade and subjugate the anti-Persian Greeks (Herodotus 6.94). The Spartans contumeliously answered the Persian threat by tossing the envoys into a well with the command that they "dig (the water and earth) out themselves" (Herodotus 7.32).

The Roman analog to Persian earth and water was fire and water, offered to a bride when she entered her new husband's home for the first time, as a symbol of his responsibility for her welfare. In Rome, "prohibition from fire and water" (*aquae et ignis interdictio*) was exile, as the state retracted its legal protection from the outcast. Fire and water were also central to the Parilia, the annual celebration of the city's founding, where citizens were purified with water and smoke (*CWW*: Chapter 8).

Water is essential to life and the smooth workings of human society. Consequently, in efforts to control this precious resource, ancient legislators oversaw fees for the use and abuse of municipal water, and they debated questions of ownership, the rights of access, and the obligations of maintaining waterways and mechanisms for drainage. Legislators sought to minimize contamination of the public and sacred water supplies while maximizing access to clean, safe water in sufficient supply. In Rome at least, usufruct rights aimed to safeguard landowners from the abuses that were suffered by the Pima in the Gila Valley in the 1890s.

2

Water Quality and Urban Planning

Introduction

Even if a population has access to water, it may not be safe to use, especially in remote or poor rural areas in the developing world. For example, in 2014, high levels of lead leached into the water supply in Flint, Michigan, resulting in a serious public health crisis.

Such concerns regarding water quality and safety are perennial, as we also see from growing anxieties over the environmental impact of increasingly violent storms (e.g., Hurricane Florence, September 13, 2018) that deposit dangerous pollutants into the watershed.[1] The questions raised by the water crisis in Flint and the challenges posed by superstorms in the twenty-first century were, likewise, addressed in antiquity. Access to clean water was (and is) paramount to human success. Medical writers and encyclopedists surveyed water traits. Infrastructure and urban planning aimed to control the water quality that supplied large populations: apparatuses were in place for "purifying" municipal water (or at least for allowing sediment to sink to the bottom of storage containers), and urban planners avoided building near toxic, vermin-infested marshes.

Water Quality

We consider first the Greco-Roman understanding of water quality. Depending on the source, water can vary by density, color, healthfulness, taste, temperature, and potability. The Greeks recognized three classes of water: for drinking (from fountains), for bathing and cleaning, and sub-potable cistern waters that were theoretically reserved for animals.[2] The qualities were discussed at length especially in the fifth-century-BCE Hippocratic *AWP* 7–9, and—on the Roman side—by Vitruvius (8.1-4; 85–20 BCE), Seneca (*NQ* 3.20-25: 4 BCE/1 CE–65 CE), and Pliny the Elder (23–79 CE).

Living in southern Italy, the Greek philosopher, Alcmaeon of Croton (fl. ca. 470 BCE), believed that health derives from the harmonious blending of elemental qualities (heat, cold, moist, dry). Perhaps the first to discuss the effects of water quality on community health, he remarked that:

> disease may also sometimes come about from external causes, from the quality of water, local environment, overwork, hardship, or something similar.[3]

The human body, it was thought, can also assume the properties of local waters. Waters that flow from under the base of a hill in Chalcis (on Euboea), for example, are stinking and clotted owing to the putrefying remains of centaurs who had been slain there, and consequently the local residents are called Ozolians ("Foul Smelling").[4] Some waters are nourishing, others not. Just as waters vary in taste, so too do their effects on the body (*AWP* 1; cf., *CWW*: Chapter 3). Some waters are fresh (and therefore drinkable), others are salty. Some promote health, others are deadly. According to Pliny (23.37), different liquids have various effects on the human body: wine promotes strength; blood, completion; milk, bone; beer, sinews; water, flesh.

Water Freshness

Freshness or quality was considered a factor primarily of taste and smell, and was thought to be determined by many variables, including the winds.[5] The Hippocratic authors promoted waters that derive from thunder and the dry upper aither as good, but those waters from violent storms or hurricanes as bad (*Epidemics* 6.4.17). They defined the lightness or digestibility of water according to the rate of vaporization and condensation: "the most digestible water is quickly heated and quickly cooled."[6] Assessment of water quality changed little, if at all, over the centuries. The Greek pharmaceutical writer, Dioscorides of Anazarbus (Asia Minor, first century CE), praised the general healthfulness of water: "on the whole, it is excellent if it is clean and sweet, free of any quality whatsoever" (5.10). In Pliny (31.37), the healthiest water "ought to be very much like air," lacking entirely in taste and scent, like the wonderfully pleasant waters in Mesopotamia where Juno supposedly bathed. Betraying his chauvinistic patriotism, moreover, Pliny commended the water from Rome's *Aqua Marcia* (a water source also lauded by Frontinus, *Aqueducts* 1.14.2), as the coolest and most wholesome. The pleasant waters in Palmyra in Turkey were famously noteworthy (5.88). Marsh, rock-spring water, and stale, melted snow were "heavy" or unpleasantly redolent.

Affecting flavor, medicinal value, texture, and temperature, several factors influence water quality: levels of contamination; whether air is reconfigured as water as it cools; whether dirt is transformed into water as it acquires moisture; and the nature of the terrain through which the water passes, together with the "juices" of plants that grow in the soil.[7] Unhealthy waters are easily detected by the presence of slime, stains, or sediments encrusted on vessels, as noted by Pliny who may be referring to the accretion of sinter (calcium carbonate or silicon) that would clog the aqueducts in Rome as limestone leached into rainwater.[8]

Conventional Greco-Roman wisdom deemed cold running water from high altitudes as the healthiest. The Hippocratics (and others) preferred sweet-smelling, light, and sparkling waters, particularly if the source emerges from deep springs that flow towards the east or northeast (*AWP* 7). Such water was palatable even if mixed with only a "little wine." Advantaged by accreting geographical and hydrological data, the Romans posited their own ideas about the best orientation of water sources. Vitruvius advocated north-facing slopes, which are cooler, shadier, and more protected from the sun, but also because "the largest rivers flow from the north" (8.1.6.2). The Roman encyclopedist, Celsus (first century CE), recommended drinking and bathing in frigid water (1.1.2, 1.2.2). Pliny often specified "cold water" as a solvent for medicinals (he considered the therapeutic use of hot water by Greek physicians as irresponsible: 29.23).[9] The culturally Greek Galen, who wrote extensively on all aspects of medicine and served as Marcus Aurelius' court physician (ruled 161–180 CE), would later revert to the Greek practice, stipulating abstention from cold water.[10]

Other waters were less pleasant. The Hippocratic author described marsh and lake water as tending to be warm, thick, and unpleasant in the summer (*AWP* 7), and water from rock springs (or hot springs), like its source, is "hard." Such waters are constipating, so it was believed, in imitation of their source.

In both Greek and Latin sources, the effects of industrial pollution on the water supply were also recognized, especially from heavy industry, including mining, where impurities include sulfur, alum, and bitumen, which tend to exacerbate muscle spasms and gout.[11] Plato and others, moreover, recommended reforesting woodlands in order to reverse the effects of soil erosion and improve water quality.[12]

Drinking Water

The Greek poet Pindar (fifth century BCE) refers frequently to water as a feature of ordinary Greek living. Water is necessary for washing, sailing, fishing, therapeutics (*Nemean* 4.4; "not even warm water soothes the body as much as song"), and even drinking (*Nemean* 3.6). Pindar, moreover, enigmatically declared that "best is water with regard to living" (*Olympian* 1.1). It is, in fact, "the source and sustenance of life."[13]

To what extent, then, did the ancients drink plain water? Plutarch, for example, expressly noted that Cato the Elder (234–149 BCE) was known to drink straight water owing to his moralizing austerity (*Cato the Elder* 1.7). Hodge observes that the practice must have been rare if drinking plain water was even worthy of mention:

> as indeed does the fact that one of our best modern authorities, in a book of 230 pages on Roman cuisine, cannot stretch out the topic of 'water' for more than half a page.[14]

Water was usually mixed with something else, such as wine. The common drink, even for children, was wine diluted with water, a practice that was credited to the Greeks (Pliny 7.200). Even small amounts of alcohol would kill contaminants, making the water safer to drink and less likely to cause "cholera," dysentery, or other water-borne diseases (see *CWW*: Chapter 5).

Unmixed water, nonetheless, may have been a more common drink than earlier scholarship implies, and Cato was surely not unique. Hippocrates, for example, had recommended water as a beverage for those with warm dispositions, e.g., athletes, soldiers, or ditch-diggers.[15] Andreas, personal physician to Ptolemy IV Philopater ("Father-loving"; ruled 221–204 BCE), furthermore, prescribed laser (distilled from silphium) as beneficial to those who drink water, perhaps owing to its digestive properties.[16] At Alexandria in 48 BCE, when Caesar's water supply was adulterated with a large quantity of seawater, the troops complained about the foul, salty water that was altogether undrinkable (*Alexandrian War* 6–8), implying that the soldiers (and residents) were drinking water straight or at least insufficiently diluted (below). At the confluence of the Ister and Istria Rivers, sailors could get a drink of water from a freshwater spring (Mela 2.63). Galen, moreover, recommended switching from wine to water in order to treat dyspepsia (heartburn).[17] Soldiers also drank water and could become sick from consuming marsh water (Vegetius 3.2).

Water Flavors

Some sources were considered more palatable than others. Rainwater, in particular, was praised. Rainwater can be sweet, light, fine, and sparkling. According to the

Hippocratics, the process of evaporation raises up only the lightest parts of the water, leaving behind salt and other heavy mineral particles. Evaporated rainwater is then "heated and boiled by the sun" and thus "sweetened" (purified) by means of heating and boiling. For, "all things always become sweeter when heated."[18] As clouds thicken and darken, however, rainwater becomes contaminated, eventually returning to the earth because of its weight. Rainwater quickly becomes rancid, and both Greek and Latin sources recommended boiling before use (*AWP* 8; Pliny 31.32, see below). Putrefaction also occurs as falling rain amasses pollutants, and as water stagnates in wells and cisterns.

Melting ice and snow also grow stale and harmful according to the Hippocratic author:

> it is never restored to its previous nature. But its bright, airy, and sweet quality is expelled and destroyed. The muddiest and heaviest part is left.
>
> *AWP* 8

This occurs presumably through the process of evaporation, which reclaims some of the water into the cycle of precipitation. Melted moisture and evaporated liquid were considered subpar and thought to lose volume, as proved by an experiment whereby water was first measured, then frozen (on a cold night), and thawed. The remaining moisture was measured:

> You will find it diminished in quantity. This is proof that the healthiest and lightest part has been removed and consumed by congealment, for the heaviest and thickest part would not be able to behave in this manner.
>
> *AWP* 8

Roman sources agreed that hail and snow are unhealthy since the "thinnest" part has been leached off. Pliny (31.33) erroneously argued, furthermore, that "every liquid becomes smaller by means of freezing." (Although most materials do contract when cooled, water, to the contrary, expands upon freezing.)

Although wealthier folk might consume snow or ice in first century CE Rome (Pliny 19.55), and the refreshing coolness of melted snow was desired, its stagnancy was not. The emperor Nero (ruled 54–68 CE) discovered a clever compromise: a glassful of boiled water could be chilled in a container of snow, thus obtaining the desired temperature while not tainting the water with any of the "injurious" qualities of the snow (Pliny 31.40).

For some peoples, the only source of fresh water was rainwater: e.g., the Vacathi who dwelt on the Nile side of the Greater Syrtis (Pliny 6.194) and on the remote and perhaps legendary Fortunate and "rainy" island of Pluvialia (Pliny

6.202). It was also recognized that a city's sewer system could taint the water supply with effluvia from latrines and baths.[19] Additional hazards include dead animals, rotting fruits and vegetables, or excrement.[20] Heavy rainfall, however, helped to clear toxins away from the soil and water courses.[21]

Pliny argued that rainwater is the "lightest" (31.32), and he expressed surprise that some "unnamed physicians" favored cistern water. Pliny and others considered running water as healthier (it is presumably made finer and more advantageous merely by the agitation of the current). Pliny nonetheless recognized that it was nearly impossible to test the "lightness" (i.e., specific gravity) of water.[22]

Thus the source could affect the quality and drinkability of water, and these understandings were guided by a philosophical understanding of how water interacted with its environment, including how quickly it was heated or cooled, or whether it derived from soft or stony sources, elevated or sea-level waterways, falling rain, or stagnant pools. Ancient thinkers also understood that water quality was affected by human factors, including pollution and industry.

Purifying Water for Human Use

Water purification was, therefore, an ancient concern. A painting in the tomb of Amenophis II at Thebes (ca. 1450 BCE), shows a convoluted water filtration system of sediment jars and syphons.[23] A simpler method involved storing water in oblong vessels. Five sweet almonds were crushed and smeared on the mouth, the water was then agitated and allowed to settle for several hours before it would be transferred to small earthen jars where it was clarified and cooled.[24] In Alexandria, Julius Caesar found a network of underground canals that conveyed the Nile's turbid water to cisterns for clarification by settlement (Aulus Hirtius [?], *Alexandrian War* 5–10). Caesar's enemies employed "painstaking machinery and the greatest exertions" to sabotage the aquifers. Water wheels were used to pump salty seawater into the hydraulic installation. The supply quickly became "undrinkable." Caesar then devoted his attention to sinking wells: "for all coasts naturally have veins of fresh water." In accord with Caesar's own ruthless efficiency, the task was completed in a single night.[25] The date of this installation is unknown.

Masking Unpleasant Odors

Clean water was desired, and one method to "purify" drinking water was by masking an unpleasant taste with a pleasant scent, just as physicians would burn fragrant woods, such as juniper or pine, to repel disease.[26] On the advice of court physicians, for example, Commodus (ruled 176–192 CE) sought refuge in Laurentum, a pleasantly shady city where the contaminated air was purified with laurels. Romans made "constant use of incense and aromatic herbs," thinking that the pleasant scent would fill their nostrils and either prevent the diseased air from entering the body or, "by its greater potency," repel the noxious fumes.[27]

Pliny cited Juba II of Mauretania (ruled 30 BCE–23 CE) as the authority for the extraction of water from fennel-like trees: bitter water is collected from trees with black bark; brighter trees yield water that is pleasant for drinking (6.203). In the desert regions near the Red Sea, *polenta* (barley-meal) would be added to nitrous, salty water to sweeten it (like a modern cook might add a potato or other starchy foodstuff to absorb excess salt in a brothy soup), and the barley would then be consumed with a meal (Pliny 31.36). Among Pliny's other proposals for freshening stale water are boiling parsley in water to make it "sweeter for drinking" (20.115), and adding pounded pennyroyal to "unwholesome" water (20.154). Consuming garlic, euphorbia peplis, laser, and lettuce, moreover, protects against "noxious" waters, according to folklore (Pliny 20.50, 68, 104, 210). Pliny and his sources acknowledged the fact that the shelf life of liquids, however, is limited. Recent authorities, Pliny reported, condemned the use of "old" hydromel (honey water), which they deemed less harmless (more harmful) than water, but as having a shorter shelf life than wine (22.112).

In addition, as recorded in the tenth century CE *Geoponika* (2.7), Diophanes of Nicea in Bithynia (fl. ca. 85–60 BCE) had specified macerated laurel for "curing bad water." The laurel should be placed in an open-aired jar and then allowed to settle. The clarified water should then be gently removed to another container without disturbing the sediment at the bottom. Bruised coral is another ingredient advocated by the *Geoponika*'s author for improving the taste of bad water. Hughes surmises that drinking water was flavored or "purified" with sour wine.[28]

Straining Out Particulates

Attempts were also made to remove particulate matter from drinking water. Water filters, as employed by many modern householders to purify tap water, are ancient in concept (attested as early as Egypt's eighteenth Dynasty). To prove that ocean salinity is caused by "the admixture of some substance," Aristotle (384–322 BCE) described the use of wax receptacles to treat salt water:

> But if someone places a wax jar into the sea, binding its mouth in such a way that it (the mouth) is not encroached by the sea. For the water entering through the waxy sides becomes drinkable. The earthy stuff and whatever makes the water salty through mixing is separated out just as through a strainer.[29]

Aristotle believed that salt would be sifted out as the seawater was strained through the sides of the wax container. Wax, however, does not provide a sufficiently constricted membrane for removing salt from water (the interface must trap the larger salt molecules while allowing water molecules to pass through). In Rome, Pliny, moreover, recounted a process to render seawater drinkable by draining it through clay which, like Aristotle's wax, is not particularly effective against tiny salt molecules and other contaminants.[30] Pliny also advocated water that naturally seeped through tufa, which has the additional advantage of "making the water cooler" (31.48). Pliny recommended building cisterns in pairs so that impurities might settle in one tank before the fluid passes through a filter into the second container (36.173). Water filters are known, especially at Nîmes where five water mains branched out from the reservoir, each fitted with a grill.[31]

Pliny also advised drawing well-water from the bottom, rather than the sides, and he seems to have favored "frequent withdrawals" to agitate the water as much as possible (31.39). Thus, in Pliny's opinion, water is kept fresh by churning, in parallel with running water (springs and rivers). But this practice does not allow particulates to settle, and it was even contrary to practical wisdom where water was brought into cities by aqueducts and then channeled through a series of settling tanks so that, after sinking to the bottom, particulates would not be conveyed into the common water supply. Aqueducts were generally covered to minimize detritus.

Boiling

Boiling was also employed as a reliable method of purification and softening, especially for rainwater, which is "composed of a mixture of so many elements" that it required further detergement after boiling.[32] Basil of Caesarea reported that, by ca. 360 CE, Greek and Roman sailors had learned how to desalinate seawater by boiling it aboard ship. They then gathered the steam in sponges and squeezed the captured steam into receptacles for further treatment or use.[33] This process of collecting vapor from boiled water may have been refined by the early Arab mariners who repeated the process of boiling and squeezing four or five times to produce potable water.[34]

Several methods were within the ancient toolkit to improve water for human use, by masking unsavory odors, removing debris, or killing toxins. Water, thus, seemed tastier, and the Greeks and Romans gained at least an apparent control over the waters that they employed for basic survival needs.

Water and Urban Planning

Waters were acclaimed or censured for their peculiarities and effects on the individual. Some waters were pleasant to drink, others were foul. Some were healthy, others fostered disease (see *CWW*: Chapter 5). The ancients, consequently, understood that the qualities of watercourses could have far-reaching effects on nearby settlements, a theme interrogated in the Hippocratic *AWP* where climate was explored as a formative principle of human character (people tend to resemble their climates[35]). Plato and Aristotle both recognized that some locales were healthier for human habitation because of the natural environment where healthy waters and pure springs are available.[36] There was a clear division between the neatly groomed habitable world of civilization and the uncultivatable, chaotic, liminal wilds.[37] This stark segregation of "cultivated" from "wild" is reflected in the principles that guided nearly every facet of administration and infrastructure planning in the Greco-Roman Mediterranean, from city-siting to selecting campsites.[38]

Thinkers from the Hippocratic authors onward viewed marshlands—liminal areas that nourished mosquitos and other noxious creatures—as unhealthy for human habitation. The Hippocratic author of *Regimen* described muddy, marshy, boggy places as insalubriously moist and warm owing to the thickness of the air and stagnation of the waters (2.37, 60). Under such conditions, marsh and lagoon vapors fill the air which, like the water, becomes putrid, especially during the hot, summer months.[39] Diseases (especially diseases of the spleen), moreover, could rise from the stench of mud, swamps, marshes, or bad waters.[40] Hippocratic environmental theory endured into the Roman era, finding expression in Latin authors. According to Varro (*Farming* 1.12.2), such moist, warm, vaporous places are breeding grounds for tiny, airborne creatures that could inflict "serious disease." Varro's expert, Agrius, advised that such places should be sold or abandoned since there is no method of preventing disease there. Vitruvius, moreover, warned against building in marshy bogs:

> For when the morning breezes come to the town with the rising sun and mists that spring forth are joined to them, and when they sprinkle the venomous

breaths of swampy creatures mixed with the mist into the bodies of the inhabitants with their breath, they render the place unwholesome.

Farming 1.4.1

The effluvia from moist mud is unhealthy, carrying disease (and unpleasant airs). In sum, marshlands are, according to Columella (1.5.6; 4–70 CE), triply treacherous and unpredictable. They emit disease-laden vapors. They breed deadly animals ("creatures armed with dangerous barbs, which fly at us in thick swarms" [mosquitos?], and "swimming and creeping things," which in turn spread their contagions to other areas). Damp conditions, finally, generate mould, which in turn causes property damage.

Some marshy areas, nonetheless, were successfully inhabited, including Aquileia and Ravenna, where the air is, surprisingly, "harmless."[41] Such places benefit from a delicate natural balance that could be enhanced by human intervention, especially for sites facing the sea on the north or northeast, and those elevated above sea level. Dykes facilitate the drainage of the rising sea into the marsh. Because of its salinity, seawater helps to control the populations of harmful fenland creatures.[42] Other boggy settlements, however, were relocated, such as Old Salpia in Apulia, whose inhabitants suffered annually from unspecified ailments (ca. 200 BCE). With the permission of the senate and people of Rome, the community was moved to a healthier spot where the lake was opened up to the sea, and the place flourished until the expanding marsh forced the population to abandon the site in the sixteenth century.[43]

War and Water

This concern for healthy waters extends to the military sphere. After being blocked into an area that lacked a water supply at Motya in western Sicily (398/397 BCE), the Carthaginians sued for peace with Dionysius, the Greek tyrant of Syracuse, according to the second-century-CE Greek source Polyaenus (*Stratagems* 6.16). In 479 BCE, the Persian general Mardonius "blocked and destroyed" the spring at Gargaphia (Plataea, south of Thebes) that supplied the entire Greek army, and the nearby Asopus River (Herodotus 9.49.2–3). Herodotus did not explain how the spring was destroyed. The Athenian general and military reformer, Iphicrates (418-353 BCE), moreover, ordered his men to fill their water casks before embarking on a two-night hike through a hot, waterless desert (Polyaenus, *Stratagems* 3.9.47), presumably in Egypt (see his first-century-BCE

Roman biographer, Cornelius Nepos, *Iphicrates* 2.4). Among Alexander of Macedon's generals, Antigonos I Monophthalmos ("one-eyed:" ruled 306–301 BCE) also laid in a supply of 10,000 casks of water in preparation for a ten-day march over a plain filled with sulfur mines and stinking bogs (Polyaenus, *Stratagems* 4.6.11). Antigonos was at war with Eumenes, another of Alexander's successors.

Julius Caesar, moreover, was always mindful of proximity to a good water supply for his troops,[44] and he aimed to cut off the water supply of his enemy:[45] "Lucius Afranius (Caesar's enemy and Pompey's client) must certainly withdraw his troops; he cannot last without water" (*BC* 1.71.4). In 32 BCE, Octavian's troops at Actium were fortunate in having access to fresh water from the Louros River in addition to supplies from Italy, while Marc Antony fortified his own sources of potable water with barriers to deprive the enemy.[46] Clean, potable water is paramount, and in Vegetius, who, in the fifth century CE codified the wisdom of centuries of Roman military science, military officers learn how best to maintain the health of their soldiers with a balance of proper sites, good water supplies, medicines, and exercise. Water sources cannot be disentangled from sites. The field general must then eschew pestilential, marshy areas (as well as arid plains that would altogether lack a water supply). Instead, the camp commander should choose temperate spots with sufficient water and shade. This principle is borne out in the archaeological record: campsites were painstakingly selected in order to avoid "unsalutary dangers."[47] In permanent camps (as, famously, at Vercovicium [Housesteads on Hadrian's Wall in northern England]), moreover, we find sanitation systems, continuously flushed latrines, bath houses, and even permanent drill halls where soldiers could train during inclement weather.[48]

Water supplies, furthermore, were weaponized by poisoning and diversion.[49] The first documented case occurs after the *polis* of Kirrha, which controlled the road between the Corinthian Gulf and Delphi (Apollo's important oracular sanctuary) appropriated sacred lands. The Amphictyonic League "of neighbors" (including Athens) was then formed in retaliation, sparking the First Sacred War (ca. 590 BCE: Pausanias 10.38.7–8). The besiegers poisoned the water supply with hellebore, a commonly prescribed emetic that causes diarrhea when taken in large doses, and the weakened Kirrhans were easily defeated (Hippocratic *Embassy* 4; Frontinus, *Stratagems* 3.7.6; Polyaenus, *Stratagems* 6.13). Some, moreover, believed that the famous epidemic that befell Athens (and other areas) in 430–427 BCE at the beginning of the Peloponnesian War, was caused by spies or Peloponnesian sympathizers who had contaminated the water supply (cf. Thucydides 2.49.5; *CWW*: Chapter 5). Caesar's troops had also fallen victim to the sabotage of the water supply (above).

Diverting waterways was another mechanism that was weaponized. Frontinus surveys Roman examples (*Stratagems* 3.7.1–4). In 143–142 BCE, for example, while campaigning against the Celtiberians in Hispania Ulterior (Further [southern] Spain), Quintus Caecilius Metellus Macedonicus diverted a river to flood an enemy camp. The enemy were easily despatched by troops anticipating the ensuing chaos of the panicking Celtiberians. Additionally, as governor of Cilicia in 78–76 BCE, Publius Servilius forced the people of Isaura to surrender owing to their thirst, after he diverted their water source. The Isaurians had supported the pirates in the eastern Mediterranean.

Water thus was naturally an important device of warfare. Capable generals aimed to ensure access for their own troops who could not succeed with clean, potable water. With the proper resources, the enemy could be denied access to their own water supplies that had been poisoned or altogether removed from within their reach.

Lead Poisoning

With permanent settlements came hydraulic infrastructure (Chapter 3). Although ceramic pipes were widely recommended[50] and used, water was routinely conveyed through lead pipes.[51] Lead poisoning is a real threat when water remains stagnant in the pipes, but the Roman aqueduct system was designed for constant flow, and any large-scale stoppage of the water flow would just as likely result in back-ups, overflows, and perhaps even flooding. Although stopcocks and taps were known in antiquity, they were used for diverting the water supply, not for halting it. Because of its constant flow, water did not remain in the pipes long enough for the lead to contaminate the supply. Sintering (the accretion of crystallized calcium carbonate) provided a further layer of protection that separated flowing water from the lead in the pipes.[52]

The Romans, nonetheless, were fully aware of the dangers of lead poisoning, and Vitruvius was explicit about the health hazards of lead, including muscular hardening and pallid complexions.[53] Dioscorides and Pliny both classified lead as a poison, whose fumes are "noxious and deadly."[54] Antidotes to lead poisoning endorsed by Celsus, Pliny, and Galen include mallow, walnut juice in wine, and hydromel, especially when mixed with oil of oenanthe.[55]

Conclusion

Clean water is essential to maintaining health and preventing disease, and we have here considered ancient control over water on two related levels: the personal and the civic. The quality of drinking water was a concern of both the state (where infrastructure was in place in order to filter out particulates, at least in theory) and the individual (who might mask unpleasant odors in drinking water with more palatable ingredients). The state of Greco-Roman technology and their (superficial) understanding of the properties of water allowed for no other solutions of improving the taste of water.

Urban planning provides another case study in ancient efforts to control water within the natural environment, where the nature of watercourses was considered when selecting sites for human settlement, marshy areas were either abandoned altogether to the wilds, or—as we shall see—they were drained in efforts to reclaim fertile land, "improving" it for human use.

3

Urban Hydraulic Engineering

Introduction

The human control, harnessing, and manipulation of water in the ancient Mediterranean world is a vast topic that has been treated elsewhere in both introductory and highly specialized platforms.[1] Here we shall survey some of the larger issues, including methods to find, store, and convey water, civic uses, and industrial applications, drawing from, as usual, a broad array of Greek and Roman sources.

Water is also a powerful source of hydroelectric power, employed by Herakles, who rerouted the Alpheius River to cleanse the stables of Augeas King of Elis, which sheltered countless heads of cattle but were never tended, and thus became disgustingly soiled. Herakles' mythic feat is a testament to Mycenaean advances in hydraulic engineering,[2] and a precursor of even more powerful (and extravagantly destructive) achievements in hydraulic mining. Hydraulic engineering was consequently employed both as a means of controlling water flow for human use, and for fueling human greed and indigence. Tempered with technology, water was a critical tool in the human conquest of the environment.

Finding, Conveying, and Storing Water

Many locales in the Mediterranean Basin are arid, thus the fabrication of systems to find, transfer, and store water has been a perennial civic priority. In the Aegean, for example, large rivers are lacking and smaller sources are unreliable in the summer. It was thus essential to employ multiple methods to ensure reliable access to fresh water.

Trial and error, coupled with careful observation, aid those looking for dependable sources of water. Nature offers many clues, both botanical and geological: figs, rosemary, and brambles require concentrated water supplies for their roots, and mosses can only grow in wet conditions;[3] moisture is indicated by yellow and orange stains above calcium carbonate or lime-rich mud and shale/sandstone deposits.[4] The Roman encyclopedist Pliny the Elder (23–79 CE), who compiled centuries' worth of human knowledge, listed many of these clues: truffles, bechion, ivy, alder, willow, rushes, a concentration of frogs, a misty steam visible in the distance before sunrise, and—as known only to the experts—moist reflections in the dirt in the middle of parched terrain.[5]

Water-divining is a valuable skill but the literary evidence is limited. The Roman sources Vitruvius (8.1.1–2, 4–6) and Pliny (31.43–49) reported procedures for water divining that likely derived from practical experience, as "magical" as they might seem. Because of its arid climate, Africa fostered a robust tradition of water diviners, which endured into late antiquity. One African diviner came to Rome looking for work in the sixth century CE with the endorsement of the Roman statesman Cassiodorus (*Letters* 3.53.1).

Noting the importance of geology, Vitruvius surveyed the best water sources and the water quality from each source. According to Vitruvius, black earth yields the best tasting water. Water from gravelly earth is unpredictable but sweet. A reliable supply with a good taste could be found in coarse gravel and sand. Red rock yields copious, good quality water "unless it slips through faults and leaks away." The best water—abundant, cold, and healthful—comes from the feet of mountains and flinty rocks. Water from ground-level springs is bitter and warm, but underground mountain spring water, especially when protected by shade, is desirable. In Vitruvius we have a detailed description of how to find water:

> Let there be dug some place no less than three feet wide, five feet deep. In it let there be placed at about the time of sunset a bronze bowl or a lead basin. From these things, whatever will be prepared, let it be rubbed inside with oil and let it be placed upside down. Let the highest part of the pit be spread with reeds or foliage and then let it be covered with earth. Then on the next day let it be opened up, and if there will be drops of moisture in the vessel, this place will have water.
>
> 8.1.4

Vitruvius also advised against sinking wells into clay, where water is in short supply or near the surface. By understanding geology and botany, it would be possible to tap into life-giving water sources.

Wells

The most basic system of accessing groundwater, and the most reliable, is the well. At their simplest, wells are just holes extending to the water table. Pliny praised wells as the best water sources, provided that they remain sheltered from the sun (31.38–39), but their quality would be affected by population density, drainage, and sewage systems. Well holes are known from the Neolithic era.[6] The technology advanced, and well shafts sunk into soft materials (such as earth or sand) would be lined (and thus stabilized) according to available materials and technologies (mud bricks or masonry). The wells of Minoan Crete have been carefully studied, including a large oval-shaped well (2 × 1.6 meters [6.5 × 5.2 feet]) lined with blocks at Palaikastro (ca. 1350–1100 BCE).[7] From the fourth century BCE, wells in the Athenian agora were lined and fitted with terracotta rings (with footholds).[8] Twenty-two groundwater wells have been excavated in Pompeii, many attached to private homes and many lined with limestone, showing the importance of wells as a water source. Nonetheless, many of the public wells seem to have been filled or privatized after the installation of public fountains and pipes, since flowing aqueduct water was preferred to stagnant well water.[9] Limestone was also used to face wells in western Europe, as with a Roman well discovered in Biddenham, England, in 1857.[10] In northern and western Europe, casks and wooden wine barrels were also employed as convenient liners for well shafts (as at Silchester in Britain[11]).

Uncovered wells posed a safety hazard as Thales learned first-hand in the sixth century BCE, when he reputedly fell into a well while gazing at the stars (*TEGP* 7: Thales was the first rational Greek philosopher, and he suggested water as the origin or catalyst of the physical world. See *CWW*: Chapter 1). Both public and private wells were commonly covered with stone or terracotta well heads, often elaborately decorated.

Wells had to be wide enough to accommodate a human worker (ca. 3 feet [1 meter]), but depth was dictated by the water table. Because of the volcanic nature of the Mediterranean landscape, it was recognized that poisonous fumes could jeopardize workers' lives, and Vitruvius advised lowering a lamp into the well shaft. If the lamp continues to burn, the air in the shaft is free from sulfur and other toxins, if not, ventilation shafts must be constructed (8.6.12–15). Many wells were sunk to a depth of between 13 feet (4 meters) and 50 feet (15 meters). In Roman Algeria, 60 meter-deep wells (ca. 200 feet)—sunk into hard rock—are common, and one Roman-Gallic well near Poitiers reached 80 meters (ca. 260 feet),[12] creating challenges for hauling up water over so great a vertical length.[13]

Depending on the depth of the well, various methods were employed for obtaining water. The simplest technology was simply to pull up by hand a water jar attached to a hemp rope, and some wells show ruts created by the friction of the ropes, such as those at Delos. Conical leather buckets were also employed, which were easy to construct and fill. The top-heavy conical bucket more easily dips into the water "avoid[ing] the embarrassment of seeing one's bucket obstinately floating on the water."[14] Deeper wells would require some mechanical advantage, such as a bucket rope suspended over a pulley above the well's center, keeping clear of projections from within the well. As the well becomes deeper, the mechanism becomes heavier (longer rope, greater distance), requiring greater effort, so more complex systems were utilized.[15] Wide well shafts, often with public or industrial application, were probably used together with mechanical water-lifting devices.[16] In large installations, wooden pots might be fitted to an iron chain strung over an angle gear drive that was powered by draught animals.[17]

Cisterns

Also ubiquitous were private and public cisterns for collecting rainwater during the rainy winter or, less commonly, for storing spring water (as at Epidaurus). They were fitted to roof tops, courtyards, or other paved areas. Cisterns were a handy complement to wells and were used where wells and springs might be lacking, but their efficacy was entirely dependent on weather patterns. The earliest known cisterns are the open, circular installations at Knossos (3500–1900 BCE), Minoan Zakro (Crete: 1750–1650 BCE), and Tylissos (1350–1100 BCE). Each was fitted with access steps, and the cistern at Tylissos incorporated a rectangular settling tank. Two cisterns at Myrtos–Pyrgos (*ca.* 1700 BCE: 5.3m in diameter [17 feet]; 3 meters deep [ca. 10 feet]) have a capacity of about 80 meters^3 (21,000 gallons^3).

To mitigate contamination and evaporation, later cisterns, such as wells, were covered. Their shapes and sizes varied, but the most common Greco-Roman variety was the bottle-shaped cistern with a wide shaft and narrow neck, known in Athens by the early-fourth century BCE.[18] In the fourth century BCE in Athens and the third century in Carthage, cisterns replaced wells in response to population stresses and environmental challenges (drought? lower water table?).[19] By the third century BCE, in the water-poor Aegean Islands, some small-scale residential pear-shaped cisterns were cut into the rock (with a capacity of ca. 10 meters^3 (2,640 gallons^3) and lined with hydraulic plaster to prevent water loss, as in Santorini and Aegina. At the same time, an elaborate central cistern (22.5 meters long × 6 meters wide [74 × 20 feet]) was built on Delos, where rainwater was channeled from the theater into an installation whose slab-cover was supported on a series of arched walls (the practice of collecting water runoff from public roofs was continued in water-poor North Africa under the Romans into the third century CE). Roofing technology limited the capacity of early built or rock-cut cisterns, and the arcaded cistern in Delos was an advance.

In Rome, the predominant style was the rectangular, barrel-vaulted cistern—which could be multi-chambered—with draw holes about 1.5 feet (50 cm) in circumference, into which terracotta pipes leading from private roofs funneled rainwater. Public cisterns also collected water from the aqueducts for distribution throughout the city. The Serino aqueduct, for example, fed the *Piscina Mirabilis* (the marvelous pool), the largest Roman public cistern in Naples with a capacity of 12,600 meters^3 (3.3 million gallons^3), a product of Roman innovation in hydraulic engineering. The Romans had discovered a mortared rubble process that made possible barrel and cross vaults that could cover greater areas. Thick walls and external reinforcements helped mitigate the pressure of the water, facilitating ground-level installations (they no longer needed to be subterranean), which in turn aided the distribution of water by gravity. Other large, multi-chambered cisterns are known in Spain and Asia Minor, but especially in northern Africa.[20] Vercovicium (Housesteads Fort on Hadrian's Britannic Wall), where six large tanks have survived, was probably watered by harvesting rainfall. Situated on a high ridge, the site is not easily accessible by conventional aqueducts (for which there is no evidence).[21]

Cisterns were often built in pairs, and the water would be "purified" with strainers that filter out contaminants, as recommended by Pliny (36.173; Chapter 2). The people of Arados in Syria obtained their drinking water both from rainwater and a freshwater spring at the bottom of the sea, which they pumped up with a complex apparatus of funnels, pipes, and bellows.[22] The water within the cisterns was protected from seepage and further contamination by a waterproof mortar. Greek waterproof mortar has been little studied. The Carthaginians preferred ash and gravel. The Romans used pink cement with crushed terracotta.[23] Pliny (36.173) recorded an alternate recipe: five parts clean coarse sand, two parts the hottest possible quicklime, and silex fragments weighing no more than a pound each. Rainwater from cisterns might be augmented by water collected in *impluvia* (pools) in the atria or peristyle gardens of Roman-style houses. Reinforcing earlier Hellenistic decrees, a Trajanic municipal decree at Roman Pergamon (second century CE, Asia Minor), moreover, emphasized the significance of domestic water collection. Private cisterns were registered in the public record, and the duties of magistrates included ensuring that such cisterns were well maintained.[24]

Aqueducts

Early Initiatives

Aqueducts, those great symbols of Roman imperial status and public benefaction, were hardly a Roman invention (Fig. 3.1). In Bronze Age Crete (ca. 2000 BCE), rudimentary aqueducts were constructed at several sites, including Knossos, Tylissos, and Malia. A combination of open channels and closed terracotta pipes, for example, conveyed water to the palace at Malia along an estimated length of ca. 2.4 km (1.5 miles). The Minoan aqueduct from the spring at Mavrokolybos to the palace at Knossos seems to have incorporated both descending and ascending channels (ca. 400 meters [ca. 1300 feet]).[25] At the Mycenaean site of Pylos, in the southwestern Peloponnese, water was conveyed from a spring via a crude 1 km-long channel (0.6 miles: partly wooden, partly stone-lined) near the palace.[26]

The Assyrians developed similar technology. Assurnarsipal II (884–859 BCE) commissioned a canal—led through a tunnel underneath a rock ridge—from the upper Zab River, in order to supply the city of Nimrud with water and irrigate its fields.[27] Assurnarsipal's canal was not unique, and many Assyrian waterworks incorporated tunnels excavated between vertical shafts. This may have inspired the development of qanats, a widespread method of carrying water from underground containers by gravity through tunnels to a lower level, sometimes miles from the source.[28]

Greek Initiatives (Sixth to Second Centuries BCE)

Greek aquifers of the Archaic era were often constructed for political advancement. The tyrant, Theagenes (seventh century BCE), commissioned fountain houses for the springs at Mégara to curry popular favor (Pausanias 1.40.1). Also wishing to gain popular support, the tyrant, Polycrates of Samos (ca. 550–522 BCE), authorized an aqueduct that extended 1.5 miles (2.5 km) through a 0.6-mile (1 km) rockhewn tunnel, known eponymously after its

Fig. 3.1 Scheme of an ideal Roman aqueduct, after Connolly and Dodge 1998: 131.

engineer, Eupalinos of Mégara (Herodotus 3.60). It was a remarkable feat of engineering. Additionally, the Peisistratids (546–510 BCE) in Athens underwrote an aqueduct that conveyed water via clay pipes and underground tunnels for nearly 5 miles (8 km) from the Ilissos Valley to public fountains in Athens. Water would be routed via underground terracotta pipes and diverted by gravity throughout the city to reservoirs and public fountain houses equipped with draw basins and waterspouts. In Athens, there is also evidence that pipes conveyed water into private homes throughout the city, through a system that was operational for centuries, indicating a systematic program of maintenance and repair.[29]

Although the arch was known to the Greeks, they did not employ it in their hydraulic engineering. Their aqueducts instead generally followed topographical contours, as water was funneled through pipes (masonry, terracotta, bronze, or lead), open channels, or rock-hewn tunnels. To increase the capacity of long-distance systems, water would flow through two or three large-gauge pipes (15–25 cm [6–10 inches]) that may not have run under full pressure. Such aqueducts supplied municipal fountain houses, which in turn provided potable water, beautified cities, and served as symbols of public benefaction. This new technology enabled a constant supply that could sustain increasing urban populations or support irrigation for the greater production of food crops.

The challenge of transporting water across deep valleys was resolved by the use of inverted siphons (pipes conveying water flowing in a gravity-operated installation underneath valleys or other depressions) (Fig. 3.2). Siphoning is an ancient method, attested in Egyptian reliefs from about 1500 BCE, but its application to hydraulic engineering first occurs in the sixth to fourth centuries at Olynthos (Macedonia), where a 5-mile (8 km) long pipeline was enhanced with a small inverted siphon (33 feet [10 meters] deep). The earliest large-scale siphon systems were constructed at Pergamon in western Turkey, where five pipelines supplied the city. The Madra Dağ aqueduct (second century BCE) dispatched water 26 miles (42 km) from the Madra Dağ hill to a basin on top of the city's citadel. Before reaching the citadel, water was siphoned through lead

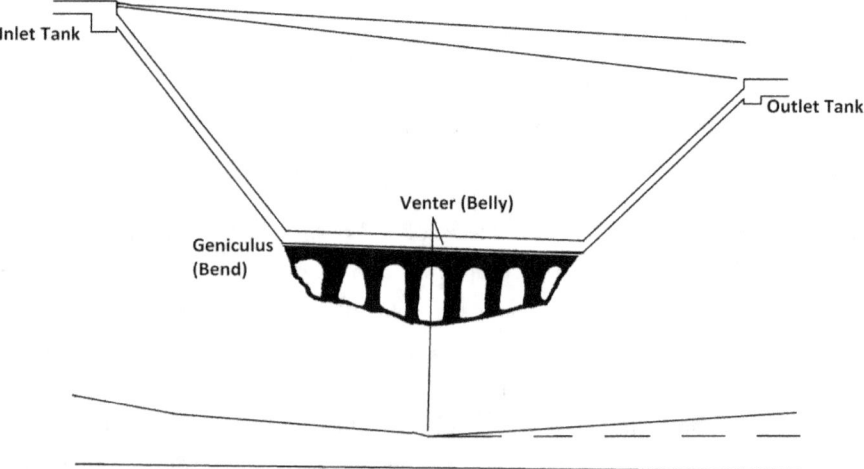

Fig. 3.2 Inverted Siphon, after Hodge 1992: 148 #102.

pipes (clay would not have been able to withstand the pressure) across a 650-foot deep (200 meters), two-mile long (3.5 km) depression.

Roman Initiatives

The Roman aqueduct system was widely praised in antiquity. The Greek historian, Strabo (64 BCE–24 CE), for example, recognized hydraulic engineering as an area neglected by the Greeks:

> The Hellenes succeeded especially well with their foundations, aiming at beauty, strength, harbors, and fertile land, but the former (the Romans) had particular foresight in what the latter (the Greeks) paid no attention to, such as the laying down of roads, the introduction of water, and sewers that are able to wash away the refuse of the city into the Tiber.
>
> 5.3.8

Pliny emphasized the ubiquity of the water supply in Rome (36.121–123), and the remarkable engineering accomplishment that the aqueducts represented: great blocks of stone were dragged along the streets above, yet the tunnels would not collapse (36.106). Frontinus proudly declared the aqueducts as a uniquely Roman accomplishment:

> May you compare the clearly idle pyramids or other works of the Greeks—sluggish but celebrated with repute—with these numerous and indispensable monuments of so much water.
>
> *Aqueducts* 1.16

Roman hydraulic engineering, nonetheless, was shaped and inspired by a long history. Vitruvius drew on Greek sources for his eighth book (on water), including a source from Pergamon.[30] The Pergamene aqueducts, although post-dating the earliest Roman ones, may have inspired the *Aqua Marcia*, the first fullscale aqueduct in Rome. Also influencing Roman initiatives in hydraulic engineering were the neighboring Etruscans, who had developed a network of subterranean drains for managing excess moisture along their westfacing coast, by diverting streams or draining swampy valleys. Roman innovations synthesized and advanced the existing technology (e.g., arches, concrete, waterproof linings, settling tanks) thus resulting in a wider, more reliable distribution.[31]

We are fortunate to have a handbook on the Roman aqueduct system, penned by Sextus Julius Frontinus, whom Nerva (ruled 96–98 CE) appointed as *curator aquarum* ("supervisor of the water-supply") in 97 CE. A distinguished Roman statesman, Frontinus served ably as governor of Britain (75–77 CE), and he also wrote on military strategy in four books (*Stratagems*). To regularize the maintenance of hydraulic infrastructure during his term as *curator aquarum*, Frontinus consulted technical reports, official plans, and senatorial decrees. This research resulted in the two-book *On the Aqueducts of the City of Rome*, a model of Roman efficiency and practicality. This handbook is rich with details on the history of the aqueduct system, its administration, and the water supply. Frontinus is an invaluable source for hydraulic legislation, water usage, fraud, and theft, as well as for aqueduct maintenance, which included regular cleaning (owing to sintering: the buildup of calcium carbonate) by a band of state-owned slaves.

Eleven aqueducts serviced Rome, and Frontinus described in detail their sources and the quality and quantity of their waters. The *Aqua Appia* was the earliest (312 BCE), followed by the *Aqua Anio Vetus* (272 BCE), both employing underground tunnels. Built in 144–140 BCE, the *Aqua Marcia* was underwritten by spoils from campaigns against Carthage and Corinth (146 BCE). Renowned for the purity of its water, the system incorporated both subterranean and elevated features. The longest of the eleven aqueducts that served the city, extending nearly 57 miles (about 90 km) from the Anio valley (northeast of Rome) to the Capitoline Hill, the *Aqua Marcia* supplied nearly 50 million gallons of water a day in the late-first century CE. Completed in 52 CE, the *Aqua Claudia* runs about 35 miles (55 km), but only the last 5 miles or so (8 km) are raised on arches. Altogether, the eleven aqueducts in Rome carried into the city an estimated daily capacity of up to 1 million cubic meters (a little over a quarter of a million gallons).[32]

Roman Provincial Initiatives

Throughout the Mediterranean and beyond (Britain and Germany, for example), aqueducts were often built by occupying Romans as tokens of imperial or private benefaction. Caesar's aqueduct in Antioch-on-the-Orontes (near Antakya, Turkey) may have been the first Roman hydraulic structure outside Italy.[33] Under Augustus and his successors, aqueducts on the Roman model were often constructed in support of Roman veteran colonies. There is also evidence of regional water planning: the 60-mile (95 km) Serino aqueduct supplied several towns in Campania.[34]

Most aqueducts were limited to a height of 70 feet (20 meters), but massive pillars, narrow arches, and double tiers were constructed to ensure stability at greater heights where elevation was mandated. The spectacular double-arched aqueduct at Segovia in Spain (first century CE?), one of the best-preserved Roman aqueducts, still brings water into the city. Because of its importance as a source of water and a civic monument, it has been restored several times. In 1929, modern concrete pipes completely replaced the Roman channel,[35] and in 2006, because of pollution and deterioration, the World Monuments Fund listed it in the World Monuments Watch. Another well-preserved Roman aqueduct, the Pont du Gard at Nîmes (ca. 40–60 CE), towering 180 feet (55 meters) above the river—the tallest surviving Roman aqueduct—features an additional tier of smaller arches on top.

Operation

Aqueducts operate on the principle of gravity. Most of an aqueduct's course is subterranean, leading water through vaulted or roofed masonry, or concrete channels that follow the terrain while avoiding destructive tree roots or intense agricultural activity. Tunnels would be employed to rout these channels through hills, and arcaded walls would support shortcut channels over valleys. In Nîmes, stone slabs above the channel (4.5 feet wide [1.3 meters], 5.5 feet deep [1.6 meters]) protect the water from sun and pollution. In Lyon, water was ferried through nine (or eleven) pipes, by means of a combination of inverted siphons and gravity channels (first century CE?). In the interest of economy and stability, inverted siphons would be employed for crossing valleys with arcaded bridges 165 feet high (50 meters) or more.[36] Near Lindum (Lincoln), in Britain, archaeological evidence suggests that a closed-pipe system conveyed water from a perennial spring into a "rising main," which then supplied the town about 1.5 miles away (2.5 km). It is possible that a manually operated force pump, of which no trace has been found, elevated water through the rising main. Engineers tried to maintain a constant gradient (about one finger's breadth for every 100 feet [30.5 meters]). Dropshafts (vertical or inclined cascading shafts) were employed where an aqueduct dropped quickly, in order to connect higher channels to older ones.[37]

Water was usually routed from open channels to closed pipes by means of headers and distribution tanks. From these distribution reservoirs, water was then circulated through an extensive system of lead pipes underneath the city streets (pipe gauges and lengths were standardized by the late-first century BCE). Pipes, moreover, were constructed to fit tightly one within the next, wider at one end and narrower at the other, creating an internal flow that helped keep the pipes clear and the water flowing. Timber-pipe aqueducts are known in northwestern Europe (where the use of wood was altogether more common), and terracotta pipes might be used in aqueducts servicing small, rural estates.[38] The subterranean distance varied according to site. In Archaic Athens, pipes were laid about 3.5 meters (11.5 feet) underground. In Olynthos in central Macedonia, the pipes were found at a depth of 3–6 meters (10–20 feet).[39]

The earliest distribution tank that has been studied (ca. 80 BCE) is from Pompeii. In its initial phase, it was an open tub into which the aqueduct debouched and from which three water mains supplied different sections of the city. The cistern was later roofed, and the mains were fitted with filters. The water mains then routed water into a dozen towers that supported satellite tanks, thus mitigating the water pressure flowing into buildings at the lowest parts of the city. Such "intermediate pressure-towers" are also found in Herculaneum but are elsewhere unattested.[40] Water supply into private homes could be controlled with a quarter turn of a stopcock (a bronze valve fitted into a bronze cylinder attached to the lead water pipe). Despite Vitruvius' principle that separate mains should supply different categories of customers (fountains, baths, private houses: 8.6.1–2), in practice, water distribution was not so stringently regulated.

Since the water flow was continuous, the supply could not be stopped. It was, however, possible to reroute the flow by taps and valves, and workers could rig temporary bypasses to conduct repairs on short sections (Frontinus, *Aqueducts* 2.124–125). Water was transferred into public and private buildings by branch pipes that radiated from main lines at intervals. Fees were charged roughly according to the diameter of the branch pipes, officially stamped, and regularly inspected. Most homes lacked running water, and householders continued to collect water at fountains or through their *impluvia*.

Unscrupulous citizens could (and did) hack into the main system (Chapter 1). Earthenware pipes were easily made in molds, sealed with putty, and then cheaply installed by unskilled laborers.[41] Cylindrical or triangular lead pipes were shaped on wooden models from rectangular sheets. Lead was more expensive, and its installation required skilled laborers. Lead pipes, however, were preferred for pressurized systems as being more reliable, despite concerns over the contamination of the drinking supply (Chapter 2).

Private Aqueducts

Finally, not all aqueducts were public. Quintus Cicero's "Baebulian" farm lost its water supply, but Quintus (102–43 BCE) was able to obtain a servitude for an aqueduct (the legal right to use waters owned by someone else).[42] Quintus' famous brother, the orator-statesman, Marcus Cicero, agreeing that Quintus should keep the property, secured an estimate for building the aqueduct (3 *denarii* a foot; the aqueduct would have extended 3,000 paces or 3 Roman miles: 2.7 miles/4.4 km).[43] A private aqueduct was erected at Pont d'Ael, Aymavilles, in Transpadana. The aqueduct bridge was inscribed *privatum* (private) by Gaius Avillius Caimus Patavinus in 23 BCE.[44] Additionally, a lengthy inscription records the private ownership of an aqueduct at Viterbo in Etruria, granted according to a senatorial decree. Mummius Niger Valerius Vegetus (suffect consul,[45] April to July 112 CE) purchased a spring from which he routed water via a private aqueduct that extended 6 miles (about 16 km) across eleven private farms as well as public roads and rights of way. Vegetus purchased the right of access from nine private owners (the decree lists the estates and their owners). Constructed with a series of arches and pipes, the aqueduct varied in width: 10 feet for the arches (ca. 3 meters); 6 feet for the pipes (ca. 2 meters). Precisely specifying the course of the aqueduct, the inscription is a visible, public record of the installation's ownership.[46]

The Roman aqueduct system was the ultimate expression of human control over the ancient watery environment. Human-built installations conveyed water miles across variable terrain to feed thirsty crops, to provide for the legitimate watery needs of the people, and to supply water to luxury installations for the comfort of the masses and indulgence of the elite.

Fountains

Public and private fountains adorned cities and grand estates. They were the gifts of imperial beneficence and symbols of authority. They were, moreover, celebrated by poets as meeting places, providing refreshing, pure drinking water and repose for travelers.[47] They were often elaborately decorated with images of water gods, including nymphs, Dionysus, Aphrodite, and some fountains were even equipped with porticos and benches.

After long years trying to return home from the Trojan War, Odysseus lands in Scheria, an island paradise whose inhabitants are renowned for their hospitality and idyllic, labor-free lifestyle. The gardens of King Alcinous' Phaeacian palace are all the more impressive for their irrigating springs and jetting fountain, operating probably by means of an open aqueduct.[48] Alcinous' fountains are likely a remnant of the cylindrical shaft step fountains that are attested at Late Minoan sites in

Crete, including the palaces at Knossos and Kato Zakro (1450–1100 BCE). These shafts, sunk into springs, provided a basin from which to draw water, and they were often equipped with steps to facilitate obtaining water as the levels changed.[49] Similar installations are noted at the mainland Mycenaean sites of Tiryns and Mycenae (1400–1200 BCE), adapted with tunnels that drew water from subterranean sources. The Mycenaean fountain in Athens (thirteenth century) was constructed to facilitate access to a natural reservoir (originally accessible only at the top through a narrow slit in the rock: 1–2.5 meters wide, 18 high, 35 long [3.2–8.2 × 59 × 115 feet]). A subterranean fountain with a stepped descent debouched into a collecting basin.[50]

Many urban residents collected their daily water from centralized fountains or spring houses, illustrated in Greek art from the Archaic period onward (from 560–460 BCE, more than 150 vases show fountain houses). On a famous black-figure Attic hydria (520 BCE), five elegantly dressed women at a Doric fountain house fill their jars with water streaming from theriomorphic spouts (four spouts are lion headed, one is donkey headed). One woman carries her hydria on her head.[51] The water flow is continuous. As on the Attic hydria, spring houses were equipped with waterspouts and roofs to protect the water from pollutants and the sun's heat, and they were erected near springs, usually just outside of the town or in prominent public spaces. Cylindrical shafts were sunk for tapping the water sources. Because of the width of the basins, water was more likely drawn than collected directly from the nozzles.[52]

Mégara

Many spring houses were underwritten by Greek tyrants in order to strengthen political standing, mitigate opponents, or demonstrate goodwill to the community.[53] The tyrant, Theagenes (sixth century BCE), is credited with underwriting the impressive fountain house at Mégara, but scholarly consensus dates construction to the early-fifth century. The spring there overflowed into two large, plastered reservoirs (260 meters^2 [ca. 2800 feet^2]) equipped with conduits and narrow basins for distribution. Up to 15 cm (ca. 6 inches), the basin was also lined with a mixture of asphalt and animal fat, applied warm, and highly polished, effecting an impermeable layer that protected the basins from the inevitable sintering evident in other hydraulic installations, a remarkable accomplishment but a unique example.[54] Also in evidence at the Mégara fountain house are traces of bronze mechanisms to control the flow. The combined capacity was 380 cubic meters (nearly 100,400 gallons) which could be filled in seven hours at the rate of 15 liters per second (4 gallons per second). The redundancy (the double basins) ensured a reliable supply in case of maintenance or water shortages.[55]

Athens

Christaki and Nastos (2018) have surveyed seventy-two fountains in Athens (into the Ottoman era). Fountain houses were a significant component of the water supply in Athens, whose hydraulic infrastructure was not altogether reliable. Peisistratos (d. 527 BCE) has been credited with the elaborate Enneakrounos fountain house (cf. Thucydides 2.15.5), but new analysis suggests that it was built during the first half of the fifth century BCE.[56] The fountain house features a portico of Doric columns and nine fountains, fitted with lion taps from which flowed cool, clean water for ritual uses (e.g., ritual cleansing before weddings). A larger, fourth-century fountain house was constructed in the Agora's southwest point in response to escalating demand. Cultically important was the fourth- century fountain house at the Dipylon ("Double Gate"). From the Dipylon Gate, initiates into the mystery rites of Demeter and Kore would begin their 22 km (14 mile) procession from Athens to Eleusis, and this would be their last water source until they reached the Kallichoron Well in Eleusis. The Dipylon fountain featured an L-shaped basin attached to the monumental gate. As with most fountain houses, users would have ladled water from the wide basin instead of collecting it directly from the spouts that would have been quite a distance from the edge. A separate shaft well in the Gate's courtyard provided water for livestock.[57] At the east/south-east side of the Pnyx hill, the Pnyx fountain was part of Peisistratos' initiative to improve the water supply. It was converted to a reservoir by the second century CE, with a square chamber with a mosaic floor that was approached by a narrow, stepped corridor. Water flowed from niches at the back of the chamber. The installations, thus, were not static, but were adapted as technology and aesthetics changed, becoming more complex as the population grew, increasing access and even distributing water into private homes.

Hellenistic Innovations

Earlier fountain houses took many layouts, from rectangular to L-shaped. From the third century onwards, square floor plans became predominant, as for example, two well-preserved porostone fountain houses in Sikyon (northern Peloponnese), built into terraces and featuring half columns that divide the basin's wall. The porostone is faced with smooth stucco (white and blue). Doric columns frame the entrance and the back wall, onto which the waterspouts were fitted, features a niche that may have displayed a small statue (it was too far removed for fountain users to offer votive gifts). A roof is likely, but does not survive.[58] Hellenistic-built communal fountain houses were also smaller, in response to increasing reliance on private water sources, but they remained an important indication of prosperity and munificence.

Free-standing Waterspouts

Water was also conveyed by pipes through open-air waterspouts attached to pillars or walls. A seventh-century-BCE lion spout (perhaps one of a pair) survived in Olympia in the northwestern Peloponnese. One Etruscan tomb fresco shows tail to tail lion waterspouts, one is debouching water into a basin (Tomb of the Bulls, 540–520 BCE). At about the same time, Apollo's sanctuary in Delphi also received a statue of a boy with a waterspout fitted into his hand, a gift of Sparta (Herodotus 1.51.5), a model that might have been duplicated at other sanctuaries, such as Asklepius' healing sanctuary at Epidaurus where a fragment of feeder pipe is preserved in a statue base.[59]

This is a type that is especially prevalent in Pompeii. Forty or so roofless street fountains debouched into stone basins, overflowing into the streets. Any home in Pompeii would have been within 50 meters (165 feet) of such a street fountain, each serving about 160 residents.[60] In Rome, nearly 600 water basins and thirty-nine ornamental fountains were fed by the aqueducts (together with seventy-five public structures), providing a convenient water source to the million or so residents of the city (Frontinus, *Aqueducts* 2.78). A great many of these fountains were commissioned, "with extraordinary mindfulness" by Marcus Vipsanius Agrippa (63–12 BCE), Augustus' friend and lieutenant (Frontinus, *Aqueducts* 1.9).

Monumental Fountains

The utilitarian fountain quickly became high art. Elaborate fountains adorned temple precincts and private estates. At Samothrace, the elegant statue of Nike (Victory) formed part of a fountain as a victory monument: water descended in front of the ship's bow, churning "up sea spray" into a bay of stone blocks that demarcated the fountain precinct and created the illusion of a ship at sea.[61] A poem preserved on papyrus describes a colorful, marble statue group (that included Queen Arsinoë II [316–270 BCE]) on a semi-circular base that also held the basin into which the water flowed. The Nike group featured several innovations that would be employed in Roman monumental fountains, including semi-circular designs and the use of colored marble.[62]

The Greek Hellenistic model was adapted into the exedra-nymphaeum at Rome (fountains with recessed seating areas), a form that was exploited by emperors usually for their private benefit. Among these are the nymphaeum of Nero (ruled 54–68), attached to his *Domus Aureum* ("Golden House"). At the palace of Domitian (ruled 81–96) in the Alban Hills outside Rome, the rectangular Doric barrel-vaulted ceiling nymphaeum (8 meters high [26 feet]) once sported large and small waterfalls fueled by cisterns and pipes installed behind the back wall and supplied by an aqueduct. Hadrian's (ruled 117–138) pied-a-terre outside Rome, Tivoli, featured many baths and fountains, including a nymphaeum with a barrel-vaulted ceiling. According to Ammianus Marcellinus, the fourth-century-CE historian from Antioch, Marcus Aurelius (ruled 161–180) was responsible for an ostentatious nymphaeum, which was a popular meeting place in the city (15.7.3).

One of the best known and best preserved Roman-era exedra-nymphaea was commissioned in Olympia by Herodes Atticus (101–177 CE), the first Greek to serve as an "ordinary" consul in Rome (143; see footnote 45). The nymphaeum features a semicircular colonnade with statues of imperial family members, including Antoninus Pius and Hadrian, situated in eleven niches.[63] Water cascades from an aqueduct through spouts into a semicircular pool, and then into a rectangular basin, and finally down to a lower rectangular pool, 100 feet long (ca. 30 meters), fitted with 83 outlet pipes.

Such installations were particularly welcome in water-poor areas, e.g., the Severan-era (193–235 CE) Nymphaeum of the Tritons in Hierapolis in Anatolia, and they were potent symbols of imperial munificence.[64] They were intended as shows of imperial success and power as (Hellenistic potentates and) Roman emperors "could dedicate aqueducts, fountains, and bath complexes, in his name and for the Roman people."[65] The benefactor thus is responsible not only for the monument but also for the life-giving water.

Baths and Showers

Water-intensive services, such as public fountains, baths, showers, and public latrines, rose in popularity in the first century CE, but such installations are ancient. What appears to be a bathtub or washbasin has been excavated at Knossos in the Queen's Megaron (1700–1400 BCE) and at the Mycenaean site of Pylos in the Palace of Nestor (1300 BCE). The 5-foot long (1.5 meter), tapered tub from Knossos has no outlet and would have been filled and emptied by hand into an outlet in the floor that connected with the main drain (see below). From about the sixth century BCE, bathing facilities were built near athletic compounds, and they utilized basins, according to evidence from vases. For example, one lost vase (fifth century BCE?) shows athletes around a washbasin inscribed ΔΗΜΟΣΙΑ (PUBLIC)[66] (Fig. 3.3). Another lost krater shows three nude women bathing around a basin as two servants hold oils and perfume.[67] Nude women are also shown showering on an archaic vase[68] (Fig. 3.4). By the second century CE, Plutarch mentioned both private and public bathing establishments for which there might be a charge (*Alexander* 76), as at Andania in the Peloponnese where the fee was 2 *chalkoi* (about a quarter of an *obol*: a tiny amount of money).[69]

Baths were popular throughout the Mediterranean, and they served as important community centers where people could socialize, bathe, read, exercise, or engage in games of chance. Vitruvius (5.10) described the ideal design, location, and construction of bath complexes: the site should face south or west (a deployment that also helps in heating the rooms); hypocaust floors (raised on an array of brick pillars under which heat was distributed) should slope towards the furnace; the pillars should be 2 feet high (0.6 meters) for the maximum

Fig. 3.3 Athletes at a public washbasin: Hamilton and Tischbein 1791: 1.plate 58.

Fig. 3.4 Women showering at a fountain house. Panofka 1843: 9, 18.

circulation of heat. Concrete construction was preferred. The sweat-room should be circular and adjacent to the warm room for efficient, even diffusion of heat, and the size of the suites should be commensurate with the size of the crowds who would use them. Rome had nearly a 1,000 public bath houses, and bathers enjoyed a series of communal cold pools, warm plunge pools, hot rooms, and saunas that were heated by nearby furnaces channeling heat from beneath raised hypocaust floors.[70] Aqueducts furnished water to the larger bath houses in Rome and other urban centers, which also maintained cisterns in reserve.[71]

In Rome the bathing habit was criticized by moralists as endemic of an overly indulgent lifestyle. Nero's tutor and speech writer, Seneca the Younger, for example, singled out the dark, old-fashioned private baths in Scipio Africanus' (235–183 BCE) villa as preferable to the bright, comfortable modern baths of his day (*Epistle* 86.8–12). Nero was particularly guilty of extravagance in all aspects of his life. According to Suetonius, the second-century-CE imperial biographer, Nero's *Domus Aureum* was equipped with sprinklers concealed in the ceiling to spray guests with a shower of perfume, and the baths were fitted with taps for both sea water and sulfur water (*Nero* 31).

Public baths featured elaborate artwork, such as the elegant Antonine-era floral and geometric pavement from the Roman baths at Blackfriars, Leicester, or the stunning Orpheus mosaic in the private bath suite at the Littlecote Roman Villa in Wiltshire. In *Hippias, Or the Bath*, the second-century-CE Greek satirist, Lucian, described an opulent, brightly-lit contemporary Roman bath house designed by Hippias with its attractive relaxation rooms, marble plunge pools, a gently warmed and oblong *tepidarium* (warm room) with apses at either side, bright massage hall, and a beautiful *caldarium* (hot room) faced with Numidian marble. In sum, Hippias' bathhouse is useful, convenient, well-proportioned, safe, and beautiful, equipped with two latrines, several exits, a water clock, and a sundial. The rich decorations and vaulted rooms were no doubt impressive, but Lucian eschewed technical analysis. Small bathing establishments were supplied with water pumped from wells by water-lifting devices.

Aquae Sulis

The famous installation at Aquae Sulis (Bath, England) provides an interesting case study.[72] Of Roman design and probably north Gallic workmanship, the sanctuary included a temple to the Romano-Celtic syncretized Sulis Minerva, baths, and, possibly, a theater, all built in masonry and richly decorated with carved reliefs. The temple complex underwent several phases of redesign to keep pace with changing aesthetics (construction was begun in the 60s CE). A circular, Greek-style *tholos* temple was added under Hadrian, a renowned Grecophile, possibly in honor of his visit to the province: common in the east, the *tholos* temple was exotic in the west. Ca. 300 CE, a new retaining-wall was erected, and a new façade incorporated a flight of steps between two flanking *sacella* (shrines), thus making the sanctuary less Roman and more Romano-Celtic in design, reflecting perhaps a cultural shift in the balance between Roman and Celtic sensibilities. In late antiquity, the open spring was completely enclosed with a vaulted timber roof, and the entire complex was re-roofed. The site's east-west axis was interrupted at the southeast corner where the spring rose, and natural fissures were surrounded by a lead-lined wall to form a watertight reservoir.

The architects were challenged with creating a visually appealing pool that was supplied by a sufficient quantity of water to feed the baths. To avoid clogging the installation with the black sand that bubbles up with the thermal water, the engineers installed a slot in the east wall that could be closed by a moveable sluice, joined to the head of a substantial drain. When closed, the water level grew high enough to feed the baths. When too much sand had accumulated, the sluice was opened, and water flushed out the detritus. The drain was also equipped with regularly-spaced manholes, allowing workers to shovel out impediments. The baths would re-open when the reservoir filled up again (two to three days). The culvert at Aquae Sulis still carries water from the spring and the Great Bath to the River Avon along a channel lined with wooden boards (*in situ*) that drains from the eastern sector of the baths.

Latrines

Another challenge of civilized life is dealing with human waste while minimizing the spread of disease (not to mention mitigating unpleasant aromas). In the late-fourth millennium BCE, the Mesopotamians developed non-flushing pit toilets, about 4.5 meters deep (ca. 15 feet), lined with ceramic cylinders (about 1 meter in diameter) over which users would squat and into which the excrement would remain. Such installations may have been reserved exclusively for liquid waste.[73] Among the technological marvels at the palace at Knossos were screened water closets fitted to conduits that flushed waste into the sewer system.[74] Such installations were only for the elite, and the technology was lost until revived by the Romans.

By the fourth century BCE, the standard latrine type seems to be the cesspit (κοπρών: *kopron*: several have been excavated in Athens). Earthenware, anatomical lavatory seats, as found in Olynthos (northwestern Greece), may have been used over the cesspits. Despite being attested in ancient texts, there are few excavated latrines, and chamber pots may have been a common method of dealing with excrement.[75] Nonetheless, both public and private latrines seem to follow a similar paradigm of marble or limestone benches with defecation holes. Public facilities were communal, accommodating tens of users at a time; some private latrines also had multiple openings, suggesting simultaneous communal use within a household (by gender?).[76] Spacing of defecation openings varies according to population density: 51 cm (20 inches) separated users of the public latrines in the Roman Agora in Athens, but users of the public lavatory in the Gymnasium of Minoa in Amorgos (one of the islands in the Cyclades) enjoyed 85 cm of distance (33 inches).[77] The floor of the Roman Agora latrine in Athens was also equipped with drainage holes for urine.[78] Many public latrines also featured small channels with continuously flowing water, perhaps for cleaning σπογγιά (*spongia*), the sponges on sticks that served as toilet paper.[79]

The Roman model resembled the Greek public installations, with defecation holes along communal benches. Rome boasted 144 public latrines, including those attached to shops in the Julian Forum. The Theater of Pompey featured a

100-seat "common bench" latrine of open design. Users had no privacy in these unisex, stall-less installations (one of the best preserved of the type is at Housesteads). Public latrines were constantly flushed with waste water from the baths. Private, single latrines, attached to houses, were most commonly established near the kitchens (for the added convenience of discarding food scraps), but were not usually attached to the sewer system, thus increasing the potential for the spread of disease in a confined environment (see *CWW*: Chapter 5).

Drainage and Sewers

The constant influx of water demands drainage, a perennial urban concern. In the Near East, sloped drains and channels of ceramic pipe, stone, or baked brick conveyed drinking water into settlements and rainwater out. Fountains were also fitted with terracotta overflow pipes, and soakaway drains removed grey water (wastewater from households that was generally free of fecal material). Mesopotamian pit toilets were also fitted with drainage tubes.[80] A sophisticated human sanitation system is in evidence in Knossos, where rainwater collected from roofs was flushed out the palace latrines into a sewer system comprised of limestone and earthenware pipes that were sunk underneath passageways in the palace.[81]

The drainage system in Athens has been ably surveyed by Chiotis and Chioti (2014). In the late-sixth or early-fifth century BCE, a stream at the Agora's western side was converted into a storm drain that drained the swampy ground and allowed for urban development. This original storm drain eventually developed into the "Great Drainage Canal" (1 square meter [about 10 feet square]), which was tiled with polygonal-shaped stone blocks and outfitted with manholes at regular intervals. The drain, covered with stone slabs that served as a street, conducted wastewater from public buildings via terracotta pipes to the Eridanos River. By the fourth century, two additional branches increased the area from which rain and wastewater were conveyed. Despite the infrastructure, even in Aristotle's day (*Athenian Constitution* 50.1), grey water was often poured onto the streets.[82] The Great Drain continues to serve its original function, draining water from the streets of modern Athens.

A feature of planned urban communities in the Roman world, integrated networks—frequently constructed of masonry—linked private drains with municipal drainage systems.[83] Grates allowed rainwater to discharge through the drainage system, and sewers received overflow from public fountains, but solid waste would accumulate, and the street drains would then be cleaned by public slaves.[84] In Rome, gutters and a sewer system (*cloaca*) beneath city streets carried overflow from baths and latrines. The *Cloaca Maxima* ("Greatest Sewer") in Rome was originally an open ditch, to which a vault was added around 100 BCE. Drainage pipes and subterranean channels extended nearly 3,000 feet long (900 meters), 14 feet high (about 4 meters), and 10.5 feet wide (about 3 meters), large enough to accommodate a small boat for Agrippa's famous inspection tour of 33 BCE (Pliny 36.104). These channels carried effluent from baths, overflow from streets, and sewage and waste from houses and streets. The *Cloaca* was flushed by overflow from public fountains into the Tiber, causing the sewer to back up when the river flooded. Few houses were connected directly to sewers. Sophisticated sewage systems were also in place in Ravenna, Colonia Claudia Ara Agrippinensium (Cologne), Timgad, Caesarea Maritima, and many Roman legionary fortresses (e.g., Eboracum [York], Novaesium [Neuss], and Vindobona [Vienna]).[85]

Initiatives to control drainage and the removal of dirty water were deliberate and mindful, organized at state level. In this way, technology was employed to control the environment by removing dirty water in order to mitigate the spread of disease and to ensure a "clean" and pleasant urban environment.

Wetland Reclamation

Wetland reclamation projects are also in evidence. In Bronze Age Boeotia, the Kopais Basin had been drained by diverting surface waters into limestone sinkholes. By Alexander's day, the system was failing, and Crates was commissioned to re-drain the Kopais (Strabo 9.2.18). Crates was never able to complete the work, but he had probably planned a shaft-and-gallery drainage tunnel, on the model of the aqueduct at Samos. There was also a proposal to drain the swampy Ptechae Lake at Eretria in Euboea (318 BCE). The contract between the city and the engineer, Chaerephanes, called for channels to collect surface water into a cistern that could then be siphoned through an outlet for drainage into the sea. A sluice gate allowed water to be reserved for irrigating fields (*IG* 9.9.19.1). The route of the drainage system was prescribed, and Chaerephanes was instructed to purchase the land required for the tunnel's outlet shafts.

In central Italy in the first century CE, Claudius commissioned a drainage tunnel through the mountains to reclaim the rich land under the Fucine Lake for agriculture. Claudius aimed to succeed where Julius Caesar and Augustus had failed. Caesar had hoped to drain the lake (or at least control it: Suetonius, *Caesar* 44.3), and Augustus refused to pursue the costly project despite requests from local residents (Suetonius, *Claudius* 33.1). Draining the Fucine was among Claudius' "most memorable accomplishments," according to Pliny (36.124–125; cf., Suetonius, *Claudius* 20.1-2; Dio 60.11.5). Inaugurated in 52 CE and requiring eleven years and the labor of 30,000 men for construction, the enterprise, however, was fraught. One tunnel had collapsed, and Claudius' freedman, Tiberius Claudius Narcissus, was apparently embezzling funds from the job. Hadrian later widened and deepened the tunnel (Dio 61.33.5). The lake was not completely drained until 1877. In Britain, nonetheless, wetlands were reclaimed in the Severn Estuary (second century CE) in Gloucestershire (where levies were raised and land parcels were walled off), and the East Anglican fens were partly drained.[86]

Not all efforts were successful. In a disastrous vanity project, Ariarathes (V?) of Cappadocia (second century BCE) attempted to manipulate the local terrain where the Melas River flowed into the Euphrates. The act of blocking the Melas River resulted in a lake in which Ariarathes "isolated certain islets as if they were the Cyclades Islands, and he created youthful amusements for himself." The infrastructure failed, and the water broke through "carrying away settlements and obliterating numerous fields, ruining no small part of the territory of the Galatians in Phrygia." Ariarathes was fined 300 talents (a great deal of money!). In Cilicia, the town of Herpa was also held responsible for damages when their efforts to block the Mallos River failed, resulting in destruction of the littoral plain (Strabo 12.2.8).

Waterways were manipulated for many purposes. Streams were dammed to control flooding, irrigate fields, power mills, and replenish municipal water supplies. Plato had advised dikes and channels for controlling run-off irrigation for his ideal state (*Laws* 6.761a–b). In northern Africa, extant cisterns and large-scale masonry dams attest efforts to control water in arid regions where irrigation was essential for agriculture, such as the Roman-built cisterns at La Malga near Carthage (second century CE). Strabo referred to the irrigation canals in Mesopotamia that helped to prevent stagnation (16.1.9–10). The elder Pliny recounted the cutting of ditches for draining marshy farmland (18.47), and his nephew, Pliny the Younger, praised the Trajanic spillway that helped drain the Tiber during flooding (*Epistle* 8.17). In the Aurès Mountains in Algeria, underground wells were tapped to supply both cities and farms.

Industrial Water Uses

Water was also employed in light and heavy industry, assisting farmers in irrigating fields, powering mills, and facilitating mining.

Water-lifting Devices and Irrigation

Developed over time, water-lifting devices were employed for water collection and for powering installations that milled grain (and even kneaded dough). The shaduf (a counter-posied weighted lever) was a simple water-lifting device that could easily jam. The Aristotelian author of *Mechanical Problems* (28.857a–b) suggested that the device's functionality would improve with the addition of a lead weight to the shaduf pole. The bucket should then be lowered slowly and deliberately. Another simple and ancient water-lifting device was the water screw ("Archimedes" screw), more reliable but with limited lift. Although the technology is much earlier, its discovery was ascribed to Archimedes whose research on spirals may have inspired improvements to the device.[87] The user would twist the spiral screw with a crank in order to lift water from a stream into a bucket at ground level. Farmers may have stepped on the screw's treads when irrigatating their fields, but they would need to lean on sturdy supports to prevent themselves from slipping (Philo of Alexandria, *On the Confusion of Tongues* 38).

Irrigation was common, even in the well-watered climate of Italy where not every farm had a reliable water source. Cicero commended the Fufidian farm that his brother Quintus had recently purchased (*To Quintus* #21 [3.1.1]). The estate was well watered, and Cicero's friend, Lucius Caesius, believed that the springs there were sufficient for irrigating 50 *iugera* of meadowland (ca. 30 acres). In the late-first century BCE, Vergil, moreover, rhapsodized about the farmer irrigating his parched crops:

> and when the scalded field boils with dying crops, behold does (the farmer) entice water from the brow of the sloping channel? He stirs a hoarse murmur, falling across polished stones, and tempers waterless fields with gushes of water.
> *Georgics* 1.107–110

Especially in parched, water-poor areas, irrigation was the lifeline that provided vital water to feed thirsty, life-giving crops, a crisp example of the technological utilization of water to control and change the envirnoment for human benefit.

Hydraulic Machines

In the third century BCE, Ctesibius of Alexandria invented a force pump whereby water could be pushed up through a discharge tube by the action of a brace of pistons, each in its own cylinder, and each fitted with separate pump handles and discharge tubes. A one-way valve ensures that water travels in the desired direction.[88] The device proved useful as a fire extinguisher because of its ability to deliver a jet of water.[89] In fire-prone Rome, such a force pump was likely standard fire-brigade equipment, fitted with nozzles and carried on carts with water reservoirs.[90] Some ships were likewise equipped with force pumps to siphon out the bilge that inevitably accumulates in the incessantly leaking wooden hulls. Water screws also served as bilge pumps.[91] As high-volume trade increased from the Hellenistic period onward, and larger ships were constructed to accommodate the increasing volume of enterprise, bilge pumps became an essential feature of nautical engineering ensuring the efficiency and safety of ships, crew, and merchandise.[92]

Waterwheels and Water Mills

By about 50 BCE, water-driven wheels were in use. Bucket wheels, operable in restricted spaces, were likely powered by animals, some of whom set a limit to the amount of work they were willing to do in a day:

> For there the cows irrigate the king's park (at Susa) with buckets raised on waterwheels, of which the number is set. For each heifer raises up one hundred buckets each day. It is not possible to exact or force more, even if you want to. But the wranglers often try to increase the quota, but the cows stop and do not move after they have delivered the pre-arranged quantity. She thus has accurately calculated and memorized the count by heart.
>
> Plutarch, *Cleverness of Animals* 974e

The cows at Susa, apparently, knew how to count.

More sophisticated were the comparted "drum wheels" and bucket chains (with greater lift but limited by the weight of chains and pots). Suited to fast streams, undershot wheels (set into flowing water) have a high torque and low speed (determined by the force of the stream).[93] Overshot wheels, driven by falling water, were employed where slower streams could be channeled down over the wheel. With a lower torque and higher adjustable speed, they were more powerful and efficient (converting up to 70 percent of the water's energy if the wheel revolves quickly enough). The less efficient undershot "Vitruvian" wheel

(Vitruvius 10.4–5: up to 22 percent efficiency at best), was, however, more common, requiring no elaborate construction and no engineering to increase the rate of flow: it worked slowly, but cheaply and reliably at low velocities. Undershot waterwheels were used to lift water and grind grain, but Ausonius (fourth century CE) referred to water-driven sawmills (*Mosella* 361–364), in evidence materially at Hierapolis (Turkey: mid-third century CE) and Jerash (Jordan: late-fifth century CE), where they are also rendered artistically.[94]

Operating on the same principle as the water-lifting wheels is the water mill, a technology that may have been developed in Anatolia where large rivers had copious supplies of water for fueling such mills. The water mill at Kabeira is the earliest one cited in our sources, perhaps constructed for the court of Mithridates VI (ruled 120–63 BCE), an exceptional installation, so Strabo observed (12.3.30), along with the palace, imperial hunting grounds, animal park, and mine.[95] In Vitruvius we have a detailed description of the underlying mechanics:

> hydrelatae (water mills) are turned on the same principle as the other water wheels, except that on one edge of the axle there is fixed a toothed drum. This, moreover, placed on the perpendicular plane, turns onto its teeth at the same time as the wheel. A second larger drum, likewise toothed, is placed horizontally. The teeth of the drum on the axle force a revoltion of the millstone by driving the flat drum. In this machine, a suspended hopper supplies the grain to the millstone and flour is yielded as it turns.
>
> 10.5.2

The largest known water-mill complex, constructed at Barbegal in southern Gaul and operational from the first to the third century CE, featured two vertical, parallel rows of eight overshot waterwheels accessed by a staircase ascending the hill into which the mills were situated (Fig. 3.5). The wheels were driven by the confluence of two aqueducts some 7.5 miles (12 km) north of Arles, connected by a sluice that allowed for control of the water supply. The mill's daily capacity may have reached nearly 5 tons (4.5 metric tons) of flour. Procopius (*Goths* 5.1 9.8–19) cited a similar concentration of water mills on the Janiculan Hill in Rome. Hydraulic and pneumatic energy was applied also to musical instruments (organs and flutes) and curiosities, including the steam-powered aeölipile.[96]

Hydraulic Mining

The Romans also practiced hushing, the process whereby water is employed to prospect for and remove valuable ores. On the paradigm of Hercules cleaning the

Fig. 3.5 Reconstructed water mill at Barbegal (with overshot wheels). After Hodge 1990: 109.

Augean stables, water is brought forcefully to the ore by diverting the surface streams that then wash the rocks towards settling tanks.[97] At Las Médulas in Hispania Tarraconensis, an aqueduct was cut into the rock to erode gold-bearing gravel. The elder Pliny, who served as procurator of the province in 74 CE, recounted this complex, expensive, dangerous, and destructive procedure, perhaps from autopsy at Las Médulas. His description (33.74–78) is a technical tour de force that details the process of diverting waterways over 100 miles away (160 km) "for the purpose of washing away the destruction of mountain ranges" (Pliny's disapproval of the extravagance is explicitly stark). Engineers must dig out steep channels so that the water can be transported from a height that results in rushing water (rather than a gentle stream). Workmen—often suspended precariously from ropes—must cut into the mountain's "pathless crags" while they mark out level lines for the channels over hard, flinty rock. At the head of this manufactured waterfall, pools are hollowed out (200 feet wide and long, 10 feet deep [60 × 3 meters]), fitted with sluices in such a way that, when the reservoir is full, the water gushes out violently so that the rock "flies out." Meanwhile, at ground level, workmen excavate stepped trenches that they line with gorse (a rough plant that holds the gold back), whose sides they confine with planks. The detritus is guided over steeply elevated channels. The debris, consequently, "slides into the sea," and the "broken mountain is washed away." Although labor intensive, the process is highly profitable, yielding large chunks of pure ore—up to approximately 10 pounds (4.5 kg)—that do not need to be "cooked." Pliny recorded annual yields of approximately 20,000 pounds of ore (9,100 kg) from the mines in Asturia, Callaecia, and Lusitania. Italy alone was spared from industrial pollution by an ancient senatorial decree: "otherwise no land would be more productive of metals." Hydraulic mining is probably the starkest and most violent example of human water use to conquer (and destroy) the natural world.

Moreover, Dolaucothi in the valley of the River Cothi, near Pumsaint, Carmarthenshire, Wales, was the site of an important ancient gold mine, exploited as early as the Bronze Age.[98] Water was brought in with aqueducts and leats (channels dug into the ground), the longest of which extends about 7 miles (11 km) from the river.[99] As with other mining procedures, hushing results in pollution, including the leaching of toxins (lead, mercury, arsenic) into the water supply.

Conclusion

It is important that a society knows how to manipulate water, by bringing it to thirsty fields to feed growing crops or by drawing it away from inhabited areas to prevent floods and the spread of water-borne diseases. Water was sought and stored in a variety of ways, from public well houses to cisterns and *impluvia* that captured rainwater in estate houses. By means of aqueducts, a technology employed in ninth century BCE Assyria but perfected as high art under the

Romans, water was conveyed into towns and cities over great distances, sometimes even against gravity, a testament to human ingenuity and initiatives to control the natural environment. Aqueducts in northern Africa could thus turn deserts into oases. Everywhere the Romans ventured, the aqueducts that followed fed the taste for water-intensive luxuries such as bath houses, public latrines, and sewers.

4

Marine Hydraulic Engineering

Introduction

Just as the ancients devised infrastructure to manage water on land (extending well beyond water collection and distribution), they also developed an extensive system of marine hydraulic engineering. Initiatives in nautical infrastructure were commercially important and often matters of "national" security. Harbors were built and maintained by dredging. Canals were cut to facilitate short cuts through water. Lighthouses were erected as navigational aids. Here we consider Greco-Roman initiatives and accomplishments in marine hydraulic engineering by which coastal waters were further controlled and harnessed for human benefit.

Anchorages and Harbors

Poets, historians, and geographical writers systematically recorded data about harbors, where they are plentiful, and where they are lacking. We shall here survey the literary and archaeological evidence to tease out the principles of harbor construction in general, before turning to two case studies, the important harbors in Athens (Piraeus) and Rome (Portus).

Travel by ship was part of the fabric of Greek mythology, and the Greek heroes at Troy had to voyage by sea in order to return to their homes on the Greek mainland or surrounding islands. Homer's *Odyssey* famously recounts the sea adventures of one of those Greek heroes, Odysseus, who made careful note of harbors before mooring: Circe's harbor in Aeaea was "fit for ships" (*Odyssey* 10.141; she who turned Odysseus' men into swine); the giant, cannibalistic Laestrygones possessed a "renowned" harbor (*Odyssey* 10.87); and in Sicily, where Odysseus encountered the one-eyed Cyclops, the harbor was "good for making-fast, where there is no need of a cable: neither to throw an anchor-stone nor to make fast at the stern" (*Odyssey* 9.136–139).[1]

Historians and geographical writers followed suit, making note of where good harbors exist. In the fourth century BCE, Pseudo-Skylax noted the Adriatic's "good (or plentiful) harbors," and even shipyards in harbors.[2] The anonymous *Periplus of the Erythraean Sea*, a practical merchant-marine guide of the Red Sea (first century CE), opened with a promise to list the harbors and markets along that waterway. Pliny, with typical Roman pride, lauded the many harbors of Italy (3.41). Josephus, a Romano-Jewish author writing in Greek, described the "immense fabricated [first century CE] defenses" around Pharos near Alexandria, where the channel is rough, and the narrow entrance treacherous, but "perfectly safe" once inside the harbor (*Jewish Wars* 4.614: Josephus would accompany the future emperor, Titus, to Rome in 71 CE). Commodious harbors also sheltered ships from southerly and easterly winds on the Black Sea in the second century CE (Arrian, *Periplus* 4.2; cf. Mela 2.3).

Authors also recorded where harbors were insufficient or lacking. The Athenian statesman-philosopher Xenophon (430–355 BCE), had complained about the lack of harbors near Salmydessus on the western coast of the Euxine (*Anabasis* 7.5.12–13).[3] This was inconvenient for Xenophon, who had been elected commander of the 10,000 mercenaries employed by Cyrus II in a bid to take the Persian throne. Writing from experience, Xenophon was trying to return his troops back to the Greek mainland across the Aegean. Strabo observed that Ostia, for example, was harborless in the early-first century CE because of siltation from the Tiber (Strabo 5.3.5). The coast of Egypt was "almost harborless" (Josephus, *Jewish Wars* 4.610), as was the dangerous, shallow Gulf of Syrtis (Mela 1.35). In Gaul, cities were sparse because the harbors there were scant (Mela 2.76), and some harbors on the Black Sea were suitable only for small vessels (Arrian, *Periplus* 5.1, 13.1, 13.5).

The literature is rich with circumstantial evidence, but handbooks on harbor construction also existed. Timosthenes, a naval commander under Ptolemy II Philadelphos ("Sister-loving:" ruled: 283–246 BCE), composed an *On Harbors*, cited by Strabo (2.1.40)[4] and Pliny (6.183), but now sadly lost. Philo's *Harbor Construction* (third century BCE) is also lost, but other harbor engineering manuals may have existed.[5] In Vitruvius (5.12), we have a tantalizing glimpse into ancient harbor construction. Although emphasizing the virtues of working with nature and topography, Vitruvius offered solutions to environmental challenges, including developing anchorages away from the mouths of rivers; construction of masonry and embankment piers; the use and construction of breakers; and building up lime and sand foundations on soft sea bottoms. Hesnard (2004) suggests that the harbor works at Massilia (Marseille) reflect the account of harbor construction in Vitruvius who may have been present there during the Caesarian siege of 49 BCE (5.12.5-6: serving under Julius Caesar during the Civil War, Vitruvius included an account of the siege of Massilia: 2.1.5, 10.16.11).

Where natural harbors did not exist but were needed to serve political, economic, or military exigencies, they were built, as at the Pireaus near Athens, Ostia near Rome, Caesarea Maritima (Judea), Lechaion (at Corinth), and Trapezous on the Black Sea (cf., Arrian, *Periplus* 16.5).[6] Harbor development, moreover, dates from early times. For example, riverside quays in Mesopotamia and Egypt were constructed as early as the second millennium BCE. Early Greek harbor works generally consisted of ashlar masonry blocks often joined with clamps (but without mortar) to form quays.[7] Quays and docks increased in size and complexity as maritime trade grew. Where sheltered bays and headlands did not occur naturally, breakwaters and moles—if aligned correctly with currents and prevailing winds—could protect exposed shores from the sea, enemy attack, and siltation.[8] Moles were often constructed by depositing rubble in long, submarine embankments, sloped to secure the loose construction against the currents. A long mole (100 yards [ca. 90 meters]) of local granite blocks (11–13 tons each) is known at Delos (eighth century BCE). A breakwater and ship sheds were built at Samos (late-sixth century), but nothing remains aside from a brief mention of the Samian "boathouses" (*neosoikoi:* οἱ νεωσοίκοι) in Herodotus (3.45.4). Ship sheds were narrow, roofed installations for protecting vessels from the weather, as famously at Piraeus. Opening towards the water, they had inclined slipways that facilitated launching or pulling ships out of the water. Remnants of the six ship sheds that served Aegina's naval harbor are still in evidence under the waterline. Aegina was also fitted with two breakwaters.[9] The original harbor dates to ca. 480 BCE, but refinements are of uncertain date. Strabo refers to the first century CE ship sheds at Cyzicus in Phrygia and Carchedon in Libya (12.8.11, 17.3.14). Herodotus, moreover, mentioned the winches that were used for landing Necho's triremes on the Erythraean Sea (2.159.1: sixth century BCE). Harbor access was controlled with slender entrances, towers, booms, or chains as at Mytilene on Lesbos (Strabo 13.2.2), Knidos in Asia Minor (14.2.15), and Tyre in Lebanon (16.2.12). Harbors were also fortified with walls as a further bulwark against land attack (as in Athens, Mégara, and Corinth).

The literary evidence is further augmented by the archaeological record that has benefited greatly from nineteenth century programs of drainage, land reclamation, and harbor modernization, together with the compilation of Admiralty Charts of the Mediterranean for the British navy. The British chart makers included valuable data such as location and layout, water sources and courses, identifying factors, and hazards.[10] In addition, the scope and nature of underwater archaeology has evolved in recent decades, diversifying the evidence and its contexts.[11]

These ancient installations have endured in large part because of the Roman innovation of remarkably durable hydraulic concrete. Hydraulic concrete was formulated with the addition of volcanic ash (pozzolana), which helps prevent the spread of fractures (cf., Vitruvius 2.6.1, 5.12), thus allowing the construction of freestanding submarine installations that were built up from timber frameworks. The infrastructure was then lowered into place (probably by crane). The procedure is in evidence from the mid-first century BCE at Cosa on Italy's western (Etrurian) coast, where five concrete pillars provide the framework of the breakwater in the harbor there. Additionally, the pozzolana used in the mortar for Herod the Great's harbor at Caesarea Maritima was imported from the eponymous region around Pozzuoli.[12] Built between 22 and 15 BCE (dedicated to Augustus in 10/9 BCE), Herod's Sebastos ("Augustus") Harbor was comprised of two pozzolana jetties, which delimited a very large artificial harbor (100,000 meters^2 [ca. 120,000 yards^2]: perhaps the largest artificial harbor in the open sea). Herod's design may have been the prototype for Claudius' Portus Harbor at Ostia.

Known to the Greeks but not fully exploited until in the hands of Roman engineers, the arch was also applied to harbor construction. The curved pier of the arched mole at Puteoli (near Naples) could more easily withstand southern storms.[13] Pliny the Younger cited yet another Roman innovation, the off-shore breakwater, as at Centumcellae (Civitavecchia), where he (Pliny) had observed the construction of the harbor: at the entrance, boats transported the huge blocks on which the piers were constructed into an arched superstructure (*Epistle* 6.31).

Owing to the nature of the winds and currents, together with the maritime needs of particular cities, some city states boasted multiple harbors. Multiple harbors accommodate ships entering or leaving port as the prevailing winds might change: e.g., Luna in northwestern Italy with "several harbors" (Strabo 5.2.5); Syracuse in southeastern Sicily, with two harbors (one of which is inside a fort) that Strabo credited in part as the source of the city state's prosperity (6.2.4); Phaselis on the Black Sea with three harbors (14.3.9); and the Great Harbor at Alexandria, deep near the shore (so it was able to accommodate "the largest ship [at] anchor at the steps"), is "cut up into a number of harbors." Multiple harbors might provide access to different waterways. Corinth, for example, had two harbors, one at each side of the isthmus in order to facilitate trade to both Italy and Asia (Strabo 8.6.19). The Black Sea city states, Amastris and Sinope, also had harbors on each side of their peninsulas (12.3.10, 11). Multiple harbors may also enable city states to maintain dedicated harbors for their fleets, as at Miletus in western Asia Minor, where only one of the four harbors was "sufficient for a fleet" (Strabo 14.1.6), Knidos (south of Miletus), where one harbor was "a naval station for 20 ships" (14.215).

Piraeus

Before the Piraeus, Athens depended on the ancient port at Phaleron on the Saronic Gulf, about 3 miles (5 km) southwest of the Acropolis. According to legend, Theseus sailed from Phaleron to Crete to destroy the Minotaur and the legendary Athenian king, Menestheus, set out from Phaleron to Troy (Pausanias 1.1.2). It was also the site of a little-known allied defeat of the Persian fleet in 490 BCE.[14]

Themistocles, however, turned his attention to improving the harbor at Piraeus, which had first been fortified in the late-sixth century by Hippias (tyrant: ca. 527–510 BCE).[15] Athens would be well-served by the limestone Piraeus peninsula (5 miles [7 km] to the southwest) with its three natural harbors. The navy utilized two circular harbors, Zea and Munychia, exclusively, and access to the military dockyard was likely restricted to enhance security.[16] Themistocles later began to fortify the Piraean peninsula with thick limestone walls to ward off enemy attack (Diodorus Siculus 2.41; completed in 479–477 BCE). The Athenians and their allies had just defeated an army of invading Persians, and Themistocles was mindful of the fact that the Persian army could more easily attack the city by water than by land (Thucydides 1.93.3–7). In 429/8, the Peloponnesian army planned an attack of Athens from the sea, since the Piraeus had been left open and unguarded, owing to Athenian arrogant confidence in their maritime prowess. But the winds delayed the Peloponnesian allies, and the Athenian forces were able to ward off the invaders. They subsequently took precautions to defend the harbor more diligently, including closing the harbors (Thucydides 2.93–94).

By the mid-fifth century, "Long Walls" joined the port to the city state of Athens, providing refuge to rural Athenian citizens who came into the *polis* during the first year of the Peloponnesian War (Thucydides 1.107.1, 2.17.3) (Fig. 4.1). If under siege, Athens could then be supplied from the water. Contrarily, enemies (or vengeful politicians) could block the importation of grain and other

Fig. 4.1 Piraeus with long walls.

supplies. In 411 BCE, petty and greedy Athenian oligarchs, who had launched a successful coup of Athens, commandeered the grain supply entering port, which they then sold (Thucydides 8.90.5).

Infrastructure was developed over time, and harbors were outfitted with open slips where triremes were moored, perhaps as early as 480 BCE. These slips were quickly replaced with a monumental ship shed, open to the sea, accessed by ramps, and roofed to protect the wooden ships against the weather. In 429/8 BCE the harbors were closed by chains that connected the towers on the moles at each harbor.[17] By the fourth year of the Peloponnesian War (428 BCE), Athens was deploying 250 triremes, the largest number in her maritime history thus far, and considerable resources were brought to bear on manning and maintaining the fleet.[18] After the Peloponnesian War, the Piraeus fell into decline as centers of power shifted away from Athens. Nonetheless, the growth in Athens from a single harbor at Phaleron to multiple harbors underscores how important maritime access was to Greek *poleis* for security, prosperity, and the reliable import of essential goods.

Portus

Rome was serviced by both Mediterranean and river ports. Wharfs and warehouses are in evidence at emporia installations along the Tiber's left bank (between the Ponte Sublicio and Ponte Testaccio: one Trajanic wharf was reinforced with an artificial concrete bank).[19] Along the Campus Martius, quays have been discovered.[20] The Severan Marble Plan (203–211 CE), furthermore, suggests harbor installations on the Tiber's east bank.[21] We focus briefly on Rome's famous port at Ostia (35 km downstream from the Tiber's estuary [ca. 22 miles]), commissioned by Claudius who utilized Caligula's obelisk-barge as part of the harbor's west mole (Pliny 16.202, 36.70; Suetonius, *Claudius* 20.3), and excavated by The British School at Rome.[22] The barge was fitted with three pozzolana moles "with the height of towers," and towed to Ostia, before it was sunk in the harbor. The moles additionally provided the base for the Portus lighthouse. Other ships were also discovered, encased in the harbor's concrete foundations. Claudius had, moreover, hoped to reduce the seasonal flooding at Rome by digging channels from the Tiber to the sea.[23] The harbor was wide (800 meters [2,600 feet]) and shallow (6–7 meters [ca. 20 feet]), and could accommodate up to 300 ships, including large seafaring vessels with a cargo capacity of 500 tons. Framing a 225-yard-wide entrance (ca. 200 meters), the breakwaters were constructed with six- to seven-ton blocks of travertine. A Neronian coin issue shows the busy harbor in aerial view: four boats are under sail, three are rowed. In the upper register, Neptune, standing on a lighthouse, wields a scepter. At the bottom, a reclining Tiber holds a rudder and a dolphin. Along the left edge the crescent-shaped pier is depicted as a portico with fourteen pillars, above which a crude figure makes a sacrifice. The right edge is balanced with a crescent of fourteen slips, above which a figure sits on a rock.[24] The design emphasizes merchant vessels (with their broad sails and swan figureheads) and may be intended to remind the Roman people of Nero's beneficence after a storm that had destroyed 200 ships in the harbor (Tacitus, *Annals* 15.18.2) and threatened the grain supply.[25] This was Nero's assurance to the Romans that the grain supply was secure.

Trajan (ruled 100–112 CE) commissioned an extension of the harbor, which resulted in a hexagonal basin that greatly increased the harbor's capacity for ships and cargo (with the annexation of more warehouses) (Fig. 4.2). An additional 200 ships could be moored in the hexagonal basin.[26] The annex took twelve years to complete. Stones covered the harbor floor, creating a gradual slope easing off to a depth of 1 meter (ca. 3 feet) at the basin's edge, so that a ship's

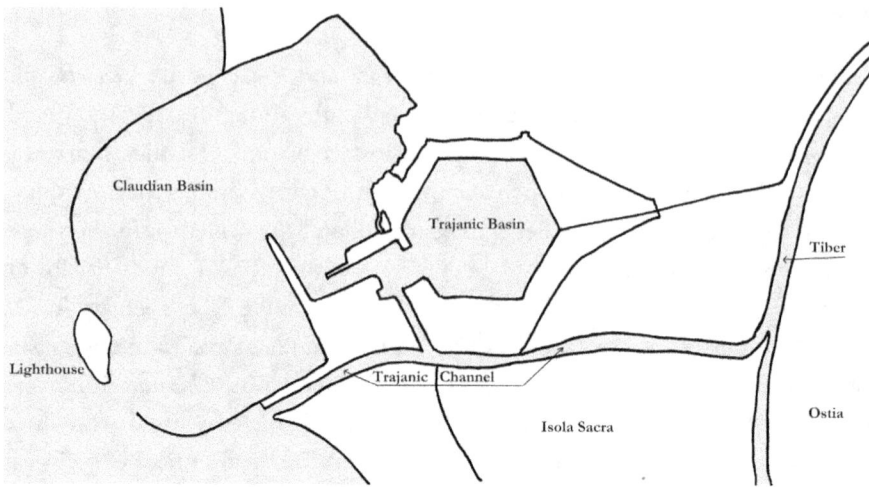

Fig. 4.2 The octogonal harbor at Portus.

deck might be level with the dock for the ease of offloading merchandise (as depicted on the Torlonia relief; Fig. 6.2).[27] Travertine blocks with mooring holes were situated about 15 meters apart (50 feet) in the brick quay face. Blackman estimates that there may have been as many as 100 pierced bollards (mooring stones) at Portus.[28] The hexagonal harbor may have been subdivided into administrative sections, as suggested by the discovery of columns inscribed with numbers. Walls with narrow doorways, five on each side of the hexagon, helped to control the flow of cargo that was offloaded by slaves. Additionally, the layout assisted with the collection of any tariffs or customs that might be due. Essentially intact, much of the harbor now lies beneath Fiumicino Airport.

The hexagonal harbor was, seemingly, unique, and it may have been designed by Apollodoros of Damascus, the architect of Trajan's Market in Rome, as a display of Roman imperial authority. The harbor is also commemorated on Trajanic coinage as a clarion declaration of imperial assurance that the import of necessary supplies (and luxuries) would not be interrupted.[29] This artificial harbor is a remarkable example of Roman engineering and human manipulation of the watery world for the sake of the comfort and prosperity of the residents of the imperial city.

Dredging

Because of the constant process of siltation and the variability of relative sea levels (the Mediterranean is geologically unstable and volcanically active), many harbors require the attention of regular dredging. When Attalus [II] Philadelphos (159–138 BCE) installed a mole at the harbor of Ephesus, siltation increased:

> The city (Ephesus) has both a dockyard and a harbor, which was made narrower by the architects, but they were deceived, along with the king who ordered it. This was Attalos [II] Philadelphos (159–138 BCE), who thought that the entrance, as well as the harbor, would be deep enough for large merchantmen—there previously were shallows because of siltation from the Kaystros—if a mole were placed alongside the mouth, which was extremely wide. He thus ordered the mole to be constructed. But the opposite happened, because the silt was held in and the entire harbor became shallower as far as the mouth. Previously the ebb and flow of the sea was sufficient to remove the silt and draw it outside.
>
> Strabo 14.1.24

Greco-Roman literature is frustratingly mute on dredging, and archaeological evidence is scant. Tacitus may imply that the port at Ephesus was dredged (*aperiendo*: "by opening") when Barrea Soranus served as proconsul of Asia in the 50s CE (*Annals* 16.23). The harbor was notorious for its silting from the alluvium of the Cayster River (see Strabo 14.1.24). Technical details, however, are lacking (*Annals* 16.23.1). Gaps, nonetheless, are evident in the stratigraphic chronological sequence, substantiating dredging activity at the Ephesian harbor into the Byzantine era.[30]

Dredging likely occurred within the basins of primitive harbors along the Nile, Euphrates, Tigris, and Indus Rivers.[31] Morhange and Marriner have examined evidence from fossilized dredging scoops, chronological inversions, and chrono-stratigraphic gaps in order to show that the coastal areas of Massilia (Marseille), Neapolis (Naples), Sidon, and Tyre have undergone regular dredging from the fourth century BCE onward.[32] The Roman harbors at Massilia and Neapolis are now both fully silted and therefore accessible to terrestrial excavations. At both sites, scoop-gouges are preserved in the harbor-floor sediment. Fossilized scour marks indicate the great force that was brought to bear on the harbor clay as well as the large quantity of sediment that they could remove. Three dredging boats—dubbed *Jules Verne 3, 4,* and *5* by the excavators—have been recovered from the Roman harbor at Massilia.[33] Dating to the first/second century CE, each barge has an open central well that could have accommodated a dredging arm. *Jules Verne 3* measures ca. 16 meters (52.5 feet) with a central well of 255 cm in length and 50 cm in width (8.5 × 1.5 feet). At Massilia, fossilized dredging scoops have been found at depths of 30–50 cm (1–1.5 feet), but the precise mechanics of the scoops are not fully understood.

The dredging arms may have been equipped with buckets to scoop up sediment or with hooks to scrape away the sediment. The process was likely tedious and time-consuming, and it remains so, even with modern equipment, as sediment is lifted from the bottom and dumped onto barges for disposal elsewhere. More commonly in antiquity, small channels were built into harbor walls to ensure that detritus could flow freely out of a basin. Additionally, breakwaters were positioned within the harbors to create currents that would wash away siltation (as at Ephesus before Attalus' "improvements"). Breakwaters also deflected silt conveyed by incoming currents, especially in harbors with two basins, as at the Phoenician harbor at Sidon. Ashlar-lined channels were installed in Mytilene's north harbor. Broad perforations were bored through a breakwater at Paphos, which in turn was protected by an additional external breakwater.[34] Both mechanical dredging and harbor design provide further examples of Roman administrative organization, efficiency, and success in altering the watery landscape to support trade and travel and to enhance the quality of life in Rome.

Canals

Labor-intensive canals were also cut to facilitate irrigation and drainage or to link navigable waterways.[35] Around 600 BCE, Necho II constructed a canal (the length of a four-day sail) that joined the Nile with the Erythraean Sea (Herodotus 2.158; Diodorus Siculus 1.33.912.). Wide enough to allow for two-way traffic, the canal was restored by Darius I (ruled 522–486 BCE), who commemorated the renovated canal with an inscription in Persian.[36] Under Ptolemy II, a canal was again dug in 270/269 CE between the Nile and the Red Sea. The Ptolemaic canal was enhanced with a navigable lock and sluices allowing vessels to travel but preventing salt water from flowing into the canal's fresh water (Pithom Stele: CM 22183). Under Trajan, a 180 km (112 miles) "Trajanic river" between the Nile to the Red Sea was either reopened (desilted) or its course was modified. The project was subsidized by a special local tax (Ptolemy, *Geography* 4.5.54).

Herodotus described another early canal, cut by Xerxes at Mount Athos (483–480 BCE), built as a display of power and memorial to himself, but also as a model of the engineering excellence of the Phoenicians. The canal was constructed in sections, each with its own work crew, including one of Phoenicians. The other work crews had excavated their trenches as deep as they were wide, and the steep sides, naturally, collapsed. But working on their own section, the Phoenicians:

dug out the upper part of the trench, making it double the required width of the channel, and reducing the width as they progressed. At the bottom their work was equal to the rest.

7.23.3

Xerxes' canal was also fitted with breakwaters of earthen mounds at both entrances to protect against flood-tides. Athenaeus (5.204c–d) implied the use of locks in a canal designed by Phoenician engineers and constructed on the orders of Ptolemy IV Philopater ("Father-loving": ruled 221–204 BCE) for the launching of a huge ship at Motya.[37] During the launching, a system of gate mechanisms allowed "the sea into all the excavated space" of Ptolemy's canal.

There have, moreover, been many efforts to facilitate the passage of ships and merchandise across the narrow isthmus at Corinth (6 km [ca. 4 miles]). Smaller ships were dispatched by means of an overland track, the *diolkos* (from the sixth century BCE). But Nero's canal project there was never completed (Suetonius, *Nero* 19; Dio 62.16). Ca. 300 BCE, Demetrios Poliorcetes ("Besieger of Cities"), had intended to construct a canal between Corinth and Aegina, but his engineers informed him (correctly) that the two sides had different sea levels and (incorrectly) that cutting through the Isthmus would flood Aegina (Strabo 1.3.11).[38] Tacitus also alluded to a canal that joined the Saône and Moselle Rivers in Gaul and Germany (*Annals* 13.53). Drusus (38–9 BCE) commissioned a canal between the Rhine and Issjel Rivers, praised by Suetonius as a "remarkable engineering feat" that saved Roman shipping lanes from the treacherous North Sea passage (Suetonius, *Claudius* 1; Tacitus, *Annals* 2.8). The canal between the Rhine and Maas Rivers (47 CE) may have been a make-work project to distract idle soldiers (Tacitus, *Annals* 11.20). In 64 CE, Nero, moreover, agreed to the construction of a navigable canal between Lake Avernus and the Tiber for no better reason than that the task would be "unendurable" (Tacitus, *Annals* 15.42). Requesting the assistance of a surveyor, Pliny the Younger proposed a canal between the Marmara Sea and a lake near Nicomedia in Bithynia, where he served as governor in 110 CE. Concerned that the proposed canal might drain the lake, Trajan responded with a recommendation that his clingy administrator employ the expert surveyors of Lower Moesia (*Epistle* 10.41–42). Along the Appian Way, mules towed boats through a canal over the marshes near Terracina (Strabo 5.3.6; cf., Horace, *Satires* 1.5.9–23), just as "mules" continue to tow ships through the locks of the old Panama Canal (constructed 1904–1914). Ongoing work at Portus, furthermore, indicates an extensive canal system associated with agriculture and other delta industries. Keay and Strutt (2014) have investigated the links between these canals and the delta's geomorphology (dune cordons, variations in the sediments across the wetland). Canals provide further examples of altering topography and the watery landscape to facilitate trade and travel.

Lighthouses

Navigation is greatly aided by both natural and purpose-built landmarks, including lighthouses, a common feature of ancient (and modern) harbors. The shield of Achilles is compared to a navigational beacon that guides sailors to safety through storm-tossed waters (*Iliad* 19.375–379). In Ovid, a mountain-top beacon at Thessaly was a sight "pleasing to exhausted craft" (*Metamorphoses* 11.393). Strabo mentioned several light towers in the Mediterranean: the Tower of Caepio at Gades (Cadiz: 3.1.9); at Neoptolemus at the mouth of the Tyros River (7.3.16: northeastern Europe); two at Aulis (in Boeotia), whose harbor can accommodate fifty ships (9.2.8; the harbor there is actually quite small); pairs of towers at Abydos and Sestos (13.1.22); the Euphrantas Tower on the Syrtis (17.3.20); and the four towers on the Cyrenaian coast above the Plynos harbor (17.3.22). Juvenal described the *pharos*-tower that marks the breakwaters at Ostia (12.75–82),[39] also featured on the Torlonia relief, as well as coins and several Ostian mosaics. On one Ostian mosaic, two ships from Sullectum (Byzacium in Africa), according to the inscription, sail into port from opposite directions. Two symmetrical dolphins guide the ships. In the background stands the two-tiered *pharos* (lighthouse), topped with a blazing light (Fig. 4.3). The ruins of one of Dover's two ancient towers can still be seen (80 feet high [24 meters]). Pliny, incidentally, recognized that a tower whose fire burns constantly might be mistaken for a star (36.83). Patterns of blinking lights thus distinguish modern light towers from celestial objects and from each other, providing a useful tool for nocturnal navigation.

Pharos

The most famous ancient lighthouse is the white limestone tower on Pharos, the island off Alexandria that was connected to the mainland by a mole in the fourth century BCE. The tower was constructed "for the safety of sailors" according to its inscription (Strabo 17.1.6).[40] One of the seven wonders of the ancient world, the Alexandrian *Pharos* was commissioned by Ptolemy II

Fig. 4.3 Dolphins lead ships into harbor past Ostia lighthouse.

Philadelphos who generously allowed the architect, Sostratos of Knidos, to inscribe his name "on the very structure of the building." Costing 800 talents (23 tons of silver), the tower warned sailors of dangerous shoals, and it marked the harbor's entrance (Pliny 36.83). Josephus claimed that its light could be seen as far away as 300 stades (or ca. 34 miles: *Jewish War* 4.10.5). If the light were visible from the distance of a day-sail (ca. 24 miles), the lighthouse may have been as tall as 100 meters (over 300 feet), as confirmed in Arabic sources who variously report a height of about 100–120 meters [ca. 340–390 feet] supported on a base of 30 meters^2 [ca. 320 feet^2]. Ibn Jubayr, who traveled through Alexandria to Mecca in 1183, claimed that the tower could be seen from 70 miles away (112 km).[41]

A beacon for navigators, lighthouses were also strategic markers of narrow straits, as Julius Caesar understood. After his defeat at Pharsalus in 48 BCE,

Pompey fled to Egypt, with Caesar in pursuit. Garrisoning his troops on Pharos Island, Caesar seized the tower to control water traffic and he thus secured access for his own supplies and reinforcements (*BC* 3.112).

The *Pharos* tower was widely praised in antiquity for both its beauty and practicality. The *Pharos* was "the greatest and most beautiful of all works, that a flame might blaze from it for sailors over much of the sea" (Lucian, *How to write History* 62), whose light "rivalled" the moon's (Statius, *Silvae* 3.5.101). Achilles Tatius (second century CE) compared the structure to a mountain, reaching the clouds, at whose summit was a second sun:

> it was a mountain lying in mid-sea, touching the clouds themselves. Water flowed underneath the structure, which stood as if suspended on the sea. At the acropolis of this mountain rose up a pilot for ships, another sun.
>
> *Leukippe and Klitophon* 5.6.3

Literary sources agree that the light on top of the tower was a bright beacon that could be seen from quite a distance. Roman imperial coins (from Domitian to Commodus: 81–192 CE) show Tritons positioned on each of the four corners of the tower's upper register, and Poseidon or Zeus on the tower's top. Coins also show Isis Pharia (her incarnation as the divine patroness of the island) holding a sail and approaching the tower.[42]

Harmonizing with Roman numismatic representations, Arabic sources include several descriptions from pilgrims on their way to Mecca,[43] the fullest coming from el-Balawi el-Andaloussi, who visited Alexandria in 1166 CE. He was unable to read the ancient inscription whose letters—large, black, and weathered—were still attached on the seaward wall. A vaulted ramp on sixteen arches led into the structure, while another ramp spiraled around the center. Draught animals conveyed fuel up this ramp to the second tier where it was then lifted up to the top by pulleys or carried up the steps by human labor. Arabic sources represent the *Pharos* lighthouse with three tapering tiers: a lower square with a central core, a middle octagon, and, at the top, a circular section (as restored after the upper tier was purloined by the Caliph al-Walid (ruled 705–715). At the apex, a mirror reflected sunlight during the day, while at night a fire was lit.[44] A Chinese source adds that the tower, with chambers for storing grain and weapons, reached a height of 200 *chang*: "four horses abreast could ascend to two-thirds of its height." By the Arab period, according to our Chinese source, the tower became even more strategic, and its mirror was used to alert the local navy of impending sea attack. A foreigner—an enemy spy—was employed to "sprinkle and sweep" for years until he found an opportunity to steal the mirror and throw it into the sea.[45]

Ibn Jubayr, who traveled through Alexandria to Mecca in 1183, linked the lighthouse to the glory of Allah:

> One of the greatest wonders that we saw in this city was the lighthouse which Great and Glorious God had erected by the hands of those who were forced to such labor as "a sign to those who take warning from examining the fate of others" (Quran 15:75) and as a guide to voyagers, for without it they could not find the true course to Alexandria. It can be seen for more than seventy miles, and is of great antiquity. It is most strongly built in all directions and competes with the skies in height. Description of it falls short, the eyes fail to comprehend it, and words are inadequate, so vast is the spectacle.
>
> We measured one of its four sides and found it to be more than fifty arms' lengths. It is said that in height it is more than 150 *qamah* [one *qamah* = a man's height]. Its interior is an awe-inspiring sight in its amplitude, with stairways and entrances and numerous apartments, so that he who penetrates and wanders through its passages may be lost. In short, words fail to give a conception of it.
>
> transl. Broadhurst 1952: 32–3

Between 320 CE and 1303, the tower was damaged by a series of earthquakes. By 1349, pilgrims were no longer able to enter the ruin (according to Ibn Battuta, traveling from Morocco that year). By the middle of the fourteenth century the tower disappeared from the historical record, and much of its construction material and adornments fell into the eastern harbor, where they still lie. By 1480 the Citadel of Qaitbay was constructed on the site, perhaps with stones recycled from the derelict tower.[46] Currently, the Egyptian government is exploring an initiative in cooperation with the UNESCO Convention on the Protection of the Underwater Cultural Heritage to construct an underwater museum for showcasing the remains, which have been included on a World Heritage List of submerged cultural sites.[47]

A Coruña

Unlike the *Pharos* of Alexandria, the Tower of Hercules, possibly built under Trajan in the second century CE, remains intact at A Coruña in northwestern Spain. The 55-meter high tower (180 feet) was designed by the architect Gaius Servius Lupus, according to the inscription at the base. The Roman tower followed the three-story Phoenician paradigm evident in the Alexandrian *Pharos*. In 1788, under Charles III of Spain, the tower was restored, a fourth story was added (21 meters high [ca. 70 feet]), and the new design conformed to neoclassical aesthetics.

Conclusion

The Greeks and Romans aimed to control, enhance, and "improve" all aspects of their world, including the seascape. Their initiatives in marine hydraulic engineering fostered travel and trade by providing regulated access between land and water with man-made harbors and canals, maintained by dredging, and purpose-built light towers that served as important navigational beacons.

Part Two

Engaging with the Watery World

5

Sailing and Navigating[1]

Introduction

Despite the dangers, humankind has always felt the call of the sea, and folk have plied the waters with many types of vessels, from plain rafts fitted with simple sails, Huckleberry-Finn style, to the mammoth nine-masted, four-decked ships of Zheng He, who traveled the Pacific in the early fifteenth century CE. Approximately 50,000 years ago, we find the earliest evidence of deep-sea sailing in Sahul (the land mass of Australia and New Guinea at a time of low sea level). Mesolithic deep-water fishing near the Franchthi Cave in the Argolid may be indicated by the remains of tuna bones and obsidian tools (7500 BCE). Fleets of monumental outrigger canoes cleft the waters of the southern Pacific, making voyages of thousands of miles over open ocean, perhaps as early as 1000 BCE.[2]

In Herodotus, Egypt was an early seafaring nation,[3] a claim supported by artistic and archaeological evidence. A statue base in Amenhotep III's tomb (ca. 1800 BCE), inscribed with Aegean toponyms, may be a record of a voyage—from Egypt across the eastern Mediterranean—that included a circumnavigation of Crete with stops in the Peloponnese, Cythera, and Troy.[4]

Generally keeping within sight of land and sailing during the day when conditions allowed for better visibility, the Phoenicians made short-haul commercial voyages, establishing commercial emporia and settlements every 25 to 30 nautical miles (29–34.5 miles [46.3–55.5 km])—the average length of a day's sail—usually in conjunction with natural harbors.[5] Furthermore, the Phoenicians were the first Mediterranean people to explore the waters beyond the Pillars of Herakles (Straits of Gibraltar). After completing his canal of the Nile River, Necho II (ca. 600 BCE) sponsored a voyage around Libya ("Africa"). He may have wanted to prove that, with the completion of his canal, Libya was in fact fully circumnavigable, as had long been suspected. A Phoenician crew sailed from Egypt through the Erythraean (Red) Sea down Libya's eastern coast. They crossed the equator, reporting the sun on their right as they sailed west in the Southern Sea (in the northern hemisphere the sun appears to the left for those facing due west). After rounding the Cape of Good Hope, they continued up the western coast, then turned east through the Pillars of Herakles. During the two-and-a-half-year-long voyage, the expedition wintered ashore, planting crops, and replenishing supplies.[6]

In the fifth century BCE, the Carthaginian "suffete" (or "king" [*basileus*]), Hanno, led an expedition through the Pillars of Herakles down the western coast of Africa. Whether Hanno intended a circumnavigation we do not know, but his team aborted their exploration when supplies ran out, reaching, perhaps, the equator off the west coast of Africa. The adventure was recorded on an inscription consecrated in a Carthaginian temple of Cronus (Saturn) and eventually translated into Greek.[7] Hanno's voyage was known to geographical writers, mentioned in pseudo-Aristotelian *On Marvelous Things Heard*, Mela, Pliny, and Arrian.[8] Hanno listed the settlements founded by the expedition up to the Lixus River (the Wadi Draa), and he recounted the adventures faced by the settlers: the friendly Lixitans, the hostile Aithiopians, and the mysterious cave dwellers, as well as broad rivers and strange animals including elephants, crocodiles, hippopotami that filled an unnamed inland river, and *gorillae* (bonabos?/chimpanzees?) whose pelts were displayed back in Carthage in the temple of Tanit, until the Romans destroyed them together with the city, according to Pliny. The Carthaginians met with many obstacles during their voyage: lack of water, burning heat, and even rushing torrents of fire falling into the sea (i.e., the eruption, perhaps, of Mt Cameroon, still an active volcano: *Periplus* 16).[9]

Necho and Hanno were among many Phoenician mariners, long recognized as excellent sailors and accomplished navigators. Because of their superior seamanship, Phoenicians were prominent in the Persian fleet at Salamis (480 BCE), and Herodotus ascribed Persian maritime success to the Phoenicians alone (3.19.3; cf., 7.96.1). Both Strabo and Pliny admired their adroitness in celestial navigation, and Pliny (7.209) credited them with inventing the art. Although they may have preferred to anchor in protected coastal areas at night, they also knew how to maintain course on open water by observing the "Phoenician Star," the brightest star in Ursa Minor, whose navigational utility they recognized long before the Greeks did.[10]

By the eighth century BCE, nonetheless, the Greeks came to rival the Phoenicians in maritime exploration and "settlement," and their seamanship was no less accomplished. Although the technology has advanced, the challenges facing the sailor remain eternal (navigation, steering, and weather). Storms may rage, or winds may blow unfavorably or not at all. Passengers suffer from seasickness, then as now. Blithely enduring a sea trip, moreover, is sometimes construed as a sign of intemperance among those who do not ordinarily sail. While the proper wife, for example, becomes ill at the mere thought of embarking on a perilous voyage by boat (even under the protection of her husband), the wanton mistress—whose stomach is steady when she travels with her lover—"dines with the crew, wanders through the ship, and rejoices to haul rough lines" (Juvenal 6.100–103).

Although handbooks exist in other areas, no ancient manuals on the art of sailing have survived. Timosthenes of Rhodes (fl. 270–240 BCE), nonetheless, discussed the wind-rose at great length, and his treatise may have served as a navigational handbook. Furthermore, Varro's "Naval Books" (first century BCE) were cited by Vegetius (late-fourth century CE) in his treatment of weather signs and navigation, and Varro may have been Vegetius' source for the winds.[11] Our evidence must be culled from iconographic and textual accounts. Limited by scale, artistic convention, and an artist's often inexact knowledge of the technical aspects of sailing, iconography—especially as rendered on stylized images on coins and intaglios—rarely shows navigational maneuvers in any realistic way.[12] Textual sources, composed for a variety of reasons, lack comprehensivity and impartiality. Such sources, nonetheless, speak to the importance of sailing in the ancient Mediterranean, and they often provide vivid details about sailing and navigation. A compelling relief in Copenhagen shows three ships maneuvering to save a boy who has fallen from his dory into a choppy sea (third century CE). Two ships are wearing (turning away from the wind), and their sails are luffing (flapping owing to the change in the relative direction of the wind), while crew on all three handle and take in sail.[13] A crew member in the ship closest to harbor reaches out desperately to the drowning boy. To the right of the ships, at the mouth of the harbor, is the distinctive two-tiered lighthouse and signal fire of Ostia (the sarcophagus' provenance). To the left, two men stand watch at the top of a tower, one man on the dock is ready to handle dock lines. Two ships are fitted with sprit sails (a square fore-and-aft sail rigged at about a 45 degree angle to the water on the Copenhagen sarcophagus, one of the earliest depictions of this rig). The Copenhagen sarcophagus may record the unfortunate, premature death of a child, but it also speaks clearly to the dangers of travel by water and human engagement with the sea.

Officers of the Deck

...Before being a rudder-man, it was necessary to be a rower, then to serve as lookout, and then to examine the winds, and only then to govern (the boat).

Aristophanes, *Knights* 542–544

In this maritime *cursus honorum* (ladder of honors) on the ancient warship, we see a rational progression of nautical responsibilities and skills: from rower to lookout to helmsman to captain who holds the greatest authority aboard ship.[14]

Kubernetes/Gubernator

The captain of a ship was the *kubernetes* in Greek ("steerer:" κυβερνήτης), *gubernator* in Latin (whence "governor"), so-called for his ability to lead a crew, rather than simply for his skill in navigating a ship (Plato, *Republic* 1.341c–d). But the term refers both to the helmsman, *per se*, and to the captain, whose prerogatives include taking the steering oar. Odysseus, for example, was at the helm when he left Aeolus' island (*Odyssey* 10.32–33). Centuries later, the third-century-CE Greek philosopher Philostratus reported a merchant ship in service between Egypt and India on which there were "many *kubernetai*, subordinate to the oldest and ablest," that is, the captain whose mates took their turns at the helm.[15] Whoever is at the helm seems to have command of the deck (e.g., Telephus, below).

Plato defined the helmsman's skill (*kubernetike techne*) as maintaining "safety while sailing."[16] The job requires experience and skill in reading the weather, knowing the winds, and handling the vessel.[17] The helmsman must work with the wind to "keep (the ship) straight."[18] Only the worst storms could disorient a talented, experienced pilot:

> When the hold was half-full with the flood, and the masts were already unsteady while the waves splashed over the other side of the stern, when experience offered no help to the white-capped helmsman, he began to strike a deal with the winds.
>
> Juvenal 12.30–34

As stormy conditions become apparent, quick action is crucial. Reacting as soon as he senses a storm, the alert skipper "draws tight the slack sails from all directions, so that not any light breeze escapes" (Ovid, *Metamorphoses* 6.231–233). "When a wind swoops down on the sea," after "making all fast, drawing tight the sails," a quick-thinking *kubernetes* must exercise "his skill, disregarding the tears and entreaties of sea-sick and fearful passengers" (Plutarch, *Pericles* 33.5). In Ovid a harrowing storm at sea destroyed Ceyx,[19] and the crew took meticulous precautions to run before the storm:

> The captain shouts: "Lower the yards, already, and close-reef the whole sail." He commands, but hostile winds ensnare his orders, nor does the din of the waves allow any voice to be heard. On their own initiative, however,

some crewmen hurry to remove the oars, some protect the hull, some deny the sails to the winds. This one bails the water into the waves, that one snatches up the spars.

Metamorphoses 11.482–487

This well-trained crew anticipates orders, and each sailor instinctively goes to his own duty station during emergencies.

A good helmsman might even win renown. Typhus, the *Argo*'s pilot during Jason's quest for the Golden Fleece, was a local hero at Tipha (Pausanias 9.32.4). In Pausanias 9.32.4, we also visit the burial site of Menelaus' helmsman, Phrontis (the "thoughtful" son of Onetor), who "surpassed the tribes of man in guiding a ship when storms raged" (the abduction of Menelaus' wife, Helen, caused the Trojan War). On the Trojan side, Aeneas' helmsman, Palinurus, was a master of the art. As the Trojans rested ashore after journeying from their war-devastated homeland, Palinurus studied all the winds, grasped the breezes with his ears, and marked all the stars gliding in the silent sky (Arcturus, the rainy Pleiades, the twin Bears, and Orion's constellation armed with gold).[20] He saw that the sky was serenely calm, and conditions were favorable for sailing. The Trojans consequently struck camp, tested the watery road (*via*), and "spread the wings" of their sails (Vergil, *Aeneid* 3.513–520). Palinurus stood at the stern when the boat embarked. Earlier in the narrative, the Trojan helmsman had found himself so disoriented in the thick of a storm that he was unable to distinguish night from day, and he was thus unable to remember the paths (*viae*) in the midst of the waves (*Aeneid* 3.201–202). In Lucan's first-century-CE versified account of the civil war between Pompey and Caesar (49–46 BCE), we meet the equally skilled Gallic helmsman, Telo (perhaps from Gallia Narbonensis), who fought in Pompey's ranks. Like Palinurus, Telo was proficient at forecasting the weather from observing the sun and moon and in handling watercraft under all weather conditions: "in an agitated sea, keels listened to no hand better than his" (*Pharsalia* 3.592–594).

Keleustes/Pausarius

After the *kubernetes*, two other deck officers were key. "The one who orders," the *keleustes* (κελευστής; *pausarius* in Latin, from the verb *pauso*: to halt/cease), gives the beat to the oarsmen, "in a piercing voice" (Seneca, *Epistle* 56.5; cf. *ILS* 2867). Oars provided the primary propulsion for ancient warships, and men would draw lots in pairs for their turn at the oars of a warship. Since it was essential for the rowers to work in unison, the *keleustes*' job was vital:

They made the men sit on rower's benches on dry land, having the same arrangements as they would on voyages; making the *keleustes* stand in the middle to teach them all to drive their striving arms back at the same time and then to stoop their thrusting arms forward, they became accustomed to begin and stop their motions at the calls of the *keleustes*.²¹

Like a coach,²² he could encourage his crew, improving morale, or he could dishearten them. If the crew were intractable, "the time to finish the same voyage" could more than double (Xenophon, *Economics* 21.3).

A strong voice is essential for the *keleustes*. Euripides recounts a version of the Trojan War in which Helen goes to Egypt, remaining faithful to her husband (a cloud replica is sent in her place). After Menelaus has retrieved his loyal wife from the clutches of the lustful Egyptian king, heading away from Egypt the rowers started their strokes as soon as they heard the *keleustes'* shout (*Helen* 1576). During a violent storm off Patrae in 429 BCE early in the Peloponnesian War between Athens and Sparta, the crew were unable to distinguish one officer from another, and they were consequently unable to follow orders (Thucydides 2.84.3). Likewise, in the thick of battle, noise and activity could obstruct communication, rendering it difficult (or impossible) to hear the *keleustes'* commands. In 413 BCE, during a naval engagement off Syracuse, Athenian and Syracusan ships were coming at each other "at the same time" with such frequency, confusion, and "din" that the crew could not hear the *keleustes* (Thucydides 7.70.6). In contrast, during stealth missions where silence was essential to success, *keleustes* would call the strokes by "clinking stones together, instead of with their voices," as when the Spartan general, Gorgopas, orchestrated a surprise attack on the Athenian navy as it returned to the Piraeus in 388 BCE during the Corinthian War (Xenophon, *Hellenika* 5.1.8). To lighten their work, rowers might accompany themselves by singing (Longus, *Daphnis and Chloe* 3.21).

Prorates/Proreta

The third watch officer, the *prorates* (πρῳράτης: "who observes from the ship's bow;" Latinized as *proreta*), keeps lookout for shallows, smaller craft, debris, and other hazards.²³ The *prorates* would even call out when he caught sight of swordfish for sport fishing (Polybius 34.3.3). As *prorates*, Danaus kept watch during his sea-borne escape from his hostile brother, Aegyptus (who had demanded marriage between his fifty sons and their fifty Danaid cousins).²⁴ While Herakles' son, Telephus, was at the helm, he commanded his *prorates* "to watch the course that the sons of Atreus take to Troy." An ally of the Trojans

Fig. 5.1 Attic black-figure hydria showing the three deck officers. Paris, Louvre E735.

during the war, Telephus had given refuge to Paris and Helen when they landed in Mysia where he reigned.[25] With a sharp, penetrating glance, Odysseus also kept lookout for Scylla (Homer, *Odyssey* 12.230; see *CWW*: Chapter 7). This deck triumvirate is depicted on an Attic black-figure hydria: the seated *kubernetes* mans the steering oar at the stern; amidships an animated *keleustes* calls the beat; and the *prorates* looks aft.[26] (Fig. 5.1)

Launching the Vessel

In calm weather, when not underway, ships were left at anchor or beached. Jason beached his ship on the shores of Colchis (*Heroides* 12.15). Also beached were the Greek ships at Troy around which the combatants fought.[27] Odysseus beached his ships on the Sicilian shore before taking sail (Ovid, *Heroides* 12.15). From the historical record, Arrian's fleet was beached at Athenai on the Black Sea when the sea "turned savage" in 130/1 CE (*Periplus* 4.4). When violent weather threatened, ships were usually pulled ashore, bows toward the sea, and hauled back into the water when the weather broke.

Launching from the beach was a labor-intensive procedure for large ships, as shown by Apollonius' description of the *Argo's* launch. First the Argonauts excavated a trough of sufficient length (from the prow to the sea's edge) and width ("of the vessel's beam"), in which they laid rollers. The crew then lashed the oars to the benches, handle-side out, and pushed her down the track as Tiphys urged on their work with his calls (Apollonius, *Argonautica* 1.367–390). The stern cables of anchored craft were cast off before getting under way, as when

Telemachus left Ithaca, searching for information about his father, Odysseus, who had left for Troy twenty years before (*Odyssey* 2.418). The skipper would wait for an opportune time to put out to sea, using coastal winds and tides to his advantage, usually departing at daybreak when land breezes would carry the ship out.[28]

Anchoring

If an adequate beach was not available, or if security demanded, ships would anchor out, if they could (e.g., historically, Caesar's fleet off Britain [55–54 BCE]: *BG* 4.23.4, 28.3, 29.2–3, 5.9.1).[29] In the annals of legend and myth, Aeneas dropped anchor near a little port beyond the dangerous island of Leukate (western Greece), where Apollo's temple was a beacon for mariners at sea (Vergil, *Aeneid* 3.274–277), and he anchored again off Cumae (6.4). When mooring in Caietas Bay, nonetheless, the sterns were beached while the prows were secured with two anchors, one on each side (*Aeneid* 6.900–901). Two anchors, naturally, could better safeguard a moored ship (Propertius 2.22.41). Sailors would tie up their vessels with additional anchors whenever sea urchins steadied themselves by clutching stones for extra ballast, a sure sign, mariners believed, of rough seas.[30] Anchors were usually secured by ropes, but chain cables were also used. Alexander had protected his anchored ships from sabotage with anchor chains (Arrian, *Anabasis* 2.21.6). Caesar reported that the Veneti, like Alexander, used iron chains, instead of cable ropes, to hitch their vessels in the rough waters of the English Channel (*BG* 3.13.5).

Divers

Divers were important to the success of many naval expeditions. They delivered messages and supplies, as in 425 BCE when swimmers smuggled poppy seed and honey to Spartans during the Athenian blockade of Pylos.[31] Occasionally an anchor would foul, and a diver would then release it. The apocryphal Massalian diver, Phoceus (from Marseille), whose name evokes *phoca*—the Greek word for "seal," another underwater acrobat—was particularly talented at "saving" his breath, so that he could search for the sandy seabed or release anchors that were too firmly fixed (Lucan 3.696–704).[32] Defending his native waters against Caesar's invading forces in 49 BCE, and prevailing in an epic underwater fight with a Roman enemy soldier, Phoceus attempted to surface but was ironically stricken on the head by the hull of a ship, an act of unintentional suicide in the horror of a destructive civil war.[33]

Anchoring, however, might also leave a fleet susceptible to sabotage, and divers also crippled enemy fleets. In 480 BCE, during the war between Persia and the allied Greek city states, the legendary diver, "Skyllis" (of Skione, on the westernmost of the three Chalcidicean headlands), severed the mooring lines of the Persian fleet at Artemisium.[34] The apocryphal "Skyllias" sought help against the Persians by means of a legendary nine-mile underwater swim (Herodotus 8.7).[35] The swimmer's name suspiciously recalls our historical "Skyllis." Cables and hawsers were sometimes cut to facilitate escaping from the enemy (Livy 28.36.11). At Byzantium in 194 CE, divers slashed the anchor ropes to commandeer enemy ships (Dio 75.12.2).

Divers furthermore repaired sabotage. In 413 BCE, during the Sicilian engagement of the Peloponnesian War, the Syracusans drove piles into their harbor to protect their own ships at anchor while also endangering the invading Athenian navy. Athenian divers shaved down the piles, as if they were reefs, thus facilitating their access to the Syracusan harbor (Thucydides 7.25.7). In 48 BCE, at war with Caesar, Pompey sent divers to open up the harbor at Epirus (northwestern Greece), whose mouth had been blocked with stone-laden ships (Dio 42.12.2). Coastal communities also baited unsuspecting mariners. False beacons in the Euboean Strait lured ships into the shallows in order to pilfer the cargo.[36]

The Greeks, moreover, were excellent swimmers, a skill that saved many Hellenic sailors at Salamis. (Not knowing how to swim, many Persians and their allies drowned as their ships were destroyed: Herodotus 8.89.) In 325 BCE, taking advantage of his skilled swimmers as he prepared to launch a surprise attack along the Tomerus River (modern Hongol in southwestern Pakistan), Alexander's admiral, Nearchus, ordered his troops to fall into a phalanx formation on the sea's bottom and then to make their charge only once the ranks were three deep (Arrian, *Indika* 24.6–7).

Propulsion

Propulsion was achieved through a combination of a main square sail (plus often an *artemon*, "fore-sail") and men at oars, often simultaneously.[37] Triremes (so-called because three men together pulled at each oar) could reach up to 10 km per hour (5.4 knots; about 6 miles per hour) at battle speed in calm seas,[38] a respectable velocity for ships under square sail. We see a ship vividly propelled by oarsmen in Lucan (3.525–528):

> On one side Caesar's ships are exalted, on the other a fleet with many a Greek oarsman; oar-stricken hulls quivered, and repeated strikes set the lofty ships in motion.

According to the principle of the lever, rowers amidships exerted the greatest impact on propelling the ship: there was simply more oar (lever) amidships than fore and aft (the greater the distance between force [rower] and fulcrum [thole-pin], the greater the weight that can be moved). It is unclear if the peripatetic author of *Mechanical Problems* assumed oars all of the same length or longer oars amidships:[39]

> Therefore the ship is moved, while the oar-blade is supported against the sea, owing to the oar's handle, being within the ship, moving forward, and the ship, attached to the thole-pin, advances where the oar's handle goes. For where the oar-blade cleaves through the most (quantity of) sea, there the ship is mostly thrust forward by necessity. And it cleaves the most water where there is the greatest portion from the oar's thole-pin. Because of this, the rowers in the middle move the ship the most. For amidships, there is the greatest length from the oar's thole-pin within the ship.
>
> *Mechanical Problems* 4

The author's explanation is not quite right. He had assumed a lever of the first class where the pivot point is located between the force and the load to be moved (e.g., a see-saw), and he assumed that the oar is driving the water backwards. The oar is instead a lever of the second class where the weight lies between power and fulcrum (like a wheelbarrow where the axle is the pivot point, and effort is exerted at the handles in order to move the weight between fulcrum and effort). For an oar, the fulcrum is the point where the blade dips into the water, the effort is inside the ship as the rowers pull on their oars, and the weight to be moved is the ship's weight at the thole pin.[40]

Although oars were preferred (as more dependable especially when the winds failed), the image of a ship under full sail is irresistible. Ariadne saw the sails of the Athenian ship that had conveyed her absent-minded lover, Theseus, away from Naxos (in the Cyclades: Ovid, *Heroides* 10.6).[41] Odysseus' ship left Circe's island under "stretched sail" (*Odyssey* 11.11). Additionally, during the funeral games honoring Aeneas' deceased father, Anchises, despite the damage she took mid-race, running aground on a reef, the *Centaur* came into port "with full sails"

(Vergil, *Aeneid* 5.281). The poets also depicted sailors at their work, hoisting "lucky sails to the mast-head" (Propertius 3.21.11–13).

The ship's wheel was a late-seventeenth-century Dutch invention, improving on the sixteenth century whipstaff that enabled steering ever larger ships. Until then, ships were steered either with a tiller attached directly to the rudder or with steering oars. The tiller is a simple lever that provides torque to turn the rudder. Our peripatetic author recognized that the length of the tiller and the force needed to move it are proportional to the size of the boat; the force required to control a rudder with a tiller increases with the size of the vessel, and the force, applied by the rudder to the sea, results in steering the ship. Because of the mechanical advantage of the lever (rudder), a tiller attached to a rudder can be controlled with a "gentle exertion." The rudder, whose fulcrum is the point where it is attached to the ship, works on a horizontal plane, "taking the sea obliquely:"

> For the fulcrum is turned to an opposite direction, the sea to the inside, the fulcrum to the outside. The ship follows the fulcrum because it is attached. Therefore the flat oar-paddle pushes the weight and, pushed in the opposite direction by it, it drives the ship straight. But the rudder, since it is positioned obliquely, causes an oblique movement, either this way or that.
>
> *Mechanical Problems* 5

The author did recognize some differences between oar (acting squarely on the water as an agent of forward propulsion) and rudder (acting obliquely as an agent of lateral propulsion). Understanding that the water is the fulcrum, Heron (first century CE) clarified the Aristotelian explanation: "the rudder rests or supports itself on the water and is so able to overpower the ship."[42] Likewise, so thought our Aristotelian author, raising the yardarm increases the speed of a ship under sail: the peripatetic argued that the mast is also a lever that is acted on by the wind; the socket to which the yard is attached is the fulcrum; and the boat is the weight to be moved by the lever. Since the same power (wind) moves the same weight (ship) more easily when the fulcrum (mast socket) is further away from that weight, raising the yardarm increases the distance between weight and fulcrum and also the vessel's speed (*Mechanical Problems* 6). Heath dismissed this explanation as "thoroughly misconceived."[43] Instead, as a sail (at right angles to the ship's deck and the wind) is raised, the bow will dip progressively, owing to the forces that act downward at the pivot point.

Larger ships were equipped with a brace of steering oars, one at each side of the stern, emerging from outboard oar boxes. Sometimes these steering oars rested in cradles and were restrained by straps. The oars were socketed to a tiller bar or made fast directly to the hull,[44] and were consequently vulnerable to sabotage from enemy divers. In the middle of a naval battle, skilled marines in small skiffs might sever the cables by which the steering gear was attached to a hull, rendering a ship rudderless and thus disabled (Vegetius 4.46). On a warship the steering oars were massive, and they might be operated by "some tiny little old manling ... whirling around the rudder by means of a delicate vine-pole tiller," like the helmsman of Lucian's *Isis* (second century CE).[45] If well balanced, the steering oars were light to the touch, and the ship would steer easily (Lucian, *Navigium* 5). Artwork usually depicts the helmsman as seated, such as Telephus (above) and the *kubernetes* at the port (left) steering oar of our Attic black-figure hydria (Fig. 5.1).

The Military Sailing Season

Like the Phoenicians before them, the Greeks and Romans preferred not to sail during the stormy winter months when reduced daylight and persistent dense cloudiness hampered visibility in the Mediterranean Basin. For merchant mariners making the trek through the Erythraean Sea to India with the south-west monsoon, the extra two or so hours of winter daylight in waters closer to the equator would have been advantageous.[46]

Sailing did occur during the winter months, especially for commercial ventures (Chapter 6), nor was it altogether uncommon.[47] But our sources emphasized the distastefulness of winter sailing and often described the point at which the seas again become "navigable."[48] The Attic orator, Andocides (fl. 440–390 BCE), for example, asked: "For what greater danger is there than a sea voyage in winter?" (*On the Mysteries* 1.137). Sources also foregrounded the sailing season that was observed by ancient navies, in parallel with the infantry's military calendar that extended roughly from March to October, taking advantage of more predictable conditions on land and sea.

The best conditions for sailing extend generally from late May to mid-September ("after the rising of the Pleiades to the rising of Arcturus:" Vegetius 4.39). A millennium earlier and with the perspective of different waters, the "lubberly" farmer, Hesiod, restricted sailing to the fifty days following the summer solstice (late-June to mid-August). But sea conditions cannot be generalized over an expanse as broad as the Mediterranean.[49] To Aristotle, sea conditions were usually more treacherous during the changing of seasons (*Meteorology* 361b30–33). After mid-September, "navigation is doubtful and more exposed to danger," according to Vegetius (4.39), since Arcturus, "a very violent star," heralds fierce equinoctial storms. The winter months (mid-November to mid-March) were characterized by violent winds and frequent storms,[50] but sea ice is virtually unknown in the Mediterranean Basin.[51] Although Casson claims that the seas were "effectively closed" in the winter months, a growing body of evidence suggests that sailing continued well into the winter months (no doubt undertaken with greater caution).[52]

Winds[53]

A ship under sail is at the mercy of the winds, and it is therefore important to understand the winds, their directions and effects, and to use them to advantage: "therefore only in extreme difficulty does he, who has diligently studied the science of the winds, endure shipwreck" (Vegetius 4.38). It is reasonable to assume that records of wind data were maintained in order to facilitate navigation.[54] To this end, Cape Sigeum, near Troy, may have been the site of a weather station, as was the spectacular octagonal Tower of the Winds (*Horologium*) built by Andronicus of Cyrrha at Athens (mid-second century BCE). Standing 13 meters high (about 43 feet), the tower was equipped with an internal water clock and, on each face, a sundial and relief of the appropriate wind. On its roof a weather vane, taking the form of a Triton or sea monster, signified the direction of the prevailing wind with his rod (Varro, *Farming* 3.5.17; Vitruvius 1.6.4–7).[55] The *Horologium* is an elegant presentation of useful solar and meteorological information.

During the summer, prevailing winds in the Mediterranean are northerly (e.g., blowing from the north), thus expediting southward travel: that is, from Italy or Greece to Egypt, Africa, Asia Minor, and Syria. Casson shows an average speed of 6 knots (11 km/7 mph) for the southerly journey from Ostia to Africa, but 4.3 knots (8 km/5 mph) for the northerly trip from Carthage to Syracuse.[56] These prevailing winds nonetheless are a substrate for numerous local winds ensuing from ever-varying topography, weather, and meandering coastlines (for more on weather, see *CWW*: Chapter 4).

Aristotle treated the winds in the *Meteorology*.[57] Most winds, he noted, are northerly or southerly. In alignment with the tenets of Aristotelian physics, winds result from the evaporation of rainfall in the watery eastern and western regions, along the course of the ecliptic (the sun's apparent transit). The trajectory of the ecliptic also accounts for the obliquity of the direction in which the winds blow. North (Etesian) winds blow continuously after the summer solstice (originating from the evaporation in the cold, moist, arctic regions), but there is no corresponding southern wind after the winter solstice (arising from warm, dry places that lack sufficient moisture to sustain a southerly breeze). There are, consequently, more northerly winds than southerly. Contrary winds, furthermore, cannot blow simultaneously, according to Aristotle.

Aristotle's student, Theophrastus, explored the individuality of both seasonal and local winds in his less speculative *On the Winds*.[58] Each wind has its proper place, and each is shaped by force, temperature, quantity, moisture, topography, elevation, and origin. According to Theophrastus, most winds are created by the agency of the sun, which provides the energy for the exhalations (16, 19–23). To a lesser degree, the moon's light can also bring about winds (17), so it was surmised. Southerly breezes are colder than those "in our climate," and the winds passing through narrow confines (such as mountain passes) are colder still and more powerful because of temperature and confinement (3). Winds that travel over some distance tend to generate more force because they propel more air as they advance, condensing air into clouds and producing rainy weather (7). Theophrastus recorded the frequency and regularity of the contrary north and south winds, blowing respectively in winter and summer (10). Individual winds assume their idiosyncrasies according to their sources and the topography over which they blow. The east-north-east wind follows a circular pattern with a skyward concavity because it "blows away from the earth" (39). The west wind is warmer than the north wind because "it blows off vaporizing water, not snow," and its origins over the water account for its gentleness and cloudiness. Contrarily, mountains and snowmelt explain the strength of the north wind (40–44). Westerly winds take on local characteristics as well. They are destructive in the Malic and Pierian Gulfs (open to the east), but—when enclosed by mountains to the west—the West Wind:

> blowing from equinoctial sundown, takes the heat from the sun, which has fallen on the mountains, and deflecting it earthwards, launches it straight into the plain and scorches it.
>
> *On the Winds*, 45, trans. Coutant and Eichenlaub 1975

With their ridges and hollows, mountain ranges, in particular, were recognized as causing air currents and producing winds (Pliny 2.115). In sum, local conditions are as important to understanding the winds as are the underpinning philosophical principles.

Twelve prevailing winds were recognized, each originating from a separate point of the compass rose, 30 degrees apart. In Homer, only four winds are mentioned, a system dismissed eight centuries later as "stupid" for its oversimplicity by Pliny (2.119). An ancient wind rose was developed by Aristotle, perfected by Timosthenes, and adopted by Varro and the Greek Stoic philosopher Posidonius, whose Roman friends included Cicero and Pompey.[59]

Varro ("a careful scholar") and Seneca both accepted the paradigm of twelve winds, where each quadrant is dominated by a primary wind with two associated, subordinate winds, because the sun rises and sets in different places according to the season. These twelve predominant winds can be linked with the sectors of the zodiac along the ecliptic (Seneca, *NQ* 5.16.3). Rejecting the Stoic explanation of winds as caused by the earth's "breath," Seneca understood that air, like fire and water, has an innate ability to move (air is never completely motionless). Winds ("flowing air") are generated as air expands through evaporation, and they abate as air is condensed on the following analogy, air : wind : lake : river.[60] Shore breezes rise when the cold, wet air—emitted from marshes, lakes, rivers, or confined in mountain passes—is warmed by the rising sun. Air currents, nonetheless, are not caused exclusively by the sun's heat, otherwise ships would not be able to sail at night (*NQ* 5.10.1). Nor do these winds blow everywhere: "for some are shut out by the earth's contours," and some are strictly local (*NQ* 5.17.1, 5). Considering this system of twelve primary winds plus twenty-four subsidiary winds as too detailed and perhaps too technical, Pliny (2.119) preferred a model where two winds prevail from each of the quadrants: east, southeast, south, southwest, west, northwest, north, and northeast.

Many of the winds are geographically named: Thrascias from Thrace, Olympias from Mt. Olympus, Skiron from Skiron's rocks in Mégara, and Kaikias from the River Kaikos in Mysia (Asia Minor). The Lips emerges in Libya,[61] and the Phoinikias (south-south-east) arises from Phoenicia (Aristotle, *Meteorology* 63a21–365a13). This seems to suggest the maritime origin of an ancient compass rose based on a central Aegean framework.[62] However, the names and quarters of these winds were the subject of disagreement, prompting Aulus Gellius (125–180 CE) to seek clarification (*Attic Nights* 2.22). The names varied, and the quarters of the winds migrated: Eurus, originally the East Wind, became the Southeast Wind by the fifth century BCE, as reported later by Timosthenes.

Without endorsement, Pliny posited four explanations for the creation of winds:

1) Winds are caused by the constant motion of the world[63] and by the contrary revolution of the stars;
2) They are generated by a tide-like, ebbing and flowing "breath" of the universe;

3) They derive from the irregular emission of rays from celestial bodies that put air into motion; and
4) They are emitted by celestial bodies.

Produced naturally according to rational laws of physics, "winds have a law of nature that is not unknown even if it is not thoroughly understood" (Pliny 2.116).

The winds have various effects. Following winds sped the *Argo* away from Thrace (Apollonius, *Argonautica* 1.953), the Greek fleet to Trojan shores (Ovid, *Metamorphoses* 12.37), Odysseus' ship to the Sirens' island (Homer, *Odyssey* 12.167), and Aeneas' fleet to Crete (Vergil, *Aeneid* 3.130). The Argonauts were compelled to row towards Ares' island after the wind had dropped in the night (Apollonius, *Argonautica* 2.1032). Winds delayed Protesilaus at Aulis before he (and the rest of the Greeks) could set sail to Troy to retrieve Helen (Ovid, *Heroides* 13.3). Winds ceased to blow when Odysseus passed the Sirens, and his men had to row (Homer, *Odyssey* 12.168–169). Likewise, Aeneas' fleet drifted to Sicily when the winds failed (Vergil, *Aeneid* 3.568). Historically, Pompey's forces nearly overcame Caesar's navy near Dyrrhachium in 48 BCE, until a

Table 5.1 The Winds and Their Cardinal Points

Cardinal point	Greek name	Latin name	Season
E	Apheliotes	Subsolanus	Equinoctial rising point
ENE	Kaekias	Euroborus	Summer solsticial rising
ESE	Eurus	Vulturnus	Winter solsticial rising
S	Notus	Auster	
SSE	Leuconotus	Albus Notus	
SSW	Libonotus	Corus	Summer solsticial setting
W	Zephyrus	Subvespertinus	
WSW	Lips	Africus	Winter solsticial rising
WNW	Iapyx	Favonius	Equinoctial setting point
N	Aparctias	Septentrio	
NNW	Thrascias	Circius	
NNE	Boreas	Aquilo	

rising south wind rescued the Caesarians (Caesar, *BC* 3.26), and the breezes off Actium favored Agrippa's fleet in 31 BCE. Northerly winds steadied and calmed the Black Sea (Arrian, *Periplus* 6.1). Arrian sailed with the winds in the Pontus (*Periplus* 3.2).

Homer and Apollonius frequently specified the wind's direction, essential knowledge for a ship under sail. Zeus sent a northerly wind as the sons of Phrixus set out to return to their ancestral Greek homeland:[64] their sails had been ripped, and their hull split; they floated on the keel of their ship to the safety of Ares' island, where the Argonauts eventually found them (Apollonius, *Argonautica* 2.1097–1020). Odysseus battled the North Wind as he escaped from the Cicones, Thracian allies of the Trojans, when the violence of the storm had also ripped his sails, which he immediately stowed to prevent further damage (*Odyssey* 9.67–73). Odysseus finally came to Calypso's island on the keel of his storm-battered ship (*Odyssey* 12.420–425); Calypso would delay Odysseus on her lonely island for seven years. Some winds were notoriously precarious. The Roman poet Ovid envisioned Jupiter as unleashing the Notus (North Wind) to facilitate a flood that would destroy humanity for their arrogance (Ovid, *Metamorphoses* 1.262), and Horace reviled the Aquilo, Africus, and Notus that might threaten his friend Vergil's journey to Greece (*Ode* 1.3.12–14). Arrian moreover distinguished the harbor at Athenai as a safe mooring where ships were protected from the Thrascian winds (*Periplus* 4.2, 18.3).

Navigation

In its earliest (legendary) form, the art of navigation was magical, and Homer noted the miraculous ships of the Phaeacians that required neither navigator nor rudder (*Odyssey* 8.557–563). On one of these self-steering ships, Odysseus was finally returned to his home in Ithaca after ten years of war and another of wandering the Mediterranean. Navigation is the art of guiding a craft safely to a particular destination, aided, until recently, with reckoning only by landmarks or by the stars (the Viking lodestone aside, magnetic compasses are not attested in Europe until the late-twelfth century CE, to say nothing of the *Global Positioning System*).[65] The Minoans may have utilized magnetic direction-finding devices to align important buildings.[66] Thales (see *CWW*: Chapter 1) also seems to have recognized the attractive power of the lodestone in the sixth century BCE.[67] But the use of magnetic devices for navigation is unlikely in antiquity. Without such aids, the successful navigator had to be prepared to handle many challenges on the water, including winds and currents, tides, reefs, and shoals, by reading the skies, coasts, and natural and human-made landmarks.

Surface currents are products of the winds, but currents and tides in the Mediterranean are too weak to have any noticeable effect except at the straits (affected by currents) and deep inlets (affected by tides).[68] Apollonius remarked that the swift currents of the Hellespont propelled the *Argo* quickly across the depths of the Black Sea (*Argonautica* 1.922). The Po's current (flowing eastward across northern Italy), moreover, was taken into account by Vitellius' engineers in 69 CE, when they were charged with constructing a pontoon bridge there: the prows of the boats comprising the bridge were turned against the current (Tacitus, *Histories* 2.34). The prevailing coastal counterclockwise current in the Euxine may explain the counterclockwise order of Arrian's *periplus*.[69] It is unlikely, nonetheless, that the average skipper possessed the conceptual and mathematical tools to consider systematically the errors caused by currents (and drift).

Latitude and Longitude

By the late-fourth century BCE, latitude was understood, and it was computed from astronomical observations. Traveling into the northern Atlantic at that time, Pytheas of Massilia (Marseille) noticed that hours of daylight increased as he traveled further north during the summer months. Observing the changing angle of the sun's elevation, he employed simple trigonometry to estimate latitudes, which he connected to the duration of a place's solstitial day, a remarkable achievement.[70] Longitude, however, could not be reckoned with any precision until the development of accurate chronometers in the eighteenth century.[71] Inability to determine longitude could lengthen a journey or throw a ship well off course. Strabo (14.6.3), for example, gave the distance from Paphos on Cyprus to Alexandria as 3,600 stades (660 km/400 miles), a distance accurate for a ship heading due south and then veering westward towards the east coast of Egypt, rather than taking a more direct southwest tack to Alexandria. Mediterranean sailors, however, lacked instruments that could record the distance traveled by a ship on the water, and Menippus of Pergamon (ca. 80–20 BCE), the author of a *Periplus of the Inner* (Mediterranean) *Sea*, attributed discrepancies of distances in geographical authors to variable wind speeds.[72] It seems likely that distances were calculated according to the average number of stades sailed during a given day, approximately 20 knots per day (about 35 km/23 miles, as Menippus implied).[73]

Bearings

Navigators were guided by coastlines and landmarks. Apollonius described the passing topography as the *Argo* headed towards Lemnos: the entire coast of Magnesia with the tomb of Dolops, the wheat fields of Pelasgia, the cliffs of Pelion, the craggy heights of Meliboia, the flat beaches of Homólé.[74] Lemnos emerged from northern Aegean, and its highest peak (Mt. Athos) cast a "shadow as far as Myrína" (80 km away [50 miles]).[75]

Mariners also found their bearings by studying the behavior of animals, especially birds, and for this reason terrestrial birds (swallows, doves, or ravens)—which cannot land on water—might be kept aboard in cages to be released for determining where (or if) to make landfall. Land birds would make for the closest shore or return to the ship if land was too distant.[76] Doves guided ships from Chalcis to Cumae, where the Chalcidians established a settlement (Velleius Paterculus 1.4.1). A swan reputedly shepherded merchant mariners safely to the east coast of the Black Sea during a "blinding storm" (Mela 1.110). The Roman poet, Silius Italicus (first century CE), envisioned an eagle leading Scipio Africanus' fleet from Sicily to Carthage in 204 BCE (*Punica* 17.52–58). Fish were also instrumental in guiding ships safely to land, and the pilot fish (*pompilos*) was considered sacred because it escorts (*propempein*) ships from the sea into harbor (Athenaeus 7.284d).

Shoals

Knowing water depth is essential for maritime activity, and the ancients collected data that facilitated safe passage by taking soundings in order to determine water depth, an essential tool in navigation, especially for determining where it is safe to sail or anchor. Ca. 60 CE, Paul's storm-savaged crew, for example, took soundings by heaving a lead line when they suspected that they might be near land. When they read 20 *orguiai* (ca. 35 meters/ca. 120 feet, they let out four anchors to avoid running afoul on a shoal (Acts 27.27–32). Several varieties of ancient lead lines have been recovered, many of which were designed to lift samples of bottom sediment, another important source of information for the navigator. Sediment quality changes at the mouths of riverbeds and other geological phenomena, and this knowledge can enable navigation according to seabed topography when visibility is impaired.[77] The anchorages around Muza on the Erythraean Sea, for example, have sandy bottoms (*Erythraean Sea* 24). These data are constantly shifting because the relationship between earth and water is perennially fluxing owing to siltation (see *CWW*: Chapter 3).

Pilots

It was common for pilots to guide merchant ships through unfamiliar waters, as pilots continue to direct large ships through congested sea roads.[78] This was less an issue in mainland Greece where the rivers tend to be small, shallow, and unnavigable, but is especially important for the large rivers of western Europe and reefy shoals near coastlines. For example, as the mythical *Argo* passed through the reef-filled waters near Cyene, Poseidon's son, Triton, came to Tiphys' aid so that the *Argo* might thread through the narrow reefs and emerge undamaged (Lycophron, *Alexandra* 877–890). The art of piloting was so important to maritime communities that the Egyptians considered it one of the seven primary occupations (Herodotus 2.164). In Latin they are *navigatores* and *navigii*: navigators; for the Greeks, *kubernetes* (κυβερνήτης) referred both to helmsmen in general and to local pilots. In the second century BCE, Polybius observed that pilots were required for outsiders navigating Lake Maeotis (Sea of Azov: 4.40.8). Pilots are also attested in Cilicia (Plutarch, *Pompey* 24.3) and the Erythraean Sea where, for example, local fishermen in Barygaza, with their deep knowledge of marine topography and the local tides guide merchant ships between the shoals in the bay and then tow them up the river (*Periplus* 43–33, 46). In the fourth century CE, an elderly local mariner was called upon to pilot a ship that had run aground through the shoals off the coast of Azarium in North Africa (Synesius, *Letters* 4.176).

Knowledge of local waterways was essential for military campaigns, especially in unfamiliar areas, as Livy observes during his account of the battle of Corycus in 191 BCE: "knowledge of the sea, lands, and winds would be a great help, while all these things would throw ignorant enemy into confusion" (36.43.7–8). Hannibal the "Rhodian," a Carthaginian commander of the First Punic War (264–241 BCE), knew the Sicilian coast so well that he was able to evade Roman blockades by threading safely through the shoals in 250 BCE (Polybius 1.46.4). But Julius Caesar conceded that his ignorance of the Amorican coast (Brittany), its shoals, harbors, and coastal islands, would make his campaigns against the Veneti all the more challenging in 56 BCE (*BG* 3.9.3).

The intimate knowledge of riverbed topography possessed by pilots is as much instinct and art as it is data, a paradigm followed much later by the riverboat pilots on the Mississippi as described by Mark Twain:

> **Twain** So I see. But it is exactly like a bluff reef. How am I ever going to tell them apart?

> **Mr. Bixby** I can't tell you. It is an instinct. By and by you will just naturally know one from the other, but you never will be able to explain why or how you know them apart.[79]

A good pilot depends on repetition and the ability to anticipate fluctuations in the unobservable changes on the river's bottom. Although acknowledging that the technology employed by river pilots on the Danube had advanced over the years, Vegetius withheld details as beyond the scope of his work, restricted, as it were, to ancient military science (4.46). Along the Rhone, nonetheless, especially at Lugudunum (Lyon), a guild of local boatmen are robustly attested in the epigraphy,[80] including on memorials for curators and pilots,[81] as well as records of gifts (three denarii) for "all who sail."[82] *Navigii* are also recorded epigraphically in Rome, Sulmo, Etruria, Numidia, and Carthage.[83] Navigation cannot be successful without deep knowledge of marine topography, weather, local flora, and fauna. Coastal and river pilots ensured the safe arrival of goods and travelers to port. They were the keepers of knowledge which they gained through long years of sailing their particular waters, wherein they learned the environment and its moods, handing down their mysteries by word of mouth and apprenticeship to the next generation of coastal and river pilots.

Navigability

Literary sources described waterways as navigable or not. Unnavigable waters are thus impossible to ford, as at the Sarus River in Cilicia (Procopius, *Buildings* 5.6.9). Navigability is important for conveying "an abundance of necessities" into a city (Plutarch, *Roman Questions* 274f), and both Cato (*Agriculture* 4) and Varro (*Farming* 6) recommended farmland conveniently situated near navigable streams. Navigability is a factor of the size of the waterway,[84] and it is of interest to the general at war,[85] the geographer, and the travel writer.[86]

Waterways are not uniformly navigable. For example, the Nile is navigable in some places, but not all, according to Mela.[87] Siltation, furthermore, varied over time. Nile floods during the Ramesside period (1295–1069 BCE) were low; but, by 650 BCE, higher floods resulted in greater siltation and reduced navigability.[88] Other factors also affected navigability. In winter, ice prevented navigation along the Rhine and Ister Rivers, which were then traversed as if solid, dry land, and the ruts of chariot wheels were sometimes seen on the surface ice.[89] In the fourth century BCE, the Ister, branching between the Pontus and Adriatic, was not navigable at those points (Aristotelian *On Marvelous Things Heard* 839b10–19).

By the sixth century CE, derelict infrastructure and the ruins of Trajan's Bridge (ruled 98–117 CE), destroyed by floods, rendered the Ister no longer navigable at that point (Procopius, *Buildings* 4.6.16).

Knowing if and where bodies of water were navigable remains important to soldiers and traders. In the early-second century CE, tributaries of the Ganges and Indus Rivers were known to be navigable (Arrian, *Indika* 4), as were those that debouche into the Black Sea (*Periplus* 7.5, 10.1, 2). At Kenchreae, Corinth's harbor, Isis' boat was launched on "navigable waters" (Apuleius, *Metamorphoses* 11.5). Pliny the Younger praised the navigability of the Clitumnus River (Umbria), which was broad enough to accommodate two boats traveling in opposite directions (*Epistle* 1.8). But Sextus Junius Brutus was disadvantaged when he campaigned in Spain in 138 BCE because guerrilla fighters were able to take full advantage of their proprietary knowledge of the navigable rivers that bounded Lusitania (Appian, *Roman History* 6.71). In order to compensate for this tactical handicap, Brutus engaged in a devastating attack on the civilian population.

Celestial Navigation

Although the Greeks and Romans preferred to sail by day when visibility was better, evidence of night sailing and celestial navigation is not lacking. Trained maritime navigators could thus sail at night out of sight of land,[90] and early on the Phoenicians had learned how to navigate by their pole star (footnote 10). Lucan explicitly cited Cynosura (Ursa Minor) as "a very certain guide" for Pompey's eastern allies as they rallied to his cause (3.218–219).

The Greeks were knowledgeable celestial navigators. An accomplished navigator in the mythic sphere, the god, Bacchus, learned to watch the standard navigational constellations: the Pleiades and the Hyades (indicators of seasons), Capella (the brightest star in Auriga, abutting the circumpolar constellations), the rainy stars of the Olenian Goat, and the circumpolar Arctic Bears. Setting off eastward from Ogygia on his makeshift raft (fitted with sails and a steering oar), as instructed by Calypso, Odysseus also kept careful watch of the Pleiades and Boötes, keeping to his left the Bear (Ursa Minor), who "alone is without a share in Ocean's bath" (i.e., she never sets).[91] Odysseus' grasp of celestial navigation may reflect Archaic advancements or the state of technology of the Bronze Age setting in which the epic cycle is ostensibly cast. Aristotle, moreover, casually referred to "nautical astronomy" (*nautike astrologia*: *Posterior Analytics* 78b34–79a6; cf., Manilius 4.279–280). Pompey, furthermore, at least in Lucan's imagination, had learned the art of celestial navigation from his skilled helmsman:

> That star guides the ships, the north pole which is not submerged in the waves. Since this one will always rise to the zenith and Ursa Minor will always rise over the yard-arm, then we face the Bosporus and the Pontus rounding the shores of Scythia. When Boötes descends from the highest yard-arm and Cynosura is borne closer to the sea, the ship makes for Syrian ports. Then follows Canopus, a star content to wander in the southern sky, fearing the North Wind: with that one on your left, proceed past Pharos; the ship will reach Syrtis in mid-ocean.
>
> Lucan 8.174–201

Eschewing the planets that "glide and slip," the unnamed helmsman relied on the "fixed" circumpolar stars. Aeneas' helmsman, Palinurus, had marked the stars in the sky before embarking (above). While sailing to his Latian camp at night, furthermore, Aeneas explained the course of the stars to his young friend and new ally, Pallas (Vergil, *Aeneid* 10.160–162). This implies a knowledge of celestial navigation that might have been transferred to nautical contexts by merchants who were able to steer by the stars when they traveled the desert by night (e.g., Strabo 17.1.45). Vegetius, finally, repeated the seasonal signs that indicate when it is safe to sail, at least for the fourth-century-CE Roman navy.[92]

Maritime Highways and "Charts"

Following Gisinger's 1937 *RE* article on *Periploi*, Casson denied that Greco-Roman navigators made and used nautical charts in the modern sense,[93] arguing that such charts were unnecessary, since most of the sailing was coastal. This is a broad assertion, contradicted in both the archaeological and literary records (Strabo 1.3.2; Chapter 6). Preliterate Polynesians, furthermore, equipped with keen observation of the environment, instinct, and oral tradition, created elaborate nautical stick charts that accurately depict winds, currents, and islands.[94] Literary contexts suggest that Greek and Roman mariners had a graphic understanding of safe and well-trafficked sailing routes as recorded by shipwrecks and maritime debris. Some routes date as far back as the Bronze Age. For example, Cilicia to the north Syrian coast and the Aegean to Egypt.[95] Most extant ancient maps are, in fact, narrative; and pictorial maps were, if not rare, at least almost always monumental.[96]

A tradition of orally disseminated knowledge is implicit. In the *Aeneid* (3.274–275), the cloudy summits of Mount Leucas are dreaded by sailors. The statement is general; the knowledge is common. Palinurus also seems to have remembered sea routes in waters that he has probably never before sailed (above).

Epic heroes were forewarned of dangers, and they were usually advised on how to avoid them. Notorious was Cape Pelorus, the narrow strait between Sicily and Italy's toe, where legend placed Scylla and Charybdis. The latter, a ferocious whirlpool, was described by Apollonius as an endless roar and upsurge in his harrowing litany of "ship-smashing horrors at the sea's crossroads" (*Argonautica* 4.921–923). Regarding the double threat of Scylla, a sailor-eating sea monster, and Charybdis, a dangerous whirlpool, Circe's instructions to Odysseus are clear: "sailing quickly past Scylla's lookout, drive close past (her), since it is best to regret (the loss of) six from your ship's crew than all at once" (*Odyssey* 12.101–110). Scylla is a threat to only a few crew members; Charybdis is certain death for all. Vergil's Helenus gave similar advice to Aeneas (*Aeneid* 3.684–686): that the Trojan hero keep clear of the breakers on the coast and thus avoid the brace of dangers posed by the monster, Scylla, and the whirlpool, Charybdis. Between Scylla and Charybdis runs a narrow path that must be strictly navigated in order to avoid the loss of any crew.

Nautical charts may have been constructed in monumental contexts. On the pillars in Aeëtes' Colchian palace, the Argonauts saw tablets on which were "all the roads and paths (πᾶσαι ὁδοὶ καὶ πείρατ': *pasai hodoi kai peirat'*) of both the wet and the dry for those who travel around," all the more amazing for Jason who has sailed with his crew into an unexplored region beyond the Mediterranean Basin. These illustrations perhaps reflect monumental charts that might have been displayed in the Hellenistic courts of Apollonius' day (but they were likely unknown in the Bronze Age Colchis that provided the ostensible setting for Jason's quest for the Golden Fleece and other adventures).[97]

The narrative accounts discussed above suggest that the ancients had a similar understanding of aquatic superhighways that they envisioned for land routes. The ancient conception of sailing channels was graphic; and it was expressed with the same terminology that was employed for land travel.[98] The ancients perceived travel on water as occurring along organized routes with a commonly known infrastructure that was all but physically visible. Greek and Roman mariners were not traveling as if on some vast, barren wasteland. Or at least they rarely did so. Helenus had warned Aeneas that a long "pathless" (*invia*) route separated the Trojans from their final destination in Italy. The journey could only be by sea. Helenus stressed that the trip would be without a road (*in-via*: Vergil, *Aeneid* 3.383) and thus remarkable. Most sea routes were not so. *Via* (path, road) designates sea roads in Vergil (above), and Apollonius employed μιξοδία (*mixodia*: a place where several paths meet), a rare compound of *hodos* (ὁδός "road"). *Ploos* (πλόος, "a sailing or voyage by sea") and its compounds come to have the meaning of a sailing route as we see in Apollonius, where the Argonauts must return from Aia by "another sea route" (πλόος ἄλλος [*ploos allos*]: Apollonius, *Argonautica* 4.259). Compounds of *ploos* (*anaploos, diaploos, paraploos*), likewise, are commonly employed in Polybius, Strabo, Appian, and other ancient sources on geography and history.

Conclusion

Travel by water could be fraught with danger, and it was not undertaken lightly. But the sea was also the great lifeline of Mediterranean communities:

> The sea became as much a bridge between people as it was a natural barrier, viewed with notoriously ambiguous feelings both as a pathway of opportunity and as a sacred domain not to be violated.[99]

Skill and teamwork were essential in conquering this realm, and a crew that could anticipate orders during emergencies could mean the difference between life and death. Ceyx's crew responded courageously and professionally; the crew of Paul's ship, however, made a cowardly attempt to abandon the vessel and her passengers. For the purpose of survival, commerce, warfare, expansion, and travel, the peoples of the ancient Mediterranean had perfected the necessary arts of sailing and navigation from their earliest days on the Mediterranean. Like Benjamin Franklin who interviewed sailors when he was collecting data in order to chart the Gulf Stream, many ancient scholars, including Eratosthenes (Strabo 2.5.24), Hipparchus (Strabo 2.1.11), and Strabo himself (2.5.8, 2.5.24), relied on sailor reports for knowledge of the interstices of land and water, localized and far-ranging weather patterns, and for glimpses of the curiosities from distant places.

6

Maritime Trade and Travel

Introduction

The philosophy of cargo shipping has remained unchanged for millennia: to transport as much merchandise as quickly as possible, as cheaply as possible.[1] From the third millennium BCE onward, Egyptian ships ventured into the Erythraean (Red) Sea, and from there to the eastern coast of central Africa for trading in exotic goods.[2] By the mid-second millennium BCE, the Phoenicians had established extensive trade routes, including with Israel (1 Kgs 5:8–9, 10:11), Egypt (Herodotus 2.9.2), Greeks *poleis* (Herodotus 3.107, 111), and others, especially along the northern coast of Africa.[3] Minoan Greeks were trading as far west as Sicily; and, by the early-sixth century BCE, Athens was trading with the Etruscans of western Italy. The evidence for the early Roman imperial era (first century CE) is particularly rich.

Despite the well-developed and carefully engineered system of Roman roads, travel and trade by water remained cheaper, was frequently more comfortable (storms aside), and was usually quicker (as it is to this day). Marble for Pericles' building projects (430s BCE) on the Acropolis could only have arrived by boat. Augustus (ruled 27 BCE–14 CE) and Caligula (ruled 37–41 CE) had both ordered the construction of special barges of unparalleled dimensions to convey obelisks from Egypt to Rome, ballasted by 120 bushels of lentils and with a capacity of 1,450 tons.[4] Maritime trade linked the distant outposts of empire, as goods were trafficked across the waters. By the end of the first century BCE, Roman wine and drinking vessels were coming into Britain, as was Spanish olive oil and fish sauces, glassware, jewelry and tableware, medical tools, board games, toilet utensils, all organized through the Gallo-Belgic hub at Camulodunum (Colchester).[5] Pliny credited maritime trade with linking distant corners of the first-century-CE Roman empire, as goods were imported into Rome from Alexandria, Gades, and elsewhere.[6] The complex network of trade that existed in the first and second centuries CE was maintained for supplying the Roman army and feeding the inhabitants of the empire's capital city.[7]

Plying the Mediterranean were freighters laden with standardized long-distance shipping containers, amphorae, with a capacity of 3 *modii* (about 7.5 gallons [28 liters]) and uniquely designed to fit tightly in the hold of a merchant ship. These ancient containers (usually ceramic) were used primarily for liquid commodities (wine, olive oil, and garum), but they might also be filled with grapes, olives, fish, grain, or other foodstuffs. The very large grain transport ships (carrying up to 3,000 amphorae [about 150/165 tons]) would have anchored at sea while smaller vessels shuttled cargo into port. The fragmentary survival of these terracotta shipping containers, in urban trash heaps[8] and shipwrecks that have settled on the seabed, contributes to the reconstruction of trade routes.[9]

Here we shall explore some of the issues that concern maritime trade and travel in the ancient Mediterranean, including the commercial sailing season, trade routes, transport of goods, and piracy.

Trends in Ancient Mercantilism

As now, in antiquity trade was practiced both locally and globally between networks of posts that dotted the littorals. Trade was "a composite process in which production, transport, and marketing processes are intrinsically linked," as goods were conveyed partly by land, partly by water.[10] This mercantile complexity strongly suggests deliberate, targeted avenues of trade. Evidence from shipwrecks (often indicating where a cargo vessel was loaded) argues for high levels of specialization among the merchant mariners, as goods were exported from production centers to hubs and then distributed to local markets. Single cargoes are evident in many wrecks: e.g., wine (the wreck off Madrague de Giens, east of Toulon was loaded with wine stored in Campanian fineware and coarseware vessels produced at San Anastasia [see below][11]), and stone (for building, decorative stone, finished artworks, or second-hand materials for re-use[12]). Some harbors were adjacent to warehouses dedicated to the import of specific wares, such as—at Rome—marble (*Marmorata*), grain (*Horrea Lolliana*), or wine (*Portus Vinarius*: *CIL* 6.9189, 9190, 37809).[13] Other harbors may have been purpose-built for the export of raw materials, e.g., the harbor at Palatia (modern Saraylar, Turkey), near the Proconnesian marble quarries.[14]

Greek Trade

Bronze Age Wrecks

Bronze Age wrecks (Uluburun, Cape Gelidonya, and Point Iria) provide insight into the goods that were traded between communities in the Bronze Age.[15]

About 6 miles [10 km] southeast of Kaş in southern Asia Minor, resting on the seabed at a depth of 44–61 meters (about 140–200 feet), the Uluburun wreck was discovered by a sponge diver in 1982. About 15 meters long (49 feet), the ship was a shell-first construction fitted with mortise-and-tenon joints (the projecting tenon locks within a mortise hole). The planks and a rudimentary keel were Lebanese cedar; the tenons were oak. The ship seems to have been en route towards the Aegean, perhaps the commercial hub at Rhodes or a mainland Mycenean palace. In the late-fourteenth century BCE, traveling in a northwesterly direction, the ship may have failed to clear the rocky promontory or may have been blown off course by a sudden southerly wind.[16] The ship's cargo capacity was at least 20 tons burden. Twenty-four stone anchors and 1 ton of cobble ballasted the ship. The cargo was varied, including raw materials and luxury items. The raw materials include copper, tin, glass ingots, Canaanite jars with Pistacia (terebinth) resin, a preservative in ancient wine; African blackwood (Egyptian ebony); ivory. Also in the cargo were Cypriot oil lamps, seashell rings, Baltic amber, beads (agate, faience, ostrich eggshell, rock crystal), jewelry (including a gold scarab inscribed "Nefertiti"), weapons (daggers, spearheads, arrowheads, maces), tools (sickles, awls, drill bits, chisels, a ploughshare, whetstones, and adzes), a green volcanic stone ceremonial axe (from Bulgaria), pan-balance weights, and foodstuffs (almonds, figs, olives, grapes, pomegranates, black cumin, safflower, coriander). The geographic origin of the objects on board was vast: from Mesopotamia to Sicily, the Baltic, Egypt. Some of the finer objects may have been intended as royal gifts.[17]

At Cape Gelidonya, off the coast of Lycia (ca. 1200 BCE), a merchant ship (Canaanite or Phoenician) scraped across a rock pinnacle before sinking as cargo spilled out. The stern eventually settled on a boulder about 50 meters (about 160 feet) to the north. Of similar construction to the Uluburun wreck, the Cape Gelidonya wreck also employed brushwood to secure cargo in the hold (cp. *Odyssey* 5.256–257 where Odysseus covers his raft with brushwood as a defense against the water). The cargo was primarily scrap for recycling and raw materials for weapons: copper, tin, stone hammerheads, stone polishers and whetstones (a craftsman may have been on board).[18]

Found off the western Argolid coast, the Point Iria ship was a small (10 meters [ca. 30 feet]) Cypriotic craft that carried a workaday cargo predominantly of Mycenaean and Cypriot pottery, large storage jars, and Cretan transport jars with some decorated Mycenaean vases. The ballast seems to have been comprised of river stones of various sizes. The hull has disintegrated and there is a perplexing absence of metal objects associated with the wreck site. The ship likely foundered owing to a sudden onset of bad weather at a dangerous point (high winds and heavy seas are typical for the spot). The wreck was found 10 meters offshore (30 feet), and the cargo was scattered haphazardly. The remains confirm mercantile relations between Cyprus, Crete, and the Argolid for the Mycenean era.[19]

Also flooding into Asia Minor in the twelfth century BCE was a uniform pottery style (gray monochrome thrown and glazed: "Minyan ware") that was probably manufactured in Argolid factories.[20] Such evidence from shipwrecks and the broad distribution of Minyan gray-ware indicates a complex network of mercantile trade and travel across the eastern Mediterranean between diverse peoples who were exchanging raw materials, finished goods, foodstuffs, and luxury items. The evidence attests the human ingenuity that imposed control over the waterways.

Athens

There is also robust evidence of broad mercantile networks throughout the Greek world from the Archaic period onwards. In the sixth century BCE, asphalt was being imported from Mesopotamia or the Caspian to Mégara.[21] In the sixth and fifth centuries BCE, Athenian pottery was "etruscanized" with regard to the shapes and imagery that was especially popular with an Etruscan market. Examples of this pottery have been found in tombs, sanctuaries, and city states throughout the large region of Etruria.[22] In turn, Etruscan bucchero pottery was imported to Sparta and Athens. Moreover, by the fifth century BCE, Athens, whose soil was not suitable for grain crops, was importing dietary staples from the Black Sea, Egypt (Plutarch, *Pericles* 37.4), and Sicily. From the distribution of amphoras, according to material, manufacturers' stamps, and find sites, it is possible to tease out the networks of the ancient wine trade.[23] It was a perennial concern for politicians to maintain a reliable supply, and Athens expressed interest in Sicilian grain production before mounting their Sicilian expedition (Thucydides 6.90: 415–413 BCE). According to Demosthenes, Athens was importing 40,000 *medimnoi* of grain from the Bosporus annually in the 350s BCE (*Against Leptines* 20.32).[24]

The Fourni Wrecks

Evidence in Aegean (Greek) trade routes and wares is also brought to bear at Fourni, an archipelago between Samos and Ikaria in the western Aegean, an area characterized by sudden, violent squalls, a rocky coastline, narrow passages, and windstorms that arise suddenly from the mountainous terrain. In 2015, a free diver discovered and mapped about forty possible wreck sites. Marine archaeologists are currently excavating fifty-eight wrecks on the seabed, the earliest of which dates to ca. 525 BCE. Ships and their cargoes are also datable to the Classical era (480–323 BCE), Hellenistic (323–31 BCE), late Roman (300–600 CE), and Medieval periods (500–1500 CE). The cargoes were comprised primarily of storage amphorae which, according to analysis of the different clays, originated from many sites across the eastern Mediterranean including Cyprus, Egypt, mainland Greece, Rome, Spain, and North Africa. Speculatively, the contents were typical: oil, wine, fish sauce, and honey. The finds have not yet been published, but will no doubt shed important light on the broad network that connected mercantile centers, not just in the Aegean but throughout the Mediterranean.[25]

Trade in the Roman Era

The *Periplus* ("Coasting Guide") of the Erythraean Sea

Fortuitously extant, the anonymous *Periplus of the Erythraean Sea* provides a vivid glimpse into the patterns of travel and maritime trade between eastern Africa and the Indian subcontinent in the first century CE[26] (Fig. 6.1). Like river pilots, most cargo captains would know their stretch of coast well, making frequent runs between their own particular harbors (*Erythraean Sea* 14), and such knowledge would be passed from one captain to his successor. Ambitious merchants might have controlled broader runs, extending along a coast or an entire region.

As valuable to the merchant as to the cargo captain, the short treatise is part sailing manual, part mercantile handbook, and part guidebook. The *Erythraean Sea* preserves a wealth of information regarding harbor conditions, imports, exports, tariffs, local peoples, governments, and the crime rates one might expect at various coastal emporia. The fish eaters dwell beyond Berenike (2), the inhabitants of Malao are peaceful (8), at the wooded bird sanctuary of Menuthias Island there are no wild animals except crocodiles who, contrary to their behavior

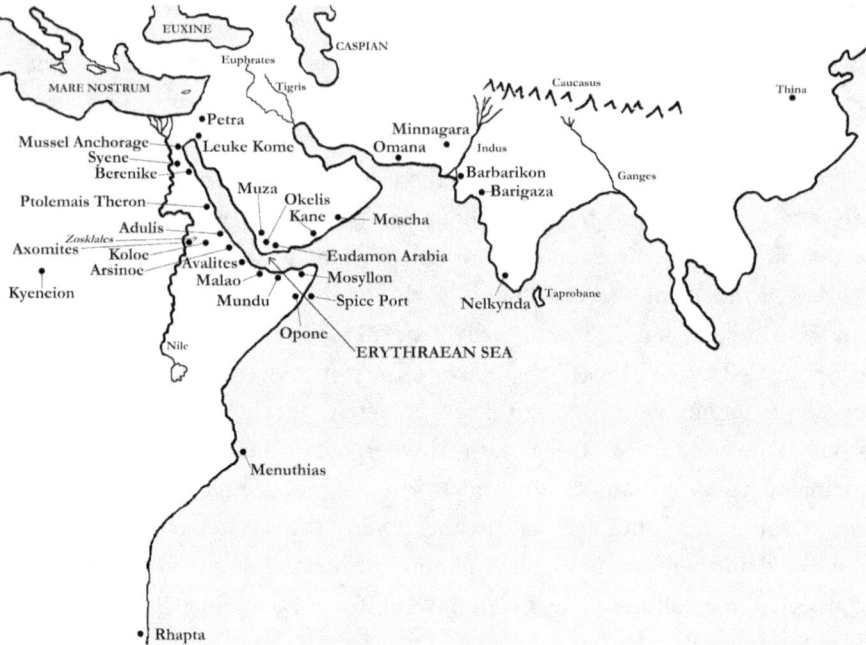

Fig. 6.1 Emporia of the Erythraean Sea.

elsewhere, "do not injure people" (15). The languages of the many tribes along the Arabian coast differ to lesser and greater degrees (20). The traveler should be aware of the warlike Bactrians (46) and the cannibalistic "Horse-Faced Peoples" (63). The sightseer might be interested in the vestiges of Alexander's expeditions at Minnagara (327–325 BCE): temples, camp foundations, and huge walls (41).

Traders would also profit from knowledge of the mercantile climate, including taxes or tariffs that might affect business. The sellers at Mundu, for example, are reputedly "hard bargainers" (9). Malichas, king of the Nabataeans, requisitioned a hefty duty of one-fourth of the wares imported into Leuke Kome, secured through a centurion (commander of a unit of 100 men), together with an armed garrison (19: Leuke Kome linked Petra with the Mediterranean). According to Strabo, increased trade between the Arabian Gulf and India, especially in high-ticket commodities, resulted in heavy tariffs by the early-first century CE:[27]

> Formerly no more than 20 ships dared to pass through the Arabian Gulf in order to have a view outside the narrows, but today large fleets are sent as far as India and the Aithiopian extremities. From these places the most valuable cargo is brought to Egypt and then sent out again to other places, so that double duties are collected, on both imports and exports, and things that are expensive have a

heavy duty. There also are monopolies, for Alexandreia is not only the receptacle for such things, for the most part, but also the supplier for those outside it.

Strabo 17.1.13

A ship's captain would also find it particularly useful to know the hotspots of sea-borne crime. Raids were commonly directed against merchants putting in to Adulis, and thus skippers preferred to anchor out instead of docking at the harbor (*Erythraean Sea* 4). The vicious Kankraitai would plunder passing ships along the Arabian coast, enslaving any shipwreck survivors (20). The fortress at Sygaros was probably intended to guard Sachalites Bay against pirates or other raiders, one assumes (30). Another pirate threat existed at India's southern cape (53).

Some ports primarily served local markets, exporting regional commodities and importing foreign goods and raw materials that were not readily available. A typical example is found along the Aithiopian coast, south of Adulis, where there was a market for "unused" clothing, wraps from Arsinoë, printed fabrics, linen, double-fringed cloth, agate-like glass ("the kind produced in Diopolis"), a little Roman money for resident foreigners, copper for cooking pans, armlets and anklets, iron for spear points (for warfare and hunting elephants), pans for cooking, axes, adzes, knives, olive oil, and wine from Laodicea and Italy. Commodities were imported especially for the king: silver- and gold-ware, "fashioned according to the local practice," Indian iron and steel, heavy unadorned cloaks, and broad cotton cloth (6). Technology and taste drove the market: during the first–second centuries CE, glass production was concentrated in Syro-Palestine (Syrian glass was exported as far west as Rome), and there is no evidence of Roman-era glass workshops south of Berenike.[28] "Genuine" tortoise shell was among the raw materials exported from Ptolemais Theron.[29] Obsidian was to be found only deep under the sandy bay of the kingdom where Zoskales ruled (5: the king was "a fine man, fluent in reading and writing Greek"). The cassia trade at Mosyllon was robust, attracting large ships that would put in at the port (10). Good quality slaves, we are told, came from the markets at Opone (13).

Trade brings prosperity,[30] and hubs would bustle—Muza, for example, was replete with "Arab ship-owners and sea-men" (21)—as wares were conveyed between ports and inland markets either over land or water. Before the first century CE, Eudaimon Arabia (Arabia Felix ["lucky"]) was the node between Egypt and India, a clearing house for merchandise from both east and west. Egyptian ships would not sail eastward, nor did Indian ships venture westward

(26). As demand increased and ship technology improved, cargo ships (in the first century CE) consequently journeyed beyond historical mercantile runs. Confirming that trading runs under the Ptolemies were far more modest, as "only a few dared to sail and trade merchandise with the Indians," Strabo (2.5.12) added that he had personally accompanied Gaius Aelius (prefect of Egypt, 26–24 BCE) up the Nile to Syene, where they observed 120 ships setting out from Mussel Anchorage (Myos Hormos, at the northwestern coast of the Erythraean Sea) to India. Adulis was a hub for the ivory trade.[31] Elephants and rhinoceroses were slaughtered inland, and their tusks were collected from "beyond the Nile" at Kyeneion and then sent to Axomites. The ivory then traveled for five days to Koloe, "the first trading post for ivory," and finally for another three days to the market at Adulis where a few pachyderms could be seen along the coasts (*Erythraean Sea* 4). Frankincense came to the port of Kane by camel and boat (27), but at a human cost: growing in unhealthy districts ("harmful to sailors and fatal to workers"), frankincense was handled by slaves and convicts "who die on account of lack of nourishment" (29). Onyx came in great quantities to Barygaza (on India's Gulf of Khambat) in wagons along "road-less routes" (51). Rivers also facilitated trade between inland and coastal emporia and ships awaiting cargo for export, such as those moored at Barbarikon Harbor (39). At Omana, large vessels would put in with their imports of raw copper and wood, taking on cargos of pearls, *madarate* (local sewn boats), and frankincense from Kane, before heading off to Arabia (36). Silk, wool, thread, and fine cloth, moreover, made their way from "Thina" ("a very large inland city") overland through Bactria to Barygaza and then along the Ganges (64) before heading west to ports at Nelkynda (56), Barbarikon at the mouth of the Indus River (39), and Barygaza on India's south-western coast (49).

For many coastal towns, prosperity came from mercantile commerce (54). Good harbors or anchorages, consequently, were invaluable. Skippers would therefore need to know the conditions of the harbor—including currents and drafts—for determining if a given port could be approached safely. On a deep bay, for example, Adulis was thus suitable for craft of any size (4). Rafts and small craft could put in at Avalites (7), whose harbor must have been shallow. In Strabo (16.3.3), traders on the Euphrates utilized rafts. Malao has an open roadstead of a harbor, sheltered by a promontory that would protect ships from windy weather (*Erythraean Sea* 8). Ships could moor safely at an island near Mundu's shore, but Mosyllon, on a beach, had a poor harbor. The periplus author failed to explain the harbor's indigence (9–10). The harbor at Spice Port could be dangerous because of its exposure to the north and, therefore, bad weather (12). The coast of Arabia was "more or less entirely dreadful" owing to its lack of harbors, poor anchorages, rocky stretches, steep coastal cliffs, and threats of piracy. The author experienced that stretch of water first-hand:

> for this reason (tricky coastal conditions), as we sailed, we set a course for the Arabian land down the middle, and we sped up all the way to Katakekaumene, straight after which are continuous lands of peaceful inhabitants, pasturing animals, and camels.
>
> 20

His relief at surviving the sail along the Arabian coast is palpable. The waters around harbor-less Muza, at least, provided a good roadstead for mooring (24). Although not a market town, Okelis nonetheless served an important function as a watering station for long-haul merchants, "the first place to put in for those sailing on" (25). The narrow gulf at Barygaza could challenge helmsmen. Consequently, the left-hand approach for vessels putting in from sea is recommended (43).

The distances between ports are also carefully recorded, given either in terms of days, "runs," or stades.[32] For example, it was three days' travel between Adulis and Koloe (4); two "runs" (*dromoi*) between Malao and Mundu (9), and 800 stades (148 km/92 miles) between Avalites and Malao (8).

Finally, the *Erythraean Sea* provides valuable hydrological data. At Spice Port, growing turbidity of the waters indicates approaching storms (12), a sign of the reciprocal effects of water and weather. It would also be crucial for the captain to have an accurate understanding of seabed topography (Chapter 5). Stretches along the territory of the Kankraitai are rocky (20). At Barake, sheer drops alternate with rocky, shallow bottoms, threatening watercraft with running aground (40). The appearance of snakes could indicate proximity to land (38). Riverine topography poses its own challenges because of unfamiliar shoals, and pilots were often employed to guide ships from the open sea into harbors, as at Barygaza where the mouth of the river debouching into the gulf was hard to find (43–44, 46).

Trade Routes

"In response to a distinct lack of evidence,"[33] scholars have long debated whether ancient maritime trade occurred either over the open sea or along the coasts according to various practices: *capotage* (from coast to coast) and *cabotage* (along "national" coasts), together with *tramping* (sailing from port to port in opportunistic searches for markets for portions of cargo), utilizing a network of primary distribution emporia linked with secondary cabotage emporia for redistribution.[34] Arnaud, however, observes that trading patterns must accord with sailing patterns, neither of which can be reduced to this simple binary model, as evident from hints in the literary, documentary, and archaeological evidence. From the time of Eratosthenes (third century BCE), trade occurred over open waters (f13Roller=Strabo 1.3.2). Trade, furthermore, fell under state control, closely regulated and taxed, with strictures regarding where and with whom trade could occur, according to international treaties (*synthekai, spondai*). Agreements between Rome and Carthage, for example, included reciprocal interdictions against marauding, trading, or founding cities in certain areas.[35]

Into the Roman era, many merchant mariners would follow the trade routes long since established by the Phoenicians, continuing to hug the coastline and taking on as return cargo "whatever they fell upon" (*Erythraean Sea* 14). Before long, nonetheless, trade routes were developed over open waters, and perhaps it always had been so (shoaley coasts pose their own dangers), depending on market demands. According to the *Erythraean Sea* author (57), Hippalos may have been the first skipper to discover open-water routes that linked the ports of trade between Kane and Eudaimon Arabia in the Arabian Sea "having grasped the disposition of emporia and the shape of the sea." In Strabo (2.3.4), Eudoxus (fourth century BCE) was credited with discovering that sea road. In Pliny, Hippalos was the name of a local wind that blew Alexander's fleet (and later sailors) safely from Patale to Sigerus.[36] Regional routes may have been recorded in other *periploi* (lost or unattested) or simply preserved orally by guilds of local pilots.[37]

Trade and sailing routes can be theoretically reconstructed by marshaling literary and archaeological evidence. Pliny (19.3–4) described a Mediterranean Sea sprinkled with ships that were importing the exotic corners of the empire to Rome: a gentle breeze could bring a ship from Pozzuoli to Alexandria in nine days, or from the Strait of Messina to Alexandria in six or seven. The route from Gades to Ostia could be completed in seven days. A ship from Narbo could reach Rome in three days, and one from Carthage in two. With favorable breezes and currents a ship could travel 4–6 knots (7.5–11 kmph/4.5–7 mph), covering up to 800 stades (148 km/92 miles) in a day.[38] But conditions were often unfavorable, and voyages could be much extended. Because of adverse winds that buffeted Posidonius' ship (first century BCE) between the Gymnesian Islands (Majorca and Minorca), Sardinia, and the coast of Libya, he spent three months "sailing upstream" (against the weather) "with difficulty" trying to reach Rome on his return from Gades (Strabo 3.2.5) (see Table 6.1).

Table 6.1 Lengths of Some Historical Sails[39]

Route	Length in days	Estimated Distances: Nautical miles/km/miles	Speed: knots/kmph/mph	source
Corinth/Puteoli	4.5	670/1240/771	6.2/11.5/7.1	Plutarch, *Marius* 8.5
Ostia/Africa	2	270/500/310	6/11/6.9	Pliny 19.4
Ostia/Gibraltar	7	935/1731/1076	5.6/10.3/6.4	Pliny 19.4
Ostia/Hispania Citerior	4	510/944/587	5.3/9.8/6	Pliny 19.4
Epidamnus/Rome	4.5	600/1111/690	5.5/10/6.3	Procopius, *Goths* 3.18.4
Ostia/Gallia Narbonensis	3	380/703/437	5.3/9.8/6	Pliny 19.4
Messina/Alexandria	7	830/1537/955	5/9.2/5.7	Pliny 19.3
Rhegium/Puteoli	1.5	175/324/201	5/9.2/5.7	Acts 28.13
Carthage/Gibraltar	7	820/1518/943	4.9/9/5.6	Pseudo-Skylax, *Erythraean Sea* 111
Puteoli/Alexandria	9	1000/1852/	4.6/11.8/5.2	Pliny 19.4
Alexandria/Ephesus	4.5	475/880/1151	4.4/8.1/5	Achilles Tatius 5.15.1, 17.1
Carthage/Syracuse	2.5	260/481/299	4.3/7.9/4.9	Procopius, *Vandals* 1.14.8
Rhodes/Alexandria	3 3.5	325/602/374	4.5/8.3/5.1 3.9/7.2/4.4	Appian, *Civil Wars* 2.89 Diodorus Siculus 3.34.7
Byzantium/Rhodes	5	445/824/512	3.7/6.8/4.2	Marcus Diaconus, *Porphyry* 55
Alexandria/Massilia	30	1500/2778/1726	2.1/3.8/2.4	Sulpicius Severus, *Dialogues* 1.1.3
Puteoli/Ostia	2.5	120/222/138	2/3.7/2.3	Philostratus, *Apollonius* 7.16
Rhodes/Byzantium	10	445/824/512	1.8/3.3/2	Marcus Diaconus, *Porphyry* 37
Caesarea Maritima/Rhodes	10	400/741/460	1.7/3.1/1.9	Marcus Diaconus, *Porphyry* 34
Alexandria/Cyprus	6.5	250/463/287	1.6/2.9/1.8	Lucian, *Navigium* 7

Major hubs in the Mediterranean included settlements founded by the Phoenicians, (Gades, Malaca, Tingis, Utica, Sidon, and Tyre), Greek centers of trade (Emporia, Massilia, Cumae, Syracuse, Cyrene, Thera, Athens, Rhodes), harbor towns along the Black Sea (Byzantium, Sinope, Chersonsesus), as well as Portus at Ostia. The Phoenicians may even have considered their routes proprietary. In order to keep the trade routes between Gades and the Kassiterides (Tin Islands) classified, a Phoenician skipper deliberately wrecked his ship in the shallows when he discovered that Roman prospectors—trying to learn the lucrative route—were trailing him (second century BCE?). The Romans followed "to the same destruction." The Phoenician skipper subsequently "received from the government the value of the cargo that he had lost" (Strabo 3.5.11).

Archaeological evidence, including debris from shipwrecks, can help reconstruct some mercantile networks. The data, however, are problematic, providing only a partial recreation, since organic commodities (grain, textiles) are rarely detectable, and remains are mostly in the form of ceramic amphorae or durable stone.[40] Although amphorae most commonly accorded with the paradigm of the large, heavy, spike-bottomed jugs,[41] other models existed, also offering valuable clues to trade routes and distribution of merchandise. Thin, flat-bottomed containers from Gaul, for example, were probably packed in straw to prevent damage during shipping. Analysis of the clay can, moreover, reveal the container's date and origin.[42] Data from shipwrecks, however useful, are fraught and must be used with caution. Known wrecks are concentrated, for example, off southern France "where it is simply easier to find" them.[43] Other wrecks, many of which were discovered in the early twentieth century by fishermen and sponge divers, are clustered off Dalmatia and Phrygia, in the shallower waters where the interplay of weather, water, coasts, shoals, and reefs pose challenges to watercraft. The wreckage of ships that were conveying perishable food items (such as grain), consequently, has likely been underestimated in so far as foodstuffs are unable to protect the wooden hulls to the same degree that non-perishable commodities are extant.[44]

The Merchant Sailing Season

According to scholarly wisdom, the Mediterranean was "closed" (*mare clausum*) to traffic during the winter because of the risks of sailing during that stormy season.[45] In 380 CE, the emperor Gratian restricted sailing from April 13 to October 15:

> The emperors Gratian, Valentinian, and Theodosius Augusti (send) a greeting to the African ship-owners ... It pleases us that in the month of November navigation will be ended, and resumed for trade in the month of April which is the month closest to the summer season. The necessity of this acceptance from the kalends of April to the kalends of October shall remain in effect. But navigation will be extended to the Ides of those same months.[46]

Gratian's prohibition likely applied only to state-controlled cargo from North Africa because of the great liability to the grain shipments.[47] But Gratian's edict was soon repeated in Vegetius, who claimed to have written from both research and personal experience. In Vegetius, sailing was prohibited during the winter and restricted to the summer months (May to October), in parallel with the Roman military campaign season that was limited to the summer months.[48] (The Greeks did engage in wintertime naval warfare in the eastern Mediterranean where winter sailing conditions were more favorable: see Thucydides 8.35.) Vegetius, however, was addressing a specifically (Roman) military audience that included army recruiters, drill sergeants, and commanding officers. The Mediterranean Sea, nonetheless, is a complex hydrological system that is far from uniform, but is instead comprised of several interlinking maritime spaces with reciprocal effects (Rougé identifies twenty-one such maritime spaces), whose behavior is transparently affected by surface currents, winds, and seasonal weather, but is also opaquely shaped by sea depth, benthic currents (at a waterway's lowest level), water temperature and density, and the shape and topography of coastlines. The western Adriatic, for example, is calmer owing to fresh water at the surface, a cold benthic zone, and a relatively smooth coastline that lacks proximity to major islands. The choppier eastern Adriatic, in contrast, is affected by jagged shores, large islands, and mountainous terrain, producing temperature variations that result in jetties and bora winds (northerly to north-easterly).[49] Evaporation in the eastern Mediterranean is higher, resulting in greater salinity and water density in calmer waters.

Furthermore, military and commercial ships were designed according to different paradigms and for different purposes. By necessity, war galleys were light, swift, and easily maneuvered, whereas the merchantman, following an ancient Phoenician paradigm, required a deeper hold for cargo. The heavier, broad-beamed, deep-hulled cargo ship could more steadily endure the exigencies of winter weather than the less-stable, shallow-drafted, long-decked war galleys.[50] Between the third century BCE and first century CE, moreover, shipbuilding technology progressed, harbor facilities were increased and improved, and hydrological knowledge deepened, thus increasing the conditions under which sailing could safely occur. Additionally, as populations became less agrarian and more urban, the demand for grain, medicinal ingredients, and luxury items was heightened.[51] Beresford imagines any moratorium on maritime trade as having "a major impact on the wider economy of the ancient Mediterranean."[52]

The sailing season, finally, was highly localized. An avid sailor, Xenocrates, victor of the chariot race at the Isthmian Games (490 BCE?), set out year-round from his home in Acragas in southern Sicily:

> Nor ever once did a swelling fair wind make him furl his sail around his friendly table, but he crossed towards Phasis in the summer, and in the winter he was sailing towards the headland of the Nile.
>
> Pindar, Isthmian 2.39–43

Xenocrates would sail eastward towards the Black Sea during the summer months when conditions were favorable there, but he would winter in the calmer waters of the south-eastern Mediterranean when the Black Sea was stormier and more treacherous. Documentary evidence also shows that in the late-fourth century BCE, merchants could continue to earn money year-round in the eastern Mediterranean on the run between Rhodes and Alexandria. They might, however, be compelled to spend a profitless winter in port if they worked out of Athens.[53]

According to customs records at Elephantine, winter sailings were not unusual in the mid-fifth century BCE, and vessels put out to sea as late as mid-December.[54] Around 60 CE, the apostle Paul was called to Rome for trial. During the winter, he embarked on a transport ship at Caesarea before transferring with other prisoners at Myra to an Egyptian grain vessel heading for Rome. The ship was famously shipwrecked off Malta.[55] The text gives no indication that the journey was exceptional or that the passengers expressed concern for their safety.[56] Heliodoros (a prefect of Egypt in the second century CE), furthermore, referred to a large Phoenician merchantman that made the run from Crete to North Africa in early spring under the threat of attack.[57] Finally, ostraca recovered from Carthage record that shipments of olive oil arrived between February and April in 373 CE.[58]

Imperial Roman authors implied that shipments came to Rome year-round. Claudius (ruled 41–54 CE) authorized winter-time grain shipments in response to the famine of 51 CE, assuring merchants that storm-related losses would be recompensed. He further enticed businessmen with exemptions from the *Lex Papia Poppaea*, provisions which, enduring into the second century, are "preserved today" (according to his early-second century CE biographer).[59] Claudius' measure may not have been exceptional.[60] Nonetheless, it would have been prudent for the emperor to encourage winter and early-spring food deliveries to Rome:

> for this was the very time of year at which the population of the capital was most likely to experience shortage and famine as supplies of grain and other foodstuffs started to run low before the new harvest could be taken in during June and July.[61]

Claudius' crisis was averted within two weeks, implying "that numerous grain freighters remained in a state of readiness throughout the winter months."[62] In February of 62 CE, furthermore, the grain fleet was at Ostia, either wintering there or waiting to be unloaded.[63] Domitian (ruled 81–96 CE) received a shipment of Egyptian roses during the winter, an extravagant, impractical gift (Martial 6.8: Martial would have preferred food crops). In the second century CE, the Greek orator Aelius Aristides attested that imports never ceased flowing into the city: "so many provisioning ships under tow arrive here from all places at all times, at every turn of season."[64] Winter trade, nonetheless, could be perilous, and creditors would protect their investments with contractual timetables that, if violated, could result in higher interest rates. Artemo and Apollodoros, for example, were threatened with an increase in interest from 22.5 percent to 30 percent if they remained in the Euxine Sea, notorious for its unpredictable winter storms beyond the middle of September.[65]

The mercantile season was clearly determined by many factors including the port, type of ship, nature of the cargo, weather, and the time of year. The *Erythraean Sea* includes timetables for trade between Egypt and India. Occasionally, ships with runs from Limyrike (on the Malabar coast) or Barygaza would have to winter at Moscha (on the Bay of Omana) because of the "lateness of the season" (32), that is to say, if the ship put out to sea during harsh winter conditions. The best time to depart from Egypt for Adulis would be September, where the trading season extended from January to September (6). July was the best time for departure from Egypt for the "far-side" ports (14; cf., Pliny 6.104) and Barygaza (49), with the favorable summertime northerly winds in the Erythraean Sea. This schedule allows for ships to arrive in India by September, avoiding the hazardous coastal conditions characteristic of August. The skipper heading eastward should leave Barbarikon around July since "the sailing with these (i.e., the Indian winds) is risky but very favorable and rather short" (39). The conditions of the summer monsoon season, moreover, would have been (and still are) challenging with their heavy swells, thick cloud cover, abundant rainfall, and poor visibility. According to Pliny, the 2,000 nautical mile voyage

(3,700 km/2,300 miles) across the bay from Cella (Okelis) to Cranganore (on the Malabar coast) took "about forty days" if the Hippalos were blowing.[66] Pliny (6.106) added that ships returning from India to Egypt should put to sea between December and mid-January, thus taking full advantage of the gentler, favorable northeasterly monsoon that starts up around mid-November. Trade routes in the Indian Ocean, consequently, are determined by the monsoons, as merchant mariners take advantage of the southwest monsoon for the voyage to India and the northwest monsoon for the return to Egypt.

The mercantile sailing season simply cannot be generalized, and the literary evidence makes clear the fact that merchant ships sailed according to demand. Local sailing conditions varied greatly from one maritime space to the next, and goods were conveyed in accord with knowledge, exigencies, and technologies that enabled tradesmen to conquer weather patterns.

Piracy

Seasonal exigencies were not the only challenges faced by merchant mariners. From Avalites to Rhapta, the so-called "far-side ports," Somalia's coast formed part of the lucrative trade route described in the *Erythraean Sea*. In the 1980s, after civil war and a power vacuum that invited foreigners to engage in illegal fishing and dump industrial waste in Somali waters, an alliance of local fisherman turned to vigilantism to protect the coastal waters and resources. But they soon turned to the lucrative practice of hijacking commercial ships. What began as a Robin-Hood-like effort to protect local communities evolved into a syndicate that threatened trade and travel through the Erythraean Sea (Gulf of Aden and Arabian Sea) and beyond. Although the international community calls them *burcad badeed* ("ocean robbers"), they prefer *badaadinta badah* ("saviors of the sea/coastguard").[67] Piracy in the waters off Somalia fell sharply by 2011.

The story is hardly new. We have already seen that local privateers would take advantage of unarmed (or lightly armed) merchantmen in the Erythraean Sea in the first century CE (above), and political careers in Rome were often made on the strength of successful campaigns against the pirates working off the Cilician coast (e.g., Pompey the Great; Chapter 10). For as long as there has been trade on the sea, there have been profiteers and sea robbers.[68] In the *Odyssey*, committed to writing in the eighth century BCE, the cyclops Polyphemus asked Odysseus and his crew:

> Strangers, who are you? From where did you sail over the watery paths? Is it for business, or do you roam about recklessly over the salt-sea like pirates, who wander hazarding their souls and bringing evil to foreigners?
>
> *Odyssey* 9.252–255

Polyphemus made a distinction between trade and "piracy" as a career choice for "energetic and adventurous men."[69] In Homeric epic, piracy was actually the less disgraceful occupation, since the pirate is more transparent in intent than traders, who were generally regarded as liars and cheats (*Odyssey* 15.415).

Piracy, which involves the forceful transfer of property usually together with violence, nonetheless threatened the legitimate trade that was expanding with Hellenic settlement. According to Thucydides, Minos, the legendary king of Crete, was the first to establish a navy with which he secured his hegemony over the "Hellenic Sea," ruling over the Cyclades and reducing piratical activity to protect his own property.[70] Historically, the Corinthians were among the first to subdue indiscriminate piracy in favor of merchant mariners, as "the procurement of wealth became a prime concern" (Thucydides 1.13.1). It is also possible that controlling piracy may have been part of the mandate of the Delian League (Chapter 9). After Alexander's death, piracy was frequent, especially in the unsettled areas of conflict between the *didachoi* (Alexander's successor generals) in the eastern Mediterranean. Additionally, coastal settlements were particularly vulnerable to sudden raids from marauders.[71] Among the most notorious buccaneers were the Tyrrhenians, Illyrians, Pamphylians, Cilicians, and Cretans with whom some larger *poleis* attempted to negotiate.[72]

Just as with the privateers who plied the Caribbean in the seventeenth century, when brigands were given license by competing countries to attack the Spanish (e.g., Sir Francis Drake [1540–1596], "corsario Ingles"), coastal raids, skirmishes, and attacks on ships could be politically retaliatory. For example, exiled Corcyran oligarchs raided their former homeland in 427 BCE, destroying the countryside and causing severe famine (Thucydides 3.85.2). In 388 BCE during the Corinthian War, a Peloponnesian flotilla attacked merchantmen setting out from Knidos (Xenophon, *Hellenika* 5.1.13–23).

Rhodes became particularly attractive to combatant oligarchs owing to its naval strength, prestige, and strategic position as a central commercial hub, earning a million drachma a year from harbor dues until ca. 167 BCE (Polybius 30.31.12). Rhodian wine, pottery, and other wares were extensively dispersed through the Mediterranean,[73] and Rhodes was an important hub for the grain trade.[74] Demetrios Poliorcetes ("Besieger of Cities") sought an alliance with the

Rhodian government against Ptolemy I Soter ("Savior:" ruled 323–282 BCE).[75] Although rejecting the coalition, the Rhodians eventually capitulated by supplying Demetrios with warships, but they refused his appeal for troops. Demetrios then demanded hostages, together with the use of the island's harbors for his military campaigns. The Rhodian rejection was met with a year-long siege (305/4 BCE) with an army comprising almost 40,000 soldiers and cavalry, 200 warships, 170 "allied vessels," plus about 1,000 cargo ships belonging to pirates and traders. "Appearing unexpectedly" there was an additional host of merchant ships of "those sailing for the purpose of plundering the land and plying the sea with regard for profit for their own sakes" (Diodorus Siculus 20.84.6). The Rhodians seized and burnt many of these craft. Demetrios clearly employed some of the pirates as allied mercenaries (20.82.4), but not all of them. Seemingly, the merchant mariners who "appeared unexpectedly" were opportunistically seeking an easy raid of a rich but vulnerable *polis* under siege. Owing to their lack of discipline and naval skill, mercenaries for hire and privateering businessmen made only a minor contribution to the campaigns of Demetrios who may have preferred their cheap price (of profits from plunder). He likely overestimated the efficacy of their great numbers.[76] Demetrios' collaboration with the pirates seems to have declined with the capture of the pirate leader (ὁ ἀρχιπειρατής: *ho archipeirates*), Timokles, whose crew—"the best of those dispatched by Demetrios"—were vanquished in a brief skirmish against the Rhodian admiral, Amyntas. The Rhodian crew overpowered three pirate ships and seized their "undecked" ships (Diodorus Siculus 20.97.5): thus, they were not proper warships. There is no further mention of pirates in service to Demetrios. They may have withdrawn their egoistic aid until conditions improved in their favor.[77]

The Rhodians remained in alliance with the Ptolemies (Diodorus Siculus 20.81), and they remained active in suppressing brigandage:

> [Rhodes] is marvelous because of its good order and its attention to political and, especially, naval matters, because of which it was master of the sea for a long time, and overthrew piracy and became a friend of the Romans and the kings who were friendly to the Romans and Hellenes.
>
> Strabo 14.2.5

In 220 BCE, the Rhodians were instrumental when the citizens of Byzantium (strategically situated at the Euxine's Hellespontian entrance) recognized the lucrative opportunity of imposing tolls on ships traveling through the Hellespont, with a detrimental effect on commerce between the Mediterranean and Euxine

Seas (Polybius 4.46.5–47.6). Declaring war, Rhodes outfitted a special class of warships and offered protection (*phylakeia:* φυλακεία) at sea to safeguard commercial interests.[78] Rhodian success, however, was limited, as they were unable to control sufficient territory to eradicate land-based pirate strongholds. Piracy continued to threaten the economic security and political stability of the coasts, forcing some settlements to move further inland. Certain Hellenistic kings, moreover, exacerbated piratical activity by providing ships in exchange for a share of the plunder.[79]

Piracy endured, and Roman efforts to eradicate piracy would likewise be met with limited success. With the power vacuum left by Rome's conquest of the Seleucid Empire in 197 BCE, marauders found the ideal combination of coast and rugged ("pathless") mountains in Rough Cilicia (Strabo 14.5.2) and elsewhere. Just like Somali brigands active in the waters of the Gulf of Aden, the Cilician marauders may at first have had legitimate grievances in response to political instability and foreign incursions into their waters. But the Cilicians were uniformly depicted as vicious and exploitative,[80] attacking ships as they happened upon them, raiding islands and coastal cities, and engaging heavily in the slave trade:

> The exportation of slaves encouraged this evil business most of all since it was especially profitable, as they were easily captured and the emporium was not very far away and was large and exceedingly wealthy. This was Delos, which could receive and send away a myriad of slaves on the same day.
>
> Strabo 14.5.2

Strabo implied that Rome, relying heavily on the slave market at Delos, was complicit in Cilician slave running, but it is more likely that piratical raids in the eastern Mediterranean had little effect on Roman interests until the first century BCE, with the eastward expansion of their political and mercantile ambitions.[81] Cilician pirates made their capital at Korakesium, a *polis* that was connected to the mainland by a narrow (defensible) isthmus, perched on a rock face that drops steeply into the sea (400–500 feet [120–150 meters]). Their numbers grew, and in 89 BCE they joined forces with Mithridates VI of Asia Minor against Rome. Mithridates' pirates were well organized and well connected. They controlled fortified roadsteads and signal stations, and they had fleets, skillful pilots, and light, speedy ships, according to Pompey's second-century-CE Greek biographer (Plutarch, *Pompey* 24.3). Plutarch's barb regarding their extravagance (gilt sails, silvered oars, awnings dyed with murex purple) is just as likely a centuries-old carry over from earlier Roman propaganda that aimed to discredit Mithridates and justify the Roman war against him.[82]

The brigands became bolder and more successful, with a fleet of more than 1,000 ships and control of 400 *poleis* (Plutarch, *Pompey* 24.4). In 75 BCE, pirates off Pharmakousa famously kidnapped Julius Caesar as he was crossing to Rhodes to study oratory. With his customary good humor, the future dictator paid their ransom, but then hunted them down and executed them with his customary ruthless efficiency (Suetonius, *Caesar* 6). In the 60s BCE, the raiders extended their activities to Italian shores, where they succeeded in kidnapping and ransoming the daughter of Marcus Antonius Creticus, Marc Antony's grandfather, who had campaigned against Cilician pirates in 102 BCE.[83] The bandits treated their captives cruelly, devising an early form of "walking the plank" for any captive who, like the apostle Paul, tried to appeal on the strength of Roman citizenship:

> Whenever some prisoner of war shouted that he was Roman and declared his name, feigning amazement and fear, the pirates would beat their thighs, and supplicate him entreating him to show leniency; and, seeing them humbled and begging, the captive was persuaded. Then some pirates would put Roman boots on his feet, and others would throw a cloak round him, so that indeed there would be no further misunderstanding. After railing thus and making sport of the man for a long time, finally tossing over a rigs-ladder in mid-ocean they cheerfully commanded him to alight and disembark; and, pushing him off, they would dunk him under the water.
> Plutarch, *Pompey* 24.7–8; cf., Paul, Acts 22.25–28

Their power, consequently, was so far-reaching that "our sea was unnavigable and became desolate to all trade" (Plutarch, *Pompey* 25.1).

Senatorial campaigns against piracy culminated in 67 BCE with a special commission under the *Lex Gabinia*, which essentially granted Gaius Pompeius Magnus (Pompey) *carte blanche* authority, including unlimited access to public funds, as well as command over the sea on "this side" of the Pillars of Hercules and the mainland to the distance of 400 stades from the sea. Within forty days, Pompey netted the pirates who were concentrated around Korakesium. They eventually surrendered but were granted clemency and relocated inland where they became farmers. Pompey's gamble would pay off in the 40s when he received their assistance against Julius Caesar. But the region remained politically unstable until Augustus restored peace in the 20s BCE (Appian, *Civil Wars* 5.130), consolidated Roman authority in the Mediterranean, and gained control over an uninterrupted stretch of coastline, "effectively restrict[ing] piracy to the margins of the Roman Empire."[84]

Ships and Cargo

Let us now turn from the routes and threats to trade, to the merchantmen that plied the seas with their cargo.

A product of Lucian's vivid imagination in the second century CE (*Navigium* 5–6), the *Isis*, a mega-freighter from Egypt whose appearance at the Piraeus in Athens has caused such a sensation that a crowd has come out to watch the top men scrambling up the shrouds and running along the yards:

> What a ship! 120 *pecheis* (180 feet/55 meters) long, the shipwright said, and the beam over a fourth of it, and from the deck to keel, 29 *pecheis* (43 feet/13 meters) through the deepest part of the bilge. How big a mast, and what a yard-arm it supports, and with what a forestay it is furnished and held together! How the curved stern crests up gently, sweeping towards its golden goose-head finial, and on the opposite end the prow rises up proportionately, and having on each side is the running nameplate of this (nautical) temple, *Isis*. As to further ornamentation, the illustrations and the fiery bright topsail sail-cloth, and more than these, the anchors and the rigging and the windlasses and the deckhouses on the stern seemed to me entirely marvelous.
>
> The crowd of sailors could be likened to an army camp, and it was said that she was carrying as much grain as would suffice as food for everyone in Attica for a year. Finally, some tiny little old manling kept all these things safe, whirling around the rudder by means of a delicate vine-pole tiller.

A prompt for a discussion of impossible wishes, the ship is hyperbolized, like the genie wishes of the three friends (to own an *Isis*-like ship filled with gold to subsidize a life of luxury; to be a world-conquering king; or to have magic powers, including invisibility). According to Casson's conversions, the *Isis* was 180 feet long (55 meters), with a beam of more than 45 feet (14 meters), and a hull depth of 45 feet, with the capacity to carry 1,200–1,300 tons of grain.[85]

In the west, ships of such a size would not be built until the mid-eighteenth century,[86] and ancient grain ships were on average a fifth the size of the *Isis*. A fragmentary papyrus preserves specifications for a fleet of grain ships operating between Ptolemais and Alexandria in 171 BCE, the smallest of which had a capacity of 225 tons, the largest could transport 450 tons, and most were carrying 250–275 tons.[87] Some mega-freighters, however, were built, such as the wine ship that foundered off the coast of Madrague de Giens (75–60 BCE, discovered in 1967, the largest Roman ship thus far excavated), with a length of more than 130 feet (40 meters), 30-foot beam (9 meters), hull depth of 15 feet (4.5 meters), carrying between 7,000 to 10,000 amphorae of wine (up to 400 tons).[88] A ship that went down on the "Herodotus Abyssal Plain" (10,000 feet underwater, between Rhodes and Alexandria) ca. 300–200 BCE, carried a cargo of 2,500 amphorae of Greek wine that had settled in a mound approximately 80 feet long (24 meters) on the seabed, preserving the ship's outline.[89] The length of the "typical" Roman merchantman averaged between 20–56 feet (about 7–17 meters), with a cargo capacity of about 75 tons. The "Herodotus Abyssal Plain" wreck was larger than average. We cannot know if her cargo was stowed properly, if she carried sufficient ballast, or if a storm arose suddenly.

Most famous, however, is the mega-freighter whose construction was supervised by the Greek mathematician and engineer, Archimedes, for Hieron II of Syracuse (ruled 270–215 BCE).[90] Materials were imported from throughout the western Mediterranean: timber from Sicily and Italy, esparto from Spain, and hemp from the Rhone for cordage. Archias of Corinth was the foreman over a crew of 300 craftsman plus their assistants. Since the vessel was so large, Archimedes designed a screw-windlass for the launching. The *Syrakousia* (which Hieron renamed *Alexandris* as she set out for ports east) had three levels of gangways, one to accommodate each deck, including the cargo hold and the crew and passenger cabins below the weather deck. The ship boasted elaborate décor, including mosaic floors that recounted the *Iliad*, a gymnasium, shaded promenades, flowering gardens, a shrine to Aphrodite (see *CWW*: Chapter 9), library, bath complex, stables, watertight fish tank, and elaborate weaponry. The largest of the ship's boats had a capacity of 78 tons. Several others could handle half that load. Plus there were an unspecified number of "small boats." In addition to the ship's boats and amenities, the *Syrakousia*'s cargo included 60,000 *modii* of grain (518 cubic meters, 137,000 gallons), 10,000 amphorae of pickled Sicilian fish, 20,000 talents of wool (660,000 kg/1,455,050 pounds), and another 20,000 talents of miscellaneous cargo. The ship was simply too large for most of the harbors on its intended route ("some would not accommodate the ship at all, others were risky"). Hieron thus gifted it to Ptolemy III, and the vessel was docked at Alexandria, where its cargo was much-welcomed relief during a grain shortage. Athenaeus did not record the ship's dimensions. But, with three decks and its elaborate siege works, it would have been extraordinarily large, expensive, and impractical, none of which could be justified by the ship's cargo capacity if it could not even approach, much less dock at, the usual ports. The ship seems to have made it safely to Alexandria, but its sheer size raises further questions about ballast and stability.

Another mega-ship was commissioned by Ptolemy IV Philopater ("Father-loving:" ruled 221–204). His τεσσαρακοντήρης (*tessarakonteres*: a ship with forty banks of oars) was 280 *pecheis* long (420 feet/128 meters), and 48 *pecheis* high

(72 feet/22 meters). She required 4,000 rowers and could accommodate 400 sailors and 3,000 marines. Ptolemy's ship, however, was difficult and dangerous to move and was thus employed as a royal exhibit (Plutarch, *Demetrios* 43.4–5).

Unfortunately, archaeology gives only a poor account of the transport of perishable items, and documentary evidence regarding imports and exports is at best inconsistent. The preservation of commodities is uneven (organic wares usually perish). Interpretation of the material can also be fraught: stones may be the commodities, or ballast, or both.[91] It has, however, been estimated that 60,000,000 *modii* of grain (135,000 tons) were imported annually from Alexandria to Rome's busy port at Ostia.[92] Other imports included wine and olive oil (from Gaul, Spain, and Greece), fish from Spain and Carthage, dried fruits, olives, and honey from Thrace and Gaul, spices from the east, medicinals and perfumes from Alexandria and the Black Sea, ceramics, copper, lead, and tin from Britain, and much more.[93]

Cargo ships were also responsible for importing wild animals for the amphitheater games at Rome, and such scenes are recorded artistically. From Portus, lions in cages are shown in relief on a ship approaching the lighthouse.[94] A Roman mosaic from Veii (Isola Farnese, Italy) shows an elephant forcefully wrangled onto a ship by eight men who tend the ropes shackled to its feet. Four men pull from the ship, another four manage the ropes behind the elephant.[95] From the Villa Romana del Casale in the Sicilian city of Piazza Armerina, the monumental "Mosaic of the Great Hunt" (third century CE) shows handlers loading captive animals, including an ostrich flapping his wings, and an ibex or gazelle also resisting its captors who pull it by the horns and restrain its hind legs.[96]

Pliny (6.101) noted that wealthy residents supported maritime trade with a cash outflow of "not less than" fifty million *sesterces* on luxury items, such as jewels, ivory, silks, and spices from Asia. Other ports, especially at Carthage, Alexandria, Antioch, and Constantinople, were just as busy importing food and goods to support their large communities. The logistics were complex. With a population of about one million people at its height, Rome developed an *annona civica*, allotting a measure of wheat and olive oil (and, later, wine) to each citizen.[97] As in the Erythraean Sea, goods would be transferred from sea-going vessels to smaller river craft that could maneuver in the narrower, shallower channels.

River Barges

Once cargo reached a hub, goods were offloaded for local use or distribution inland. Artistic sources also depict stevedores loading and offloading ships in port. On a mosaic from Ostia, a lone man transfers cargo from one ship to

Fig. 6.2 Torlonia Relief, Museo Torlonia, #430.

another, braced together stern to stern.[98] The busy Torlonia relief (Fig. 6.2) displays a cargo of wine amphorae being offloaded across a gangway. Another relief from Portus illustrates two men offloading amphorae down a gangway to the quay where three seated men wait. One receives a token (or cargo list or import tariff), while the other two make notes in wax tablets.[99]

It was cheaper and quicker to redistribute cargo by water, and the rivers of western Europe served this purpose well. River barges were an important component of the trade network, especially in the west where large, navigable rivers connected camps and settlements to coastal hubs. A nearly intact river barge, dating to 66–70 CE, is now on display at the Musée Départemental Arles Antique. Discovered in 2004, and salvaged from the mud of the Rhone River in 2011, the barge measures 101 feet in length (31 meters), about 10 feet across the beam (3 meters), with a capacity of about 21 tons.[100] At the confluence of two important roads—Caesar had built pontoon bridges over both the Saône (Arar) and the Rhine (*BG* 4.17–19)—Arelate (Arles) was an important maritime hub that linked other riverine and land routes for cargo and communications eastward from Rome to as far west as Britain.[101] The harbor at Arles may have been protected by a life-size marble Neptune, found near the barge. Such barges were likely a common feature of local river traffic, as shown on a second-/third-century-Romano-Gallic bas-relief, where a riverboat, laden with freight, is hauled upstream by teams of men (Fig. 6.3).[102]

Fig. 6.3 Romano-Gallic bas-relief showing a river barge, second/third-century. Musée Lapidaire d'Avignon, Fondation Calvet.

Thus, cargo ships and river barges conveyed a variety of materials into and out of hubs and distribution centers, including both the basic necessities of life and the frivolities for the delight of the wealthy and the entertainment of the masses. Trade by water in the ancient world guaranteed a certain quality of life for residents in every social strata, including grain for the grain dole and animals for the amphitheater shows.

"Passenger" Ships

Cargo ships and war vessels were familiar, and travel, especially from the first century onward, was not altogether uncommon. Ferries (πορθμεῖα: *porthmeia* in Greek; *traiectus* in Latin) that made quick runs between short distances were also fairly common. Famously, Charon ferried the souls of the deceased across the River Styx (cf. Lucian, *Charon*), and the centaur, Nessos, conveyed passengers across his river (to Herakles' dismay).[103] Ferries are reported across the Cimmerian Strait in Scythia (Herodotus 4.12.1, 45.2), across the Rhone to Caballio (Strabo 14.1.11), and across the Nile to Philae (Strabo 17.1.50). Rhegium was a ferry hub, with one run to Messana (Dionysius of Halicarnassus 20.5), and another to Bruttium (Plutarch, *Timolean* 19). It is likely that a regular ferry conveyed passengers across the Hellespont (*Greek Anthology* 5.293). A small six-oared ferry operated from Tanagra across the Asopus River (Aelian, *NA* 13.21). Lucilius cited a fee of a quadrans (a very small amount) for conveyance by raft, presumably across the Tiber (1162).

It is, however, not clear to what extent dedicated long-haul passenger ships might have been in use. Lucian implied that such a route existed between Corinth and Methyma, across the Aegean Sea (*Dialogues of the Sea Gods*: 5: Poseidon and the Dolphins). One could, nonetheless, book passage on a cargo ship (just as some modern shipping companies continue to offer transit to paying passengers).[104] In the fourth century BCE, for example, 40 minae would secure passage from Athens to Ace, on the Phoenician coast (Demosthenes, *Against Kallipos* 52.20). Ships were also chartered (*Digest* 4.9.3.1). In 407 CE, Synesius, future bishop of Ptolemais, was probably sailing on a small ship—making a coastal run between Ptolemais and Cyrene—that also took on passengers. As a storm approached and the skipper put out to sea as a precaution against a threatening storm, Synesius observed:

> then a vigorous southerly wind rises up by which we were obscured from the sight of land. We are sailing quickly alongside merchantmen fitted with double sails, that did not have the same need of Libya (and safety) as we had; but they were sailing another road.
>
> *Epistle* 4.7

Roman officials usually took navy transports. Cicero was booked on an open-decked oar-powered Rhodian armored cataphract for his two-week journey between Athens and Ephesus (western Asia Minor) in 51 BCE.[105] In 70 CE, during Titus' siege of Jerusalem, Vespasian voyaged from Alexandria to Rhodes on a merchantman, whether by choice or necessity, we do not know. From Rhodes, the emperor made his way to Apulia by trireme (Josephus, *Jewish Wars* 7.21–22). For his governorship of Bithynia-Pontus, Pliny the Younger was conveyed partly by ships "under oar" (*orariis navibus*: *Epistle* 10.15–17).

Civilians would make ad hoc arrangements for transport. In the third century CE, waterfront heralds would announce when ships were making ready, confirmed by the shouts of the crew weighing anchor (Philostratus, *Apollonius of Tyana* 8.14). Libanius (314–392/3 CE) inquired about vessels sailing for Athens from Constantinople's great harbor (*Oration* 1.31). The star-crossed lovers, Leukippe and Klitophon, arrived at Berytus (Beirut) just as a ship was casting loose:

> asking nothing about where she was sailing, we changed our outlook from land to sea, and there was only a short window of opportunity before putting out. She was sailing to Alexandria, the great city of the Nile.
>
> Achilles Tatius, *Leukippe and Klitophon* 2.31

Although the characters and their sea voyage are fictional, there is nothing of the marvelous in Achilles Tatius' (exaggerated) second-century-CE account of their embarkation (or of the storm that plagued them: 3.1–5). In Egypt travel was highly regulated because of the province's importance to the grain trade. Outbound travelers would thus first have to secure exit visas, authorized by the governor and then granted by the port authorities for a fee. The fees varied according to rank and status. In 90 CE, passengers traveling from Alexandria to the Erythraean Sea would pay as follows: 8 drachmas for merchant-ship captains, 5 for the ships' carpenters, 20 for soldiers' common-law wives, and 108 drachmas for prostitutes (a considerable sum in comparison with the more modest fees charged to other passengers).[106]

Travelers were at the mercy of both weather and cargo schedules, departing as meteorology and sea conditions permitted, and then enduring maritime exigencies. Josephus recalled that his ship, as it crossed to Rome in 71 CE with 600 passengers, was caught in a storm on the Adriatic Sea.[107] Synesius complained bitterly about putting out to sea instead of making for land when a storm arose (*Epistle* 4.8). But ships' boats could serve as life rafts, although they might lack the capacity to save both crew and passengers (Paul, Acts 27.30).

Since the ships were working vessels, amenities for traveling passengers were limited. Voyagers might travel with their own servants, and they were expected to bring their own food.[108] Drinking water was often supplied,[109] ports were equipped with cisterns or aqueducts for resupplying ships,[110] and galleys might be furnished with hearths for hot meals. Taking an island-hopping route (Ceos, Gyaros, Syros, Delos, Samos), Cicero was almost always assured of a hot shoreside meal each day.[111] In Lucian, some passengers sat on the quarterdeck next to the captain (*Jupiter Rants* 48). Achilles Tatius described passengers as "setting up their tents" near each other on deck (*Leukippe and Klitophon* 2.33).

The mutual obligations between captain and passengers, moreover, were legislated by Roman law: the skipper would be responsible for the behavior of his crew towards passengers, but only while underway, and passengers were indemnified against loss, damage, or theft of their property, but only if such damage occurred on the ship. Passengers, however, might be asked to forfeit that protection, thus safeguarding the skipper against liability (*Digest* 4.9.7).

Conclusion

Although ancient authors dismissed the sea (the Mediterranean and other waterways) as "pathless," the seas were teeming with traffic. Commodities were transported in deep, broad-hulled merchantmen along complex networks of sea roads that were shaped by winds, currents, coastlines, migrating fishes, and national tastes. Established by the Phoenicians, these sea roads were well-worn by Greek and Roman traders who conveyed "every kind of commodity" between emporia: grain, raw materials, tuna, salted fish, olive oil, luxury items, and even animals for the beast hunts and paying human passengers. During the imperial period, Roman hegemony over the Mediterranean was consolidated, and along these ancient sea roads, every corner of the empire was brought to the capital city (and other urban centers), supporting all aspects of ancient life throughout the year. By establishing and using these sea roads, those who sailed mapped the seas, staked their ownership over the trade-routes, and conquered "Our Sea."

7

Harvesting the "Barren" Sea[1]

Introduction

We now consider the sea as a source of wealth that was harvested and widely trafficked. Homer and others might have called the seas "barren" for dramatic effect (e.g., *Iliad* 1.316), but waterways were not merely conduits of merchandise, they were also rich sources of food and luxury items. Some communities foregrounded local associations with maritime mercantilism on their coinage (e.g., the octopods of Eretria, the mussels of Cumae, and the tuna of Acragas: Chapter 8). Watermen dived for mollusks, including scallops and oysters,[2] seaweed-of-the-ocean (Theophrastus, *HP* 4.6.4), murex purple (Pliny 9.131), pearls, of which the finest came from the Persian Gulf,[3] and sponges, used medicinally and for the toilet.[4]

Here we survey some methods of obtaining and processing watery wealth: fishing and fish farming, salting (including the production of garum), and murex purple dye.

Fishing

Fishing was an important industry that permeated Greek and Roman culture, literature, and art produced for an elite audience. From the Minoan era, fishing scenes have graced walls, pots, and mosaic floors. A famous fresco from Akrotiri (ca. 1600 BCE) shows a fisherman with a catch of blue and yellow dolphin fish ("mahi mahi:" *Coryphaena hippurus*) hanging from two lines.[5] Aristotle employed 135 names to describe 109 individual fish species, applying multiple names for some species, as well as specifying particular stages of development or patterns of behavior with dedicated terminology.[6] Pliny listed 74 species of fish in book 9 of his *Natural History*, while Oppian identified 122 fishes and 17 mollusks in his *On Fishing*.

The elite looked down on fishermen (and all working-class people) as belonging to the lower class, and Plato deemed fishing a passive act of laziness where the equipment does the work, but hunting was a sport engaged in by courageous ("gentlemen") athletes who actively subdued their prey by running, striking, and shooting (*Laws* 823d–824b; cp. Plutarch, *Cleverness of Animals* 9.965f–966a). Fishing weakens the soul, but hunting improves it. Plato, nonetheless, tacitly implied that the upper classes engaged in recreational fishing. The literary evidence for the Roman era is more robust, and suggests the popularity of fishing as a leisure activity for the wealthy.[7] Ca. 40 BCE, Marc Antony suffered bad luck while fishing on the Nile and, in order to impress Cleopatra, he ordered his fishermen to dive into the river and attach some of their catches to Antony's hook (Plutarch, *Antony* 29.3-4). Pliny the Younger (61–113 CE) considered fishing (and hunting) a suitable activity for time spent at the villa (*Epistle* 2.8), and he enjoyed pole fishing from his bedroom window at Lake Como (*Epistle* 9.7.4). Antoninus Pius (ruled 138–161 CE) also enjoyed fishing (Fronto, *On Wild Salt-Animals* 3.1).

Comparative ethnography can shed light on the interpretation of ancient evidence from literary, documentary, and iconographic sources.[8] The ethnic *ichthyphagoi* ("fish-eaters"), as it was applied to the folk who dwelt on the banks of the Erythraean Sea, suggests the significant role that fish played in the economic and nutritional life of coastal peoples.

A number of implements were devised for trapping a variety of fishes.[9] Aelian inventoried the necessary equipment: rope, white and black fishing line, galingale cord; straight pine wood poles with esparto line; wicker traps, anchored or weighed down with stones, and then lined with seaweed or rush and cypress leaves; and lures of crimson or sea-purple wool, feathers, cork, or wood. Four methods were commonly employed, each with its own moral currency: net, pole, weel (wicker trap), and hook: "some fish are caught by one device, others by another" (*NA* 12.43). Net fishing is the most lucrative, comparable to military campaigns; pole fishing is the manliest, requiring great strength, a small boat, and eager oarsmen with upper-body strength; fishing with wicker traps may be crafty, but it is a less suitable activity for free men; fishing with a hook requires the most skill and is therefore fitting for free men. Some fishermen compounded methodologies to increase their catches. For example, an encircling net together with a seabed net would create a bag that could trap large numbers of fish (especially tuna) in an underwater "death chamber."[10] Six-oared vessels are recommended for net fishing and small skiffs for weel fishing.

Among material remains are fishhooks, weights of lead, stone, and floats of terracotta, wood, and cork,[11] as well as scraps of nets made from hemp, flax, or the phloem of tree bark. Barrier nets and wicker and wood traps would be employed in estuaries, lagoons, and rivers where fishermen took advantage of natural migratory patterns.[12] The *pelamydes* (tuna) migrating from the Maiotis are "mature enough for catching and salting" by the time they reach Sinope (Strabo 7.6.2). Weirs were also used for harvesting sardines, shad, and mackerel (e.g., Oppian, *Fishing* 3.398–413). Fishermen used weirs—baited with tamarisk or arbutus and placed near sandy shores—to lure cuttlefish (Oppian, *Fishing* 4.147–171) and to harvest murex mussels (Pliny 9.132). At the Menuthias Islands in the first century CE, fishermen in sewn boats and dugout canoes lowered baskets or wicker traps instead of nets to catch turtles and fish (*Erythraean Sea* 15). Organized whaling also occurred, conducted with purpose-made equipment as "a response to take advantage of an opportunity" rather than a "systematic operation."[13]

Watermen also engaged in tricks to attract larger catches. Understanding the value of the right light, fishermen preferred to fish just before sunrise and immediately after sunset (Aristotle, *HA* 8.19.602a; Pliny 9.56). They utilized torches while fishing at night,[14] but the bright green reflection of *smaragdus* (chalcedony or mother-of-emerald) frightened the tuna of Cyprus (Pliny 37.66). Aelian (*NA* 12.43) advocated dying the nets a dark camouflaging color (blue-grey or sea-purple). Fishermen would also use sound to attract or deter fish. They might whistle, for instance, in imitation of serpents to lure the murena eel (Pliny 32.14: murenas were thought to have been conceived from serpents), or they would beat the water in order to drive the schools from one area to another.[15] Fishermen also used other creatures as bait or accomplices. Female cuttlefish, for example, were trailed in the water on a line to attract and bait other (male?) cuttlefish (Aristotle, *HA* 4.1.524b–525a; cf., Oppian, *Fishing* 4.1–126). The "wolf-fish" (pike) of the Maeotic Sea would corral other fishes into the spreading nets, afterwards tearing at the nets "unless they had received their share of the catch" (Pliny 10.23). One fisherman in the Cheldoniae ("Swallow") Islands near Mount Taurus in Asia Minor had malevolently caught and sold a fish that was friendly to his business associate, by enticing catches to the latter's net. The victimized partner recognized the poor fish in the market, and he brought suit, successfully winning damages (Pliny 9.182; cf., *CWW*: Chapter 6 for dolphins who assist fishermen). Thus fishermen not only relied on their knowledge of their prey and the environment, but they also forged alliances with the animals who could very directly manipulate the creatures dwelling in the fishing grounds.

Many of these methods are recorded artistically. Classical vases show fishermen with rods, as on a red-figure kylix (ca. 510–500 BCE by the Ambrosios Painter) where a fisherman hunches over on a cliff; in his right hand he holds a rod with line and hook—a fish has taken the bait—and a basket in his left hand. Below the water, fish swim around a wicker trap, and an octopod hides in a crevice beneath the cliff.[16] One "bountiful sea" mosaic from North Africa illustrates fishermen on the beach hauling on a seine (net).[17] Another mosaic shows two pairs of fishermen in two boats. In each boat, one man rows at the stern. A wicker trap is lowered from one boat. In the other, a man stands ready with his trident.[18]

Fish and Foodies

Epic heroes seemed to avoid fish, preferring red meat, and Plato observed that the consumption of fish is missing from the *Iliad* (*Republic* 404b–405a). In addition, Odysseus' crew were unable to sate their hunger when they languished for a month on Helios' Island when the winds did not blow (*Odyssey* 12.331–334), and they seemed to prefer beef. Aristarchus argued that Odysseus' men ate fish only under duress.[19] Fish, nonetheless, was a popular protein among the Greeks. The Rhodian dogfish (a type of shark), came highly recommended: "if they won't sell it to you, take it by force!" so declared Archestratus, a Greek poet from Gela of Syracuse.[20] Boarfish from the Ambracian Gulf was also prized:

> But if you go to the prosperous land of Ambracia and happen to see the boarfish, buy it! Even if it costs it weight in gold, don't leave without it; lest the dread vengeance of the deathless ones breathe down on you; for this fish is the flower of nectar.
>
> Archestratus, fr. 15; trans. Davidson 1997: 4

The Sicilian Greeks were quite enthusiastic about fish, referring to the sea as "sweet" because they enjoyed its harvest so much, painting detailed marine bountiful harvest scenes (early-fourth chapter BCE: Athenaeus 6.273c), and producing the earliest known cookbooks. One of the first preserved recipes is for wrasse, washed and sliced, and finished with oil and cheese (preserved by Athenaeus, quoting an earlier source: 12.518c).[21] Plentiful references to fish in the fragments of Athenian Middle and New Comedy (ca. 400–260 BCE) provide an intriguing look into culinary preferences and prejudices. Little fishes (anchovies or sprats?) were suitable only for the lowest members of society (beggars, peasants, and folks who simply don't know any better). Delicacies for the elite included eels and tuna.[22]

Tuna

Of the many species that were fished, farmed, and consumed in antiquity, tuna was the most popular. The tuna industry was lucrative, enriching coastal *poleis* and Roman officials.[23] Tuna was also an expensive delicacy that could drive folks who overindulged to bankruptcy (Athenaeus 7.304). Although the correlation between ancient and modern nomenclature is uncertain, Bonito, Albacore, and Bluefin seem to have been hunted.[24] Very popular with ancient foodies, tuna was praised, for example, by the Sicilian epicure, Archestratus (Athenaeus 3.116f–17b, 7.249a). Men would look out for the tuna from bluffs or "tuna towers" (fishermen at Parion had leased the rights to such an installation).[25] When a school had been sighted, the *thunnoskopos* ("tuna-watcher:" θυννόσκοπος) sent signals to the fishermen in their boats who then wrangled their prey with the aid of currents and skinny inlets (see *CWW*: Chapter 9). *Thunnoskopoi* reputedly could estimate the number of the tuna in the school with great accuracy.[26]

Fishermen took advantage of tuna migrations through the narrow Hellespont near Cyzicus on the Propontis,[27] the Strait of Messina between Sicily and Italy (Aelian, *NA* 15.6), and the Strait of Herakles connecting the Atlantic to the Mediterranean where the tuna, preferring warm salty waters, would come to spawn.[28] The Phoenicians may have sailed for four days from Gades to reach the open waters of the Atlantic Ocean, which were rich with tuna.[29]

Fish Farms

Not all fishes were harvested in open waters. Fish farms followed many models: from the small boutique eel ponds kept on private estates to the industrial enterprises that supplemented traditional fishing on a larger scale and aimed to meet market demands. There is, moreover, extensive evidence for fish farms throughout the Mediterranean. Plato described fish corrals in Egypt (*Politics* 264b–c). Aristotle discussed eel tanks (*HA* 8.2.592a2–8). Hieron II of Syracuse (ruled 270–215 BCE) equipped his mega yacht, the *Syrakousia*, with watertight fish tanks (Chapter 6). Notorious for his cruelty, the Roman equestrian Vedius Pollo (first century BCE) would feed his lampreys with those slaves who had incurred his wrath.[30] This method of execution was particularly painful as the eels would ingest the blood of their victims by boring holes into the flesh.[31] Like many Roman aristocrats of his day, Pollo was cultivating eels in an artificial, farmed environment.

Successful fish farms relied on careful engineering, requiring the ability to control water levels and manipulate ponds independently by the use of sluices and screens in order to regulate the movement of fish at various stages of development.[32] The installations were both artificial and natural. For raising mullet and eels, river estuaries were blocked with barriers. In Egypt, fish were hemmed in by barriers in rivers (Plato, *Politics* 264b–c). Eels were bred in engineered basins along Lake Copais in Boeotia and the Strymon River in Thrace.[33] At Agrigentum, Hieron's prisoners of war built a capacious artificial enclosure (with a circumference of 7 stades [1.3 km/0.8 miles]) intended for large-scale aquaculture.[34] Coastal bays ("the very seas and Neptune") were also encased for the easy harvesting of "wolf-fish," a sign of increasing decadence, gluttony, and greed in Rome, according to Columella (8.16.3).

Eels were a culinary delicacy, prized throughout the Mediterranean, and they were also big business. Eels were transported live from Lake Copais, quite probably in tanks, with the scanty water supply described by Aristotle.[35] By the first century CE, eels were trapped in the rivers of the Po Delta "by the thousands" in the autumn (Pliny 9.75). In addition, they were cultivated, and their care was meticulously recorded. Among the recommendations, eels should be kept in water whose clarity is maintained by flowing onto flat slabs and then streaming off again, otherwise the eels would "choke."[36] The constant movement of the water oxygenates it, a major concern for breeding fish in artificial environments.[37] Ancient sources, furthermore, understood the paramount importance of changing the water in tanks where eels have been confined for extended periods.

Other fishes were also raised on both small and large scales. We've already seen that fish farming (ἰχθυοτρόφιον: *ichthyotrophion*) is attested in Greek literary sources, and fish were maintained in artificial environments in cultic contexts:[38] eels sacred to Artemis Ortygia at Syracuse's Arethsusan Spring (Diodorus Siculus 5.3.3); the jeweled fish of Zeus at Labraunda in Caria (Aelian (*NA* 12.30; *CWW*: Chapter 8). Apollo's fish at Limyra and Myra, both in Lycia, may have been oracular, revealing the will of the gods by means of their movements (Pliny 31.22, 32.17). But archaeological evidence is lacking, aside from rock-cut fishponds in Crete and Cyprus that are difficult to date and may be Roman.[39] The *pelamydes* fisheries at Sinope in Asia Minor, admired by Strabo (12.3.10), are probably also Roman.

In Rome, the evidence is both commercial and private. Many elite Roman villas (e.g., the House of Meleager in Pompeii, insula 6.9.2) included freshwater *piscinae*,[40] in which fish were cultivated as a ready supply of food.[41] These *piscinae* were built, stocked, and maintained "at great cost" (Varro, *Farming* 3.17.2). Expenses included the extravagant food that was fed to the fish. Private ponds (and public lakes) were, furthermore, stocked with spawn from the sea (Columella 8.16.2), no doubt increasing the costs. Profitable fishponds could even raise property values. Quintus Hortensius (114–50 BCE) took great care over his fishponds at Bauli, feeding the creatures with his own hands, employing many attendants for them, and taking as much care over sick specimens as he did for ailing slaves (Varro, *Farming* 3.17.5–9). The daughter of L. Domitius Ahenobarbus (98–48 BCE) conscientiously maintained and improved the fishponds on her estate at Baiae on the Bay of Naples (Tacitus, *Annals* 13.21.6). Nero (ruled 54–68 CE) coveted the fishponds that belonged to his aunt Domitia on the Bay of Naples (Suetonius, *Nero* 34.5).

More elaborate ponds were compartmentalized and customized to suit the needs of particular fish species.[42] Lucius Lucullus (118–56 BCE) went so far as to provide his beloved fish with access to cooler waters in hot weather by cutting through a mountain in order to allow a stream of aerated, ebbing and flowing seawater into his ponds (Varro, *Farming* 3.17.9). Hortensius had, moreover, criticized Lucius' brother, Marcus Lucullus, for the stagnant water in his fishponds.

Different fishes require specific environments and care, and in Columella (8.16.7–17.15) we have recommendations for cultivating various fishes in marine fishponds purpose-designed according to species. Some littorals are more suitable for certain types. Sole, turbot, flounder, and mollusks, including the murex purple, thrive on muddy seabeds. Clear, sandy sea bottoms are suitable for gilt-heads but not shellfish. Craggy sea bottoms are excellent for rock-fish. Generally speaking, imported fish cannot survive in foreign waters. Tiberius' (ruled 14–37 CE) freedman and prefect of the fleet Optatus, however, had imported the prized *scarus* (parrot wrasse) into the waters between Ostia and Campania. After a five-year fishing moratorium, the wrasse thrived, and "they are found crowding around at the Italian shores where previously they had not been caught" (Pliny 9.63). Eschewing further comment on the effects of the invasive species,[43] Pliny also observed that the wrasse was, uniquely, a grazing fish that does not feed on other fishes (wrasses are, however, carnivorous).

Columella advised situating the marine fishpond so that tidal behavior would flush out the enclosures, and thus prevent stagnation. The artificial compartments, either plaster-built (commonly employed), or (more rarely) rock-cut (as at Crete and Cyprus), should offer recesses in which fish can retreat from the sun's heat (as Lucullus had done for his fishes). Ponds should be surrounded with channels to facilitate the outtake of stagnant water. Brass gratings fixed to the channels could prevent the fish from escaping. Columella also sympathetically observed that the fish might be more comfortable (and feel less like "captives") if, space allowing, the enclosures were decorated with rocks covered in seaweed imitating the appearance of the open sea "so that, although shut-in, they perceive their confinement as little as possible" (in alignment with modern zoo-keeping principles where enclosures for the animals on display are as realistic as possible). Surely the fish were not fooled. Overcrowding was to be avoided, and feed should be tailored according to species: leftover chards of salted fish for flat fishes who, lacking teeth, must lap up their food or gulp it down whole; for toothed fishes, fresh small fishes, and, in winter, course bread and dried fruits, especially figs. Columella warned his readers that the fish must be fed by their owners: they cannot sustain themselves for long periods of time in artificial environments, and the market value of unfed or underfed stock depreciated.

Aristotle recognized the importance of aerating water in which marine species were tended, and infrastructure suggests efforts to this end. Marine fishponds, sometimes sunk into the seabed, were usually situated near freshwater channels, providing an oxygen-rich environment (fish are sensitive to oxygen levels, which decrease as salinity increases). Given the concentration of fish in industrial-scale farms, tidal activity was insufficient (the fish would "choke"). Increased levels of oxygen (and fresh water) allowed for higher yields, moderated temperatures, and diluted toxins (e.g., ammonia from piscine excrement). Completely submerged, these installations now provide important information regarding environmental change and shifts in sea level.[44] The fish depend on their human keepers for their survival, and the fish are "improved" for human consumption by this relationship.

Fundamentally, fish farmers took control of their own economic fate, manipulating (and conquering) the watery environment to ensure adequate or predictable harvests.

Mollusks: Oysters and Mussels

Shellfish also supplemented the Mediterranean diet, and oysters—served in extravagant quantity by Trimalchio among many elaborate dishes (Petronius, *Satyricon* 70.4; mid-first century CE)—were popular with the Rome elite.[45]

Ancient authors believed that oysters arose spontaneously from the mud, and that the phases of the moon affected their growth, as did the sun and wind.[46] Moreover, the qualities and tastes of oysters from different regions were well known to the connoisseur.[47] Oysters were also farmed, especially at Baiae, Puteoli, and Brundisium, where the best oysters reputedly grew.[48] Oysters from Brundisium were transported across land to be cultivated in the Lucrine Lake.[49] Strabo praised the quantity of oysters at Lusitania (3.3.1), the Rhone River (4.1.8), and in the Lokrinos Gulf (5.4.6), and he recorded that fish-eaters built their homes from oyster shells (15.2.2).

Oyster farming has changed little over the centuries. Mature oysters are gathered so that their larval spawn can be tended until the spats are large enough for harvesting. The spawn are either attached to ropes suspended in brackish water, where they reach maturity, or they are fitted to a clutch of roofing tiles that are placed on the seabed where they grow. Oystermen would transfer the maturing spats to areas with a "current," where they would fatten. In the stretch of water between Lesbos and the narrow strait at Chios, although their numbers did not multiply, oysters "increased greatly in size." Clay oyster pots were tossed into the waters off Rhodes, where oysters would congregate while mud accreted around the pots (Aristotle, *GA* 3.11.763a31–34).

Oysters were, likewise, popular in the provinces. Juvenal (4.141) singled out British oysters (from Richborough in Kent) as highly regarded in Rome, but no physical evidence of oyster cultivation has been found there. Crustacean shells, however, are commonly discovered in archaeological remains throughout Britain, and timber-lined tanks have been recovered from Essex.[50] British oysters may also have been imported to Rome in the first century CE (Pliny 9.169). Among the remains recovered from the Roman villa at Brixton Deverill, Wiltshire in 2015 are a number of oyster and whelk shells, delicacies that would have been transported some 45 miles away (70 km) in barrels of salt water at considerable expense.[51]

Bivalve mollusks (mussels; Lucrine oysters) are featured on the obverses of Greek coins at Cumae in southern Italy, dating throughout the fifth century BCE[52] and into the late fourth century,[53] occasionally with crabs[54] or dog-headed, spiked-tailed Scyllas.[55] Strabo praised the mussels of the External Sea (3.2.7). Trogodyte women wore necklaces of mussel shells as amulets (Strabo 16.4.17). For pains of the liver, Pliny (32.93) recommended the flesh of the long mussel (sea snail) taken in honey wine with an equal quantity of water or hydromel. Several varieties are known: fan mussel (Pliny 9.142); piddick (finger mussel: Pliny 9.184); myax mussel, sea mussels, giant mussels (Pliny 1.32). Martial (3.60) referred to mussels (*mituli*) as inferior banqueting fare on whose shells guests might cut their mouths. The bivalves were also recognized as laxatives and diuretics.[56]

Guilds, Dues, and Access to the Waters

Limited documentary evidence suggests the existence of guilds, particularly in the western provinces, just as there were associations of river men, especially along the Rhone.[57] In fulfilment of a vow, the *piscatores* (fishermen) of Liguria collectively made a dedication to Neptune commemorated on a sculpted memorial. The stone shows the god standing on a boat, leaning on his trident.[58] Fishermen and divers were also organized into a guild in Rome,[59] where the leader of the fishermen's guild (*princeps piscator*) erected a funerary inscription for his young wife.[60] Under Septimius Severus (ruled 193–211 CE), a guild of fishermen and divers enjoyed exclusive rights to debris on the Tiber's bed.[61] Partnerships of fishermen existed throughout the empire, as at the Cheldoniae islands, Cyzicus and Parius.[62] Documentary evidence also suggests that customs duties were imposed on fish caught in the sea, but not in lagoons, at least at Caunus in Lycia.[63] But Juvenal, *Satire* 4 also comes to mind, where Crispinus has just purchased an impressively large fish—presumably escaped from imperial waters—and a council must be summoned to decide how to prepare the beast for Domitian's table.

Salt, Fish Salting, and Garum

Salt was etymologically and culturally significant, the root of the term "salary," and a widely-used relish. Pliny declared, "by Hercules, civilized life is not possible without salt!" (31.88–92). In coastal lagoons, salt works would consist of shallow basins with sluices to control the flow of seawater, thus expediting the evaporation of moisture in the hot, summer sun. The remaining salt would be collected. Where topography was flat, mechanical devices (Archimedes' screw or tympanum) would lift the water into the salt pans, increasing efficiency. According to Vitruvius (10.4.1–2), "an abundance of water is thus made available for irrigating gardens or for managing salt works" (cp., Cassiodorus, *Varia* 12.24.6). As the water evaporated, brine would appear first, then salt crystals, together with impurities including hydrogen sulfide (the scent of rotten eggs that Pliny characterized as "unpleasing:" *ingrato*). The salt was then purified, either by carefully raking the salt from the pans, filtering it through fresh water, or by leaching it with rainwater and then drying it on the beach, as in Hispania Baetica (Pliny 31.83). On sandy beaches with high tides, extraction by artificial heating was also possible. During low tides, salt crystals deposited on the sand would be collected into a *briquetage* (course, ceramic bin), washed in fresh water, and heated from underneath to boil away the moisture, leaving the salt. By this method, about 17 kiloliters of seawater (4,500 gallons) would be required to yield 1 kilogram of salt (ca. 2 pounds).

Endorsing the salt from Salamis on Cyprus, Mégara, and then Sicilian and Libyan salt,[64] the Greek pharmacological writer, Dioscorides (first century CE; from Arnazabus in Asia Minor), preferred salt from mines, in particular, the Ammonian salt—white, translucent, clear, dense, and smooth—quarried near the temple of Jupiter Ammon in Libya. Strabo was also impressed by the salt at the Ammon temple, remarkable for its large salt fountains that "spring up to some height" (1.3.4; cf., Pliny 31.78). A mountain of salt in Karmania was a source of regional wealth on the eastern coast of the Indian Ocean (Strabo 15.2.14). Strabo also reported that salt was quarried at Tourdetania (3.2.6: Iberia), the Caucasus (11.5.6), Arabia, and Aithiopia (17.2.2). The promontory of Gerrha, somewhere on the Persian Gulf, was so salty that houses were built of the stuff. Residents would sprinkle their walls with moisture to keep them solid (Strabo 16.3.3). According to Strabo (3.3.8), at Lusitania (a major fish-salting center), purple salt would become white when beaten. According to Pliny (31.83), salt obtained from sardines was the most pleasant. Salt was not commonly traded long distances because of its sheer weight and ready availability.[65] Strabo (3.5.11), nonetheless, remarked that salt was imported to the Kassiterides, and Plutarch (*Table Talk* 5.685d) mentioned salt-transport ships. It is not unreasonable to suppose that small amounts of highly-valued regional salt were imported to medical centers, Rome, and other urban centers.

Fish preservation was an ancient and important industry, and its technology was probably developed concurrently across the Mediterranean and beyond (Punic installations, for example, are known at Gades) in efforts to preserve the meat and get it to market before it spoiled.[66] Fish and meat were dried, smoked, and salted for export to supplement the diets of Mediterranean inhabitants for centuries, including Roman soldiers and administrators.[67] Salted fish was listed among the provisions for the Egyptian prefect and his staff in 145–147 CE,[68] and among the table offerings for Caracalla's visit to Alexandria (215–216 CE).[69] Preserved/salted fish was one of the largest exports of Africa Proconsularis.[70] It was employed also in human and veterinary medicines for treating internal diseases, including ulcers and dysentery.[71]

Fish-salting factories (*taricheiai/cetaria*) arose in coastal areas near tuna fishing waters and where salt could be collected in large quantities, such as at Baelo Claudia in Hispania Baetica (Strabo 3.1.8), Cotta and Lixus in Morocco, Troia in Lusitania, and Porto Palo in southern Sicily. The lake at the eponymous *Taricheai* ("salting factory") produced "nice fish for salting" (Strabo 16.2.45). At Troia, the vat capacity has been estimated at between 850 cubic meters (1,112 cubic yards) and 1,013 cubic meters (1,325 cubic yards).[72]

The same facilities and methods were often utilized for salting proteins according to season (pork was especially popular).[73] Fishermen performed the initial cleaning on the beach (Manilius 5.656–81), as borne out in the material evidence. Few fish bones have been discovered in the processing facilities,[74] but a large pit of tuna bones has been excavated at the Baelo Claudia beach.[75] Vats of waterproofed *opus caementicum* (Roman concrete), usually erected in pairs, were filled with alternating layers of fish and salt, until the top was reached. Pressure was applied with weights to facilitate dehydration of the fresh fish and the osmosis of the salt into the flesh. The shelf life of the final product depends on the species of fish and the time allotted for the salting process. Methods of processing, furthermore, varied. Some fish were preserved fully or partially, others were salted with or without scales, and some were cut according to specific shapes.[76]

The by-products of the fish-salting process include garum, liquamen, and allec. A recipe for garum (similar to Vietnamese *nuoc cham*) is recorded on a seventeenth-century-BCE Babylonian clay tablet.[77] Garum was the standard sauce for seasoning food in the Ancient Mediterranean, and there were several varieties. Garum was usually produced from tuna (but also mackerel),[78] and the best-documented production site was at Pompeii.[79] Pliny singled out Clazomene, Pompeii, and Leptis as renowned for their garum. For the less expensive muria sauce, Pliny endorsed Antipolis, Thurii, and Dalmatia (31.94). In some recipes, tuna was mixed with modest quantities of smaller fish, piglets, ovines, and mollusks.[80] According to Strabo (3.4.6), the best garum, and the only garum he would recommend, was produced from scomber (*Scomber scombrus*, Linnaeus: Atlantic mackerel) caught off the Island of Herakles (called "Skombroaria" after the fish). Simply put, fish or fish entrails were mixed with salt at particular ratios, covered, occasionally stirred, and fermented for up to three months. During the fermentation process, most of the solids would dissolve, leaving two layers in the vat, the top liquid layer (garum) and the undissolved sediment (allec).[81] Garum was then decanted into amphorae and shipped throughout the Mediterranean. Needless to say, the unpleasant smells associated with the process elicited negative comment. For Pliny (31.93), garum was "the liquor of decay"; in Seneca (*Epistle* 95.25) it was "the costly extract of toxic fish"; and Artemidorus of Ephesus (fl. ca. 100 CE) characterized the sauce as "putrefaction" (*On the Interpretation of Dreams* 1.66).[82]

Murex Purple Dye

Although the sea and other waterways were valuable sources of protein, they also yielded personal commodities, including pearls (for jewelry) and sponges (for hygiene). Pliny equated pearls, serving no practical function, with Rome's descent into vice and luxury (9.105). The rich wore pearls on their sandals (9.114); and those who could not afford them desired them, wearing instead cheap, glass beads. Some merchants (*magaritarii*) specialized in pearls (Latin: *margarita*).[83] The sea produces other valued commodities, including coral and sea silk, whose use by the Greeks and Romans is disputed. Prized in India (Pliny 32.21–24), coral is a medicinal ingredient in Dioscorides (5.121–122: for cicatrization, lumps, and scarred eyes) and Pliny (for bladder complaints, kidney stones, fever). Pliny noted the rarity and expense of coral in his day, valued as an amulet and used to decorate Gallic armaments. Sea silk, produced from the *pinna nobilis*, was a rare, luxury item. One example survives in the fourth-century-CE tomb of a wealthy woman.[84] Another aquatic luxury product was the highly coveted purple dye.

According to a Phoenician myth, the famous purple murex dye was discovered by a royal pet. After chewing on some of the mollusks that had washed up on the beach, the dog's snout turned "purple." Tyros (keeper of the dog and mistress of Tyre's patron deity, Melqart) liked the color so much that she asked for a garment of the same hue, and the industry was born.[85] Tyrian purple is attested in Akkadian and Ugaritic texts from the fourteenth century BCE, and it was famously produced at Tyre in Phoenicia and at Crete, in whose waters the mollusk lived. The cloth became such a successful export that some historians connected the toponym etymologically with the color (*phoinos* ["dark red"]), a reference to the dye that may derive from the Akkadian word for Canaan, *kinahhu*.[86]

According to Pliny (9.125–142), several varieties of shellfish were suitable for generating the highly coveted "purple" dye, each yielding different colors and, because of their diverse habitats, each with different methods of harvesting.

Thais haemastoma is a shallow-water snail dwelling on rocks at less than 1.5 meters deep (5 feet): Pliny's trumpet-shelled whelks, collected from the crags. *Murex trunculus Linnaeus* could be found at depths between 1.5–12 meters (5–40 feet): Pliny's "purple." And *murex brandaris Linnaeus* dwells in the mud at a depth of 10–150 meters (33–500 feet): Pliny's "mud-purple."[87]

Pliny also described the methods of harvesting and dye production. Deepwater purples were harvested in wicker traps baited with cockles. Harvesting should occur before spring or just after the rising of Sirius, the hottest part of the summer. The dye was produced from the "flower" (gland) in the middle of the live snail's throat: "because it emits the (purple) sap with its life." The snails were crushed with stones or hammers, or holes would be pierced into the shell above the gland. Removed by hand (Vitruvius 7.13.3), the extracted glands were steeped for three days in open-air vats where sunlight could initiate the photochemical process of creating the dye (50 pounds [22.5 kg] of glands to 6 gallons of water: 22.5 liters), and fresh salt was added as a preservative (1 *sextarius* for every 100 pounds [45 kg] of murex glands). For nine days the concoction simmered in a lead cauldron while the liquid emerged, first colorless, then yellow, then red-purple. The strained dye was transferred to stone or metal containers and again heated "until there is sufficient hope" (of the proper color: Pliny 9.133). The dye was finally tested. The procedure was time-consuming, complex, and expensive. In Pliny, 8,000 pounds (3,628 kg) of murex glands are reduced to only 500 pounds (228 kg) of dye. Recent experiments confirm the exorbitant quantity of mollusks necessary for producing small amounts of dye: e.g., 590 milliliters (20 ounces) of dye were produced from 100–160 murex glands, sufficient to color four 15 x 20 centimeter swaths of wool (less than 6 x 8 inches).[88] Demand for this luxury preparation may have depleted the population (especially around Delos). In turn, high demand may have precipitated fees for the purchase of fishing rights or taxes on the sale of shells to dye producers,[89] or customs duties on live murex delivered to the factories.[90] The industry was, moreover, notorious for the stench of the workshops (Strabo 16.2.23; Pliny 9.127).

The final hue could range from pale pink to bluish-purple-black. Vitruvius (7.13) connected the pigments to astronomical phenomena, geographical regions, and climactic conditions. Dye produced from the mollusks of Pontus and Gaul is black "because these regions are closest to the north" (where it is darker for a longer time). Dye from the north and west "becomes a bruised-blue" (*lividum*). Dye from equinoctial regions has "a violet-wine color" (*violacio*). Southern mollusks (from Rhodes, for example)—"nearest the sun's course"—yield red dyes, owing to the intensity of the sun, which was thought to affect the hue. For Pliny (9.135), the most desirable tint resembled "clotted blood" (*in colore sanguinis concreti*). True Tyrian purple, the most expensive color, was obtained by lengthy soaking and double-dipping the cloth. Many workshops produced quality purple dye, including the factories on the "Purple Islands," perhaps off the Atlantic coast of Mauretania, which produced Gaetulian purple dye for Juba II (ruled 30 BCE–23 CE).[91] The most famous factories were at Tyre (Strabo 16.2.23) and Sidon.[92] Furthermore, the dye was colorfast. At Hermione, Alexander the Great (256–323 BCE) reputedly came across 5,000 talents' weight of purple dye which retained its vibrant pigment despite being in storage for two centuries, according to his second-century-CE biographer (Plutarch, *Alexander* 36.2–3). The color was preserved by the honey that had been added to the original preparation. Finally, a purple-bordered *palla* that Servius Tullius (ruled 575–533 BCE) had bestowed on a statue of Fortuna was still brilliant in Pliny's day, six centuries later (8.197).

Because of its expense, "purple" was associated with royalty and extravagance, and it was reserved for the elite. Purple cloth appears on the lists of tribute owed by the people of Tyre to Assyrian kings in the ninth and eighth centuries BCE. The color was also symbolic of Roman authority. In Rome, togas with purple borders (*togae praetextae*) were reserved for members of the senatorial class and aristocratic children when they came of age (Pliny 9.127; Suetonius, *Augustus* 38). Purple soon came to represent luxury. Cleopatra's barge was notoriously equipped with purple sails (Plutarch, *Antony* 26). For Pliny (9.139–140), purple dye was yet another example of contemporary exorbitance in light of the abandonment of traditional Roman self-sufficiency, frugality, and morality. The dye became subject to sumptuary laws in Rome when Nero aimed to reserve the best colors, Tyrian red and amethyst, for his exclusive use. Nero had "secretly sent an agent to sell a few ounces on a market day, and then he closed the shops."[93] By the reign of Alexander Severus (222–235 CE), the purple-dye industry came under the strict control of the state,[94] with a penalty of death for illegal dye production (*Codex* 4.40.1). Twelve types are attested in Diocletian's Price Edict (ruled 284–305 CE), ranging in price from 150,000 denarii to 300 denarii per pound.[95]

Conclusion

The sea was an important source of wealth, providing food for large urban populations and luxury items for the elite. Fishing was a lucrative business (and popular sport) in coastal communities, and tuna was especially prized by ancient foodies. Garum and liquamen, the byproducts of the salt-fishing industry, were ubiquitous in Roman cuisine, and the highly coveted murex purple dye was processed from aquatic snails. Wealth from the sea quickly became symbolic of extravagance, possibly because of the added difficulties and dangers of obtaining these rare and expensive commodities from the murky depths.

Those involved in watery industries manipulated that environment in service to the gods (artificial ponds for cultic fishes) or for personal gain (fish farms, salt-fishing factories). The ancients altered the essence of marine produce for human use, extending the shelf life of protein for consumption by salting it, and chemically processing the glands of shellfish to create beautiful dyes for luxury products. The ancients, through human technology, thus "improved" the sea's bounty for human use.

Part Three

The Sea and "National" Identity: The Political Manipulation of the Watery World

8

Minoan Thalassocracy, Archaic Expansion, and Maritime Iconography

Introduction

Here we explore how water molds "national" identity. The Mediterranean Sea and the fruits of the Inner Sea profoundly shaped how communities viewed and presented themselves, and many *poleis* (city states) defined themselves according to prominent bodies of water within their territories (e.g., the Arethusa Spring at Ortygia). The politicization of waterways is yet another expression of human conquest and control over the natural world that begins with the earliest literature. Water provides an artificial but convenient means of demarking bodies of land (continents; see *CWW* Fig. 1.1), and of separating the realm of the dead from the living (*CWW*: Chapter 8). In this regard, water is both a line on a map and a powerful symbol of cultural identity.

We do not aim to provide a full investigation of Mediterranean maritime activity. Navies, naval warfare, and individual sea battles have been treated extensively elsewhere.[1] Here, however, we shall consider the importance of maritime activities to "national" identity and empire building by focusing on visual and literary evidence in the eastern Mediterranean, starting with the Minoans.

Minos and Minoan Thalassocracy

According to legend, the mythical king of Crete, Minos, was the first to establish a navy and a thalassocracy in the eastern Aegean. With his navy, Minos quelled the piracy that threatened free trade, he facilitated maritime communication, and he established a network of "Minoan" settlements on the islands of the eastern Aegean, over which he appointed his own sons as governors (Thucydides 1.4.1, 1.8.2). Communication and commerce could occur between these island nations only by sea. Thus, owing to the very geography of the Mediterranean, which is dominated by the sea, water was central to the cultural and political identity of eastern Mediterranean peoples who inhabited island and coastal kingdoms during the Minoan era, so-called after the legendary king of Crete (ca. 1900–1100 BCE).

Reliance on the waterways is reflected in monumental contexts. At Akrotiri in Santorini, fragments of three monumental frescoes from the House of the Admiral (ca. 1500 BCE) show a maritime journey (and they may represent an early map[2]). On the longest fresco (4 meters [13 feet]), a fleet sails from one port town (left) to another (right) across an intervening body of water. A second fresco (3.5 meters [11.5 feet]) shows the plan of a river with the contours of the waterway, hillocks, clumps of vegetation, and native fauna. Included in the group are porpoising dolphins, stylized waves, reeds, and other marsh vegetation. Six boats are powered by oarsmen, and deckhouses protect either rowers or passengers from the sun. Two boats are rigged, and one is under sail. Helmsmen stand at their steering oars, and the captains sit at their sterns (see Chapter 5). At least three dinghies are being rowed within the harbor. The frescoes may show Libya[3] or a military expedition, perhaps to North Africa.[4]

The centrality of water and the sea's bounty is, furthermore, expressed in the artwork. Most famously, the elegant dolphin fresco—adorning the Queen's Megaron at the Palace of Knossos (fourteenth–thirteenth century BCE)—reveals a sensitive, realistic treatment of marine creatures and their habitats. In the on-site reconstruction, five dolphins swim in two registers (three to the right in the upper register; two to the left below). Interspersed among the dolphins are several species of fish. Clumps of sea urchins (or "spiky coral"[5]) frame the fresco. The dolphin fresco may have been part of a larger artistic group that also included a marinescape floor,[6] as was common in Egyptian and Mycenaean palaces, especially at Tiryns and Pylos on the Greek mainland. The highly fragmentary Knossos marinescape fresco features octopods, dolphins, and fishes.[7] Belonging, perhaps, originally to the "Treasury," the marinescape fresco may stem from a tradition of depositing seashells and beach pebbles in shrines, a handy offering for folk who live by the sea.[8] Additionally, the Temple Repository included several marine objects in faience: stylized argonauts (*Argonatuta Argo* L.), cardium shells, a dolphin, and a flying fish, nearly intact when Evans found it.[9] Knossos' dependence on the sea is thus reflected in the sacred landscape of temples and the votives that worshippers had offered.

It is no surprise that island kingdoms such as Knossos and Akrotiri would feature marinescapes prominently in monumental artwork. And the nautical theme carries over to the richly engraved bezels of sealstones, which were usually worn as jewelry (rings, pins, pendants, bracelets). Made of gemstones, bone, and even clay, sealstones could indicate rank, and they frequently served as amulets of protection. Fundamentally, however, their impressions (into wax or clay) were intended as proof of property ownership.[10] Because of the tiny scale of the oval bezels (as small as three centimeters in length [a little over an inch]), the details must be chosen with care and are therefore significant cultural indicators.

Nautical imagery is prevalent on Minoan sealstones, and the iconography provides an important indication of values and anxieties. About twenty examples of Minoan sealstones with ship imagery are catalogued by Casson. However crudely, these sealstones offer indispensable clues in reconstructing the size, rigging, and function of Minoan vessels,[11] and they are invaluable aids in deducing the evolution of Minoan vessels from flat-bottomed hulls with rounded prows and bifurcated sternposts (ca. 2000 BCE),[12] to more rounded craft (1600–1200 BCE).[13] Sealstones show oars and steering oars, rigs without sail, ships under full sail,[14] ships with partially furled sails (Fig. 8.1).[15] Among the remarkable details are cross hatchings for reef lines, booms, ornaments on the prows, masts and rigging, keels, steering oars, and carefully executed oars.[16] Two gold rings show ships entering or leaving port.[17] Some ships feature deckhouses to accommodate passengers,[18] and dolphins gambol at the prows of others.[19] A trident is carved into one early intaglio.[20] A large stylized horse is superimposed over a ship under oar on a Cretan clay seal (ca. 1400 BCE),[21] perhaps, as Evans suggested, depicting the transport of an aristocratic horse (Fig. 8.2). Evans interpreted a doglike sea-creature that seems to attack a sailor on a clay seal impression from Knossos (ca. 1600) as a prototype of Scylla, the sea monster from whose belly emanates the torsos of six dogs ("sharks").[22] This intaglio may have been worn apotropaically by a sailor or traveler concerned with the dangers of journeys by sea (Fig. 8.3).

"Cult" boats are also represented, helmed by bare-breasted women in Minoan skirts. The so-called gold signet "Ring of Minos" illustrates a complex scene of a ship with a seahorse[23] prow and two on-deck altars with horns of consecration. The ship is rowed along a bay of cross-hatchings (stylized rippling waves?) between two symmetrical brick structures (possibly city walls?). On the left wall a nude female seems to dance, another bare-breasted female sits on the right-hand building. The incommensurately large scale of the women suggests that they might be divine.[24] Later evidence attests onboard altars and seaside shrines to maritime deities (see Chapter 5; CWW: Chapter 8), and this gold ring may be an early indication of that practice.

Fig. 8.1 Minoan Seals: Casson 1995 #37, 40.

Fig. 8.2 Stylized aristocratic horse: Evans 1935: 827 #805 = Casson 1995: #52.

Fig. 8.3 Scylla intaglio: Evans 1935: 952, #921.

We cannot know who owned these artifacts or what personal significance they might have conveyed. But their themes, prominent in both public and private contexts, were culturally resonant. Such intaglios may have been used by ships' captains, merchants, or priests of maritime cults, all likely to have been prominent members of Minoan society who sailed the waters of the eastern Mediterranean for business, politics, or pleasure.

Homer's Catalogue of Ships and "National" Identity

As often, we now turn to Homer. In book two of the *Iliad*, we find the famous catalogue of ships, listing 20 contingents, 44 leaders, 175 towns and places, and 1,186 ships.[25] Wilcock and others have observed that the catalogue was a later insertion. Contingents with minor roles in the epic were foregrounded in the catalogue;[26] the poet employed the imperfect tense, indicating ongoing action in the not-too-distant past ("they were leaders of...," not "they had been"); and the emphasis on the number of ships is more appropriate to the beginning of the war, when the navies mustered at Aulis, than to the land battles that were the focus of the *Iliad*, a single episode that occurred during the war's tenth year.

Scholars have interpreted the catalogue in many ways. As essential to the *Iliad*'s structure;[27] a literary construction;[28] the product of a Boeotian traveling minstrel;[29] or a reflection of Hellenic naval power in the Heroic Age.[30] Wilcock defends the catalogue's geography as roughly preserving late-Mycenaean cartography. The catalogue may also loosely record contemporary (Archaic) maritime strength.[31] It is, nonetheless, a powerful statement of the importance of the sea and maritime travel in the Ancient Mediterranean, and it echoes the importance of naval activity to self-identity of Hellenic *poleis*. All the Greek contingents who came to Troy had access to the sea, ports, and harbors, and they supplied their own ships, with the exception of the sixty ships under Menelaus' command, "armed from a distance," presumably borrowed from his brother, Agamemnon (*Iliad* 2.587). Ironically, the seat of Menelaus' power was the landlocked *polis* of Sparta, Helen's homeland.

At Troy, ships provided protection and enabled warfare, and they facilitated trade and the establishment of new settlements. Hector aimed to set fire to the ship(s) that Ajax defended,[32] perhaps the "grand, fast-running, seafaring" ship that brought Protesilaus to Trojan shores (*Iliad* 15.705) and would hopefully return his Thracian soldiers safely home. Like a diplomatic embassy, the ships were, essentially, the *poleis* that they represented (anticipating Themistocles' declaration: Chapter 9). Hector's capture of a ship's stern was an invasion of the Thracian *poleis*, whose men set out under Protesilaus (*Iliad* 15.715–717). Ajax urged on the Greek heroes, emphasizing their vulnerability and the great distance from their homelands:

> We hold position in this plain of the close-armored Trojans, bent back against the sea, and far from the land of our fathers.
>
> *Iliad* 15.739–741

The ships must be protected. Their destruction would signal humiliating, irrevocable defeat and marooning in a hostile land. The ships, in fact, were symbolic representations of the Achaean homelands across the Aegean.

Archaic Settlement: Migration and the Development of Naval Power and Coinage[33]

Return of the Heroes

The Greek epic cycle, which may have been codified and committed to writing in the eighth century BCE, records robust efforts to settle Italy by the returning heroes in their *nostoi* (homeward journeys). Because of their brutal war crimes against the gods at Troy (the theft of Athena's Trojan cult statue, the Palladium; the lesser Ajax's brutal rape of Cassandra in Athena's temple; and Neoptolemus' murder of King Priam's youngest son, Polites, at Zeus' altar), many of the Greek heroes were punished by the gods. The retributions are formulaic. The epic warriors returned to troubled homes and were expelled. Escaping from enemies they sought refuge abroad and founded *apoikiai* ("homes away [from home]") where they were worshiped as heroes when they died.[34] Diomedes, for example, had wounded Aphrodite (and Ares) in battle (*Iliad* 5.330–380, 855–861), and Aphrodite devised a suitable retaliation. Returning to Argos, Diomedes found an unfaithful wife who tried to kill him (replicating the *nostos* of Agamemnon who was slain by his jealous wife, Clytemnestra, at Mycenae). Diomedes escaped to Apulia in southern Italy where he founded cities and was worshiped with a hero cult.[35] Philoctetes also sailed to southern Italy, becoming a founder of cities after his expulsion from Thessaly (perhaps he was the victim of a power shift from aristocratic to more democratic governance).[36] Like Diomedes and Agamemnon, Idomeneus was the victim of an unfaithful wife whose lover expelled the rightful king (Lycophron, *Alexandra* 1214–1225). Idomeneus consequently sailed to southern Italy where he also established cities and was honored with hero cults in southern Italy.[37] These legendary characters, like their historical counterparts, traveled by water to found settlements and expand the remit of Greek culture.

Exploration and Expansion

As Hellenic peoples spilled onto the islands of the Aegean and beyond, they were connected by the Mediterranean's sea roads, which were employed for warfare, trade, and settlement.[38] The "Middle Sea" facilitated and encouraged this mobility as people traveled in search of raw materials, exploited marine resources, and developed social networks of information and technology.[39] The people thus became skilled mariners in their quest for wealth, resources, new lands, and adventure. From the eleventh to ninth centuries BCE, the Aegean islands and western Anatolia were Hellenized by the establishment of *apoikiai* and trade routes from the "mainland" (including Attica).[40] The Phoenicians had been exploring and settling the western Mediterranean as early as the tenth century BCE, venturing into the Aegean by the ninth century BCE.[41] Hanno's expedition was undertaken for the purpose of founding *apoikiai* along Africa's western coast.[42] Settlement was a maritime activity, as opportunistic sailors from the Greek mainland established their claim on the islands of the Aegean and then the western coast of Turkey.

The duties of the colony founder (*oikistes*) are explained in the *Odyssey* (6.7–10). As the one responsible for the success of a new *apoikia*, the *oikistes* brought settlers from the mother city (*metropolis*) to the site, laid out defenses, established sanctuaries to the relevant deities, and assigned plots of land to the settlers. Even though *apoikiai* were legally independent, a close network of trade and cultural exchange often existed between *metropoleis* and their *apoikiai*. On the strength of the enduring *metropolis-apoikia* bond, Aristagoras, tyrant of Miletus, appealed to the people of Athens, Miletus' *metropolis*, for help against the Persians in 499 BCE. Aristagoras aimed to protect the autonomy of his *polis* as the Persians vied to augment their own empire by land and sea. The Athenians complied with twenty ships (Herodotus 5.97). The sea roads connected an *apoikia* with its *metropolis*, and it was by the sea that a *metropolis* could protect its *apoikia*.

Initiatives to establish Greek *apoikiai*, spreading Greek culture and commerce as far west as Spain and east as Colchis, advanced hand in glove with the development of the *polis* system of government and the growth of Greek naval power, as surveyed by Thucydides.[43] The first *triremes* were built at Corinth and the first maritime battle occurred ca. 640 BCE, when Corinth (one of the greatest

Hellenic naval powers in Thucydides' day) was defeated by its *apoikia*, Corcyra (Corfu).[44]

Corcyra was reputedly the first Greek *apoikia* in the Adriatic Sea, founded by the Corinthians Chersikrates and Archias in 733 BCE when "barbarians" were ousted (settlement was no doubt a violent enterprise that frequently resulted in the elimination of indigenous residents by banishment or extermination).[45] With a good natural harbor, Corcyra was a logical choice given its key position on the main trade route between Greece and Italy. Its major settlement overlooked the narrow coastal strait.[46] The Euboean *poleis*, Eretria and Chalcis, were significant hubs of settlement and communication, facilitating maritime contact between Hellas and the Near East from 1200–800 BCE. By the early-eighth century, both *poleis* had founded Italian *apoikiai*: Pithekoussai, Cumae (757 BCE), Naxos (734 BCE), Leontini (729 BCE), Catana (729 BCE), and Rhegium (712 BCE). Founded by 600 BCE, Massilia (Marseille) quickly became an important center of maritime trade between the Hellenes and Celts. The Ionian *metropoleis* looked also to the Black Sea, and ancient sources credit Miletus as establishing more than seventy *apoikiai*, including Cyzicus (675 BCE), strategically positioned on the Sea of Marmara beyond the Hellespont.[47]

It was owing to their command of the sea that Greek *poleis* were able to venture from the Greek mainland to expand their access to resources and their ability to extend cultural influence. This was a factor of Mediterranean geography where the sea shapes the people, and the people wield control over the sea roads.

Aquatic Themes on Archaic Greek Coins[48]

Despite formal alliances and unspoken obligations between *metropolis* and *apoikia*, *poleis* remained autonomous, adopting their own iconography, which was broadcast on local money. Although impossible to pinpoint, coin money seems to have come into common use by the end of the period of Archaic settlement (late-seventh/early-sixth century BCE) in Ionia, an important nexus of contact with the east.[49] The earliest coins showed only crude, unrecognizable shapes that were punched into the metal (the "flip" side mirrored the abstract punch). As the technology advanced, the obverse image depicted a local hero, deity, or some resonant symbol. Athena and her owl were standard on Athenian coinage. Eventually, most mints adopted a fully reverse coin type whose designs rendered important imagery of local identity or resources, such as barley on the coinage of Metapontum and Leontini; horses on Thessalian coins; cattle for

Sybaris and Macedon. The obverse would depict a patron deity (Apollo's lyre on Delian coins; Zeus' head and eagle on those of Elis) or a foundation myth (Pegasus at Corinth). Larger mints might employ a standard type that changed slightly—if at all—over centuries (such as Athena's perennial owls at Athens or ships at Phaselis). Coins were also stamped with control marks to identify batches or magistrates, but their chronology and meanings cannot always be disentangled. Legends on early coins (usually stamped on the obverse) sometimes preserve the name of the issuer or artist responsible for the design. Given their tiny size (4–25 mm [1/10 to about an inch]), images and legends were (and continue to be) chosen with great care to underscore "national" ideology or events. Small and portable, resonating with compelling messages of power and accomplishment, and uniquely valuable because of their integration of text and image,[50] coins were thus the ideal medium of propaganda, promoting the culture of an invading state, sharing news of victory or regime change, or codifying "national" identity. We can only speculate on the process and purpose of coin design. Who selected the images? What messages were those images meant to convey? Maritime themes are foregrounded on many designs, reflecting the economic and cultural importance of waterways, travel by ship, and aquatic animals that populated coastal waters. We now turn to a few representative case studies.

Dolphins

In *polis* coinage, dolphins symbolize the sea from which city states obtained commercial wealth, military success, or their very existence (through settlement from *metropoleis*). A familiar sight to mariners and coastal towns, dolphins were popular in Mediterranean art, myth, and numismatics. Spanning from the sixth to second centuries BCE and numbering in the thousands, dolphin coins take several designs. Images of dolphins appear on coinage throughout the Mediterranean, including *poleis* in Sicily, Argos, Asia Minor, the Aegean islands, and along the Black Sea.[51] Coinage from the island *polis* of Thera naturally included swimming dolphins, recalling the Bronze Age dolphin frescoes of the Queen's Megaron at Knossos.[52] A coastal *polis* in Asia Minor, Side—a harbor town that remained important into the first century CE (Pliny 5.96)—struck coins that depicted dolphins leaping beneath pomegranates, an image that appears on the obverse of all fifth-century-BCE Sidetan coins (*side* means "pomegranate"). The pomegranate was a popular symbol of fertility and eternal life because of its association with Persephone, Queen of the Underworld. It may also evoke the symbolic resonance of the sea as a conduit to the world of the dead (*CWW*: Chapter 8). Its inclusion on Sidetan coinage, nonetheless, may simply be a pun or a reference to a long-forgotten foundation legend.[53]

On coins from Sicily, dolphins were associated with maritime infrastructure, water nymphs, and sailor gods. A drachma (standard silver coin) from Zankle (later, Messana—celebrating that city's distinctive natural sickle-shaped harbor (ζάγκλον [*zangklon*] means "sickle")—shows an incommensurably large dolphin leaping into the air.[54] Several Syracusan drachma issues have dolphins swimming around the head of the Nereid Arethusa, patroness of the eponymous local freshwater spring on Ortygia.[55] As a water goddess, Arethusa would have ensured the purity and availability of an important source of fresh water. Obverses feature Nikes (winged Victories) floating above quadriga and (often) dolphins in the lower register. These types come into use with Syracuse's rise to preeminence under the tyrant, Gelon, who transferred his capital from Gela (on the southwestern coast), partly because of the excellent twin Syracusean harbors. From there, Gelon was able to consolidate his eastern Sicilian kingdom (Herodotus 7.155–156; Thucydides 6.5.3). The quadriga evokes the familiar image of Poseidon in his chariot drawn by two or four sea horses, and shows that Gelon has conquered the waters and brought the two harbors under his control. Accordingly, because of his mastery over the harbors (and the water), Gelon could regulate commercial and military access to eastern Sicily. Dolphin imagery on Gelon's coins underscores the importance of the harbors, where dolphins are, of course, frequently observed.

Dolphins with and without riders were widely included on coin issues, especially in Magna Graecia (southern Italy, heavily settled by Greeks) and Sicily, and they perhaps comprise the largest numismatic category of marine animals. There is a tradition that Phalanthos,[56] the Spartan founder of Taras, was shipwrecked off Delphi (Pausanias 10.13.10). His prayers to Apollo for safety were answered with a dolphin that conveyed the *oikistes* to southern Italy (Apollo was a god of "colonization," and proclamations from Delphi frequently included the founding of new Hellenic settlements). This founding legend was widely evoked on Tarentine coinage.[57] Other famous dolphin riders are also noted on coinage, including Ino's son, Melicertes, at Corinth,[58] and Arion whose story may have been viewed as an explanation for the prevalence of dolphin riders on coins.[59]

We turn now to the dolphin rider of Taras (Tarentum under the Romans) (Fig. 8.4). Dolphin rider scenes are enhanced with scallop shells,[60] stylized waves,[61] or hippocamps,[62] all typical features of a seascape. Dolphin riders are

Fig. 8.4 Dolphin-rider coin, Taras: Kraay 1976: #673–7.

sometimes shown hunting with tridents,[63] clutching or spearing octopods,[64] or spearing tuna.[65] Such hunts represent active, masculine attempts to conquer the watery world, in stark contrast with the more passive activity of fishing with a rod or weel (Chapter 7). Also associated with the Tarentine dolphin riders are individual mounted horsemen.[66] By the third century BCE, the obverses of Tarentine coins show dolphin riders with the Dioscuri, the twin brothers of Helen of Troy, Castor and Pollux, who protect sailors at sea (*CWW*: Chapter 9). The Dioscuri were associated with the zodiac sign Gemini ("twins"), an important tool in celestial navigation, and they were often merely hinted at on the coinage, with the emblematic inclusion of the two predominate stars of the constellation that symbolizes their epiphany.[67] The combination of dolphin (the sea animal that rescued the ship-wrecked founder) with the Dioscuri (who come to the aid of sailors in distress) is a powerful message of safety at sea. Taras, thus, as proclaimed numismatically, was a safe haven for mariners.

The Pinnipeds (Seals) of Phocaea

Other aquatic animals were rendered on coinage, including the noisome, but common, monk seal (see *CWW*: Chapter 6). By ca. 600 BCE, seals adorned the coinage of Phocaea—the northernmost Ionian settlement in Asia Minor— whose toponym is cognate with *phoke* (φώκη), the Greek word for seal. Another tradition links Phocaea etymologically to Phokis in central Greece, from where the *oikistes* came.[68] The seals nonetheless are featured on early coins showing the animals individually,[69] in braces,[70] or triads.[71] According to Herodotus (1.163.1), the Phocaeans were the first Greeks to sail long distances, establishing trade with the peoples of Egypt (Naukratis) and Spain (Tartessus), and settling *apoikiai* in Corsica, Rhegium, and Elea.[72] Their legendary success was inextricably connected with the sea roads that linked them to distant emporia to the south (Egypt) and west (Spain). Any deeper connection of seal imagery with the *polis* is lost. We do not know if the seal was part of Phocaea's foundation myth, but the early moneyers probably enjoyed the wordplay that connected the city's toponym with a familiar marine animal that had a reputation for ferocity (*CWW*: Chapter 6). Seals, nonetheless, were appropriate avatars for a *polis* that employed the often fierce sea roads for its mercantile existence.

The Tuna of Cyzicus

Also widely represented, tuna coins are concentrated at Cyzicus, Sicily,[73] and Gades (and other Iberian sites: e.g., Ilipoula, Baesuri, Sexi, Abdera), where the tuna industry especially flourished (see Chapter 7). At Cyzicus, thousands of tuna coins were minted over a span of three centuries (600–300 BCE).[74] Coin designs from Cyzicus take many forms. An elegant coin, ca. 600 BCE, shows the fish on the obverse framed by two symmetrical flourishes that might be lilies.[75] Cyzicene tuna are linked with terrestrial predators, including boars,[76] cats (a lioness pouncing on a fish;[77] a panther standing on a tuna),[78] and birds (Zeus' eagle with a fish in its talons).[79] The images of these excellent hunters may be intended to represent the prowess, courage, and strength of the men who harvested the fish. Deities are also shown with tuna: Zeus with his thunderbolt,[80] Apollo,[81] the local hero, Attis,[82] Persephone (linking the sea with death: above),[83] Tritons,[84] and a trident-wielding Poseidon,[85] perhaps in demonstration of divine patronage over the local fishing industry. Cyzicene coins, furthermore, link the tuna to aqueous mythological creatures, including Poseidon's ill-fated lover, Medusa,[86] and to ships' prows, a fitting reminder that watercraft are an essential aspect of the fishing industry.[87]

The Octopods of Eretria

Into the 440s BCE, coin issues of Eretria, a coastal town in Euboea, show the octopus.[88] The type was also noted numismatically at Dikaia, Eretria's Macedonian *apoikia*. In addition, the octopus is featured on the coins of Croton,[89] founded from Rhypes (Achaea: Strabo 6.1.12; 6.2.4), and Syracuse,[90] a Corinthian *apoikia* (Thucydides 6.3). The octopus was a culinary delicacy[91] with therapeutic applications,[92] notorious for its powerful bite (Aelian, *NA* 5.44) and for its tentacles that could strangle its enemies.[93] Themistocles (524–459 BCE) unkindly accused the Eretrians of having octopod-like pouches where their hearts should have been, perhaps a response to the city's emblem and its lively mascot (Plutarch, *Themistocles* 11). (*CWW*: Fig. 6.3) Octopods were a familiar sight in coastal waters and they were an important culinary product of the Mediterranean fishing industry. Their appearance on the coinage links the *poleis* that featured them to their watery realm and even, in some cases, to their distant *metropoleis*.

The Turtles of Aegina

In the Saronic Gulf, southwest of Athens, Aegina with its two harbors was a robust center of trade from the late-eighth century, extending to Egypt and Etruria,[94] and it was famous for its "kitschy" goods (Herodotus 2.178.3; 4.152.3). The *polis* also had connections to the slave trade and piracy.[95] Furthermore, their sea turtle (probably loggerheads) and land tortoise coinage (early-sixth to late-fourth centuries BCE) became a standard monetary medium, recognized in the Peloponnese, central Greece, and the Cyclades.[96] The ingots' convex shape resembles the shells of turtles and tortoises, but the choice of turtle as the *polis* emblem may stem from dependence on the sea (arable land was limited). The island, however, was not particularly known for its turtles. Sea turtles were a source of meat,[97] but were not popular among foodies.[98] Their livers, brains, blood, and bile were, nonetheless, employed as ingredients in medical recipes.[99] Hunting was especially good in the teeming waters of the Azanian and Phoenician Seas, where crowds of turtles would beach once a year at the Eleutherus River in Lebanon, nearly 750 miles away (ca. 1200 km) (Pliny 6.172, 9.36). Turtles were gathered by swimmers who turned the amphibious reptiles on their backs before dragging them ashore.[100] The prodigiously large shells of Indian sea turtles were used as boats[101] and roofs (single shells "of so great a size:" Pliny 6.91, 9.36). Turtle shell was imported to Rome for use as decorative furniture inlays (Pliny 9.39), but their symbolic resonance to Aegina remains a mystery.

The Crabs of Acragas

Crabs were likewise not particularly popular with Greco-Roman banqueters.[102] But they were plentiful in coastal and harbor waters, and their behavior and medicinal applications are well documented.[103] Crabs are rendered numismatically at Acragas throughout the fifth century BCE. The types are detailed and realistic, showing the defensive spines within the two claws and four pairs of quadripartite legs that include three sets of sharply pointed legs for walking and, at the rear, a brace of swimming legs with flat paddles. Especially fine is a late-sixth century BCE *didrachma* from Acragas.[104] The Acragean crabs are shown with other marine fauna: tuna;[105] eels with shrimp or crayfish;[106] octopods with conch shells.[107] Zeus is a common obverse type as are his eagles, clutching eels, or fish, including tuna.[108] Eagles are deft fishers in their own right, observed diving into bodies of water to catch seafood.

Crabs are also featured on the coinage of Cyrenaica, Cos, Motya (a Phoenician commercial settlement off Sicily's north-western coast),[109] and Terina (Magna Graecia).[110] The crab may be symbolic of Apollonia, a (claw-shaped?) port of Cyrene.[111] At Cos, crabs appear on the reverse of *tridrachmas*, on the obverse occasionally with Herakles[112] (who killed the giant crab that assisted the Lernean hydra, a poisonous, multi-headed water snake, one of the monsters that Herakles was tasked with killing).[113] Cos was an important center of worship for Asklepius, the Greek god of medicine, son of Apollo, and the crustacean was employed medicinally as an ingredient in therapies for aquaphobia (rabies), fissures, ulcers, and poisonous bites (Dioscorides 2.10).[114]

The crab's function as a talisman may explain its usage on coinage. Models in glass, marble, and bronze were consequently buried beneath the *Pharos* at Alexandria "as if their stout shells would assure its stability."[115] Created by Poseidon (Manilius 2.221), crabs are also the natural emblems of maritime deities, to whom they were sacrificed. The obverse of a small bronze coin from Bruttium shows Amphitrite with a crab-head-dress,[116] and Okeanos is depicted with crab-claw horns, as on the sixth-century Attic drinking cup by Sophilos that shows guests arriving at the wedding of Peleus and Thetis (Fig. 8.5).[117] Numismatic crabs thus link their *poleis* with their divine maritime protectors.

The Ships of Phaselis

Just as nautical creatures were foregrounded on *polis* coinage, so too was the technology that allows for travel, warfare, and the exploitation of marine fauna.

Fig. 8.5 Oceanus with crab-claw headdress. Marriage of Peleus and Thetis, drinking cup by Sophilos: BM 1971,1101.1.

Although not a common numismatic motif until after the Persian Wars, from the late-sixth century ships are rendered on the coinage of Phaselis, whose maritime heritage was significant. Phaselis was a Rhodian *apoikia* on a promontory of the Lycian coast, praised by Strabo for its three harbors.[118] The *polis* was a hub for trade between the Greek mainland, Syria, and Egypt. Phaselis would become a member of the Delian League (Chapter 9). Phaselis also served as the base of operations for Cilician pirates.[119] Early Phaselian coins feature ships' prows and sterns.[120] The type endured at Phaselis well into the Roman era, e.g., one example shows a wreath-wielding Nike fluttering over a ship's prow.[121] The moneyers of Phaselis thus chose to foreground the technology that fueled maritime trade and protected their seaward borders from naval attack.

Conclusion

Just as water shaped Greco-Roman conceptions of the origins of life, the physical cosmos, mythology, and ritual, water also profoundly informed expressions of the "national" identity that distinguished coastal and island *poleis* from each other. Sources of commercial wealth were celebrated on coinage (tuna at

Cyzicus), as were winsome marine animals that populated familiar waters (dolphins). Some *poleis* broadcast their nautical foundations, others uplifted aqueous sources of subsistence or wealth, or even naval technology. The concise, efficient maritime imagery on these portable symbols of *polis* identity underscores the watery associations and origins of many Greek *poleis* that were founded on the waves, or became wealthy and powerful because of the sea.

9

Hellenic and Hellenistic Thalassocracies

Introduction

We turn now to the initiatives that gave rise to thalassocracies in the eastern Mediterranean, with a focus on the Battle of Salamis, the growth of Athenian naval power under the "Delian League," and the maritime ambitions of Alexander's successors, especially Demetrios Poliorcetes who, like others, aimed to legitimize thalassocratic ambitions by association with gods of the sea.

Fifth-century Naval Strength

The Battle of Salamis

In the early years of the fifth century BCE, tensions mounted between the Ionian Greek *poleis* and the expanding Persian Empire under King Darius and his son, Xerxes. The Persians soon encroached onto the Greek mainland, eventually fomenting war (490–479 BCE). Athenian success at Marathon in 490 boosted waning Hellenic morale,[1] and Athens began to build up her navy under Themistocles' leadership, with profits from the Laurium silver mines.[2] A fraught oracle advised the Athenians to find their safety in a "wooden wall," which some Attic leaders had taken as indicating the ships.[3]

Ten years later, the Athenian fleet, together with the navies of the confederated *poleis*, would meet the Persian forces in the narrows off Salamis in the Saronic Gulf, to which the damaged allied ships retreated after a naval battle in the Strait of Artemisium off Thermopylae.[4] The battle at Salamis was a stunning Hellenic maritime victory over Xerxes' troops.

Detailing the naval strength that was brought to bear on both sides at Salamis,[5] Herodotus emphasized the superiority of Hellenic naval prowess, a connective motif introduced at the beginning of the *Histories,* where Greek naval preëminence had deterred Croesus (ruled 560–546 BCE) from attacking islands in the Aegean (1.27). Themistocles recognized that the outcome of this war "hangs on the ships" (8.62). For this reason—Hellenic naval acumen—Artemisia, queen of Halicarnassus, advised Xerxes not to engage the Greek fleet at Salamis (8.67). Despite Artemisia's counsel (and her warnings against greed), Xerxes sent his navy against the Greeks. The allied Hellenic fleet, although "bottled up" in the Strait, succeeded in disabling most of Xerxes' ships. The Persian ships were heavier and less maneuverable than the Greek triremes, and the invading sailors failed to hold position, making "no moves that might have followed a sensible plan" (8.86). The Greek captains, in contrast, knew the watery "terrain" and were thus able to avoid the shoals. The allied Greeks were advantaged by their knowledge of a challenging watery region, and their equipment was more suitable to that environment. At Salamis, the allied Greek *poleis* defeated a foreign superpower with a considerable navy, they preserved the autonomy of individual Greek *poleis*, at least in the short term, and they established their own control over the waters that separated the Greek mainland from their eastern enemy, as *polis* navies continued to evolve and grow in strength.[6]

With the victory at Salamis, ships, *polis* navies, and maritime power become all the more prominent in the evolving sense of *polis* identity, especially at Athens. Like the Hellenic (Athenian) land victory at Marathon in 490 BCE, the Battle of Salamis became an important cog in the Athenian propaganda machine, and a catchphrase for Athenian freedom and democracy. In a warning against arrogance, Pindar evoked the victory at Salamis:

> And now Ajax' city Salamis could have borne witness that she was saved by sailors in Ares (battle) in Zeus' destructive storm, like hail with the murder of countless men.[7]

For Herodotus (book eight), Salamis was the keystone of the Athenian salvation of Greece. For Thucydides, Salamis justified Athenian imperialism.[8]

The Corinthians supplied forty ships for the naval battle (Herodotus 8.43). They buried their dead near the city, and they erected an epitaph claiming to have saved the allied Hellenes:

> O foreigner, we inhabited the well-watered city of Corinth, but now Salamis holds us, Ajax' island. In this place, after seizing Phoenician and Persian and Medean ships, we plucked sacred Hellas (from defeat).
>
> Plutarch, *Malice of Herodotus* 39

The cenotaph that was dedicated on the isthmus is even more explicit:

For all of Hellas standing on the razor's edge, we lay dead, having plucked her out (of danger) with our lives.

For the Greeks, Salamis was a defining victory, a maritime analog to the land triumph at Marathon ten years earlier. In Plato, Clinias, the statesman from Knossos, proclaimed:

But, friend, we Cretans, at any rate, declare that the sea battle at Salamis—waged by the Hellenes against the barbarians—saved Hellas.

Laws 707b

Aeschylus, furthermore, who had fought at both Marathon and Salamis,[9] penned the earliest account of the naval battle at Salamis, envisioned from a Persian perspective and replete with technical detail (*Persians* 447–471). Celebrating Athenian "freedom," condemning *hubris*, and extolling Athenian pride in maritime success, the *Persians* was produced in 472 BCE (under Pericles as *choregos*), and it is, remarkably, the only surviving Greek tragedy on a mythologized historical topic and the earliest extant Greek tragedy.[10]

After Salamis, the Hellenic fleet remained under Spartan command, but the tactless Spartan admiral, Pausanias (who had replaced Eurybiades after the naval engagement), was accused of medizing ("going Persian"), thereby alienating his Hellenic allies (in particular the Ionians who had endured the full brunt of Persian oppression).[11] Athens accepted the allied plea to assume maritime leadership of the war-time Hellenic confederacy. The balance of power in the Aegean shifted, consequently, towards Athens.

The "Delian League" and the Peloponnesian War

In 478/7 BCE, in the aftermath of Salamis and Plataea, as many as 150 *poleis* entered into a formal agreement, the so-called "Delian League" (a modern label), secured with binding oaths "to have the same enemy and friend."[12] The confederacy was intended to check further Persian expansion, to seek financial recompense for wartime damages, and to exact "simple revenge."[13] Patrolling and controlling the waters of the Aegean would stave off further attack and help ensure the autonomy of the city states that had just recently defended themselves against Persian aggression.

The alliance, instead, quickly put Athens on the path to thalassocracy.[14] Athens would take military responsibility (e.g., leading potential military expeditions) in exchange for ships or money, while they ostensibly respected the political autonomy of their allies. Athens exercised ultimate control over the treasury, which was initially deposited on the island of Delos (hence, the modern tag). By 454 BCE, the tribute was transferred directly to Athens,[15] where the "Athenian Tribute Lists"—inscribed stelae erected on the Acropolis—listed the one-sixtieth portions of the tribute allotted directly to Athena Polias, the incarnation in which this important goddess protected the *polis* of Athens.[16] These funds were probably employed for Pericles' building program.[17] Tensions grew as Athens began to demand money instead of ships (by the beginning of the Peloponnesian War only Chios and Lesbos were still supplying ships), and, as early as 471 BCE, member states tried to withdraw.[18]

Under the umbrella of the "Delian League," Athens consolidated control of the Aegean as the allies continued to campaign against the Persian Empire.[19] They were successful both in expelling Persians from Thrace and in eradicating pirates from Skyros, a rocky island east of Euboea, where the Attic hero, Theseus, the legendary founder of Athenian democracy, was reputed to have died, and where the Attic leader Cimon also organized a "successful" search for the hero's bones (Plutarch, *Theseus* 36: Theseus was the son of Poseidon who ventured across the Aegean to kill the Minotaur). By 467 BCE, Cimon had soundly defeated the Persian navy at the Eurymedon River (Pamphylia, Asia Minor), capturing 80 Persian ships and destroying 200 ships that belonged to their Phoenician allies. The Athenians suffered a devastating defeat in 460 BCE when Pericles, hoping to weaken Persia with an attack on Cyprus, lost the greater part of his forces, thus giving the Athenians an excuse, however feeble, to transfer the Delian treasury from the Aegean to Attica. In 451 BCE, the Delian (Attic) and Persian fleets again came into conflict over Cyprus. The Greeks were victorious, and Cimon's son-in-law, Callias, crafted a peace treaty (either formally or unofficially) with Persia, thus ending armed engagements in 450/49 BCE,[20] while the Athenians turned their attentions to mounting stresses with Sparta, which soon escalated into war.[21]

It was with her navy that Athens had hoped to defeat Sparta, playing up her own strengths and the weakness of her land-locked enemy, a *polis* without a significant navy. The enemies of Athens, moreover, recognized the *polis'* maritime supremacy, and they aimed to undermine it, the effects of which would have disrupted the grain supply into the city (Attica was suitable for growing grapes and olives, but needed to import grain from abroad in order to feed her citizens). At a meeting of Spartan allies, the Corinthians opined hopefully that Athens could be overwhelmed with a single defeat at sea and that there would be time to build up a fleet for the Peloponnesian League (Thucydides 1.121–122).

Meanwhile in Athens, seeking to persuade the Athenians to go to war with Sparta, Pericles acclaimed the superior seamanship of his fellow citizens in contrast with their lubberly enemies. There were more native-born sailors in Attica than the rest of Hellas combined, so he claimed, and the Spartans lacked seamanship (1.142.7–9). Appreciating the importance of the sea to national security and commerce (cp., Thucydides 1.4–5), Pericles also threatened "if they (the Spartans) march against our land on foot, we shall sail against theirs." Recognizing that "the power of the sea is great...," he advised his compatriots "to keep a guard over the city and the sea" (1.143.4–5). We compare Pericles' eulogy over those Athenian soldiers who died during the first year of the war (431 BCE), exhorting that the Athenians would "prevail" if they attended to naval matters (τὸ ναυτικόν: *to nautikon*), restrained themselves from further empire-building, and did not endanger the *polis* (Thucydides 2.65.7) since they already held dominion over the sea. They merely needed to preserve their maritime authority.[22] This foreshadows the Athenian argument that the island of Melos should succumb to *Athenian* authority because they (the Athenians) were "masters by ship."[23] Athenian power and subsequent aggression was fueled by a relationship with the sea and mastery of nautical technology.

The Ship of State

In conjunction with a growing navy and heightened harbor security, ships became a significant feature of popular culture. Nautical and hydrological metaphors were casually sprinkled throughout the literature: e.g., "reef of justice" (Aeschylus, *Eumenides* 564); sea (of troubles: Sophocles, *Oedipus at Colonus* 1746). The "ship of state" was a popular image, dating back at least to Alcaeus (621–550 BCE) who used the imagery as a metaphor for the troubled political climate in his hometown of Mytilene on Lesbos (f6, f208 Campbell). At Athens, an oracle came to Solon (630–560 BCE) from Delphi, comparing the city to a ship and the citizens to oarsmen: "guiding straight the helmsman's work, be seated amidships. Many allies you'll have at Athens" (Plutarch, *Solon* 14.4). For Herodotus' Themistocles (8.61.2), the ships represented the state while the Persians occupied the territory of Attica. When the Persians reproached Themistocles for being a man without a city, he retorted "that he had a city and a land greater than theirs so long as they had 200 ships filled (with men)."

The image endured, finding expression in poets and philosophers. Aeschylus' *Seven Against Thebes* opens by evoking the ship of state:

> Citizens of Cadmus, it is necessary that whoever oversees the matters of the city on the stern, he should speak timely things while he guides the tiller.
>
> 1–3

Aristophanes' Sosias dreamt "about the whole hull (ship) of state" (*Wasps* 29); and in Sophocles a priest of Zeus compared Corinth with a storm-tossed ship: "for the city rolls and can scarcely lift its head from the depths" (*Oedipus Rex* 22–24). In 415 BCE, as the Athenian fleet was about to embark for Sicily, Thucydides' Nicias described the city as on the high sea.[24] Plato employed the allegory of a ship with a mutinous crew in defense of his argument that states (like ships) are best captained by benevolent, omnipotent philosopher-kings who have true knowledge of the Forms (*Republic* 488a–489d). Plato compared the general population to the ship's owner who lacks experience of seafaring; politicians were likened to inexperienced and unknowledgeable sailors who vied for power and favor from the owner while they dismissed the navigator (*kubernetes*, philosopher king) as a mere "star-gazer" although he, the true helmsman, was the only man with the knowledge to steer the ship (or guide the state).[25] Thus the image was part of the fabric of Athenian culture and metaphor, revealing how the people (at least the elite, literate folk) identified themselves and their state as dependent on power by sea. Athens looked to the sea for military success as well as for its very survival. Grain, the staple of the ancient Mediterranean diet, could be imported only over water.

The Panathenaic Festival

The metaphor also found expression in Athens at the Panathenaic festival, which spanned several days in the month of Hekatombaion (mid-summer). According to tradition,[26] the festival was established by Theseus or Erichthonius, and it marked the city's birthday by honoring its eponymous divine patroness, Athena (with the presentation of a freshly woven robe/*peplos* to the ancient wooden cult statue[27]). Athena was credited in myth with building the first ship (the *Argo*), making the ship a cultically resonant symbol for Athens. Because of her links with shipbuilding and navigation,[28] at Mégara and elsewhere, Athena was worshiped as Aithyia (αἴθυια: "diver" or shearwater, metaphorically a ship).[29] By the mid-sixth century, Peisistratos (d. 527 BCE) added athletic competitions, which might have included ship races.[30] The final day was celebrated with a procession—advancing from the ancient cemetery, the Kerameikos, through the Agora towards the Acropolis—that featured a ship on wheels, from whose yardarm Athena's new *peplos* was suspended like a sail.³¹ Philostratus (ca 170–250 CE) praised the sight: "attached to the ship, the *peplos* was more tasteful than a painting, with a fold in a fair wind" (*Lives of the Sophists* 2.550). Philostratus tantalizingly added that Herodes Atticus constructed subterranean machines for the Panathenaia of 138/9 CE, a seemingly unique installation: "the ship ran not with draught animals driving it, but it slipped along on underground mechanisms."

There is robust debate on when the ship was introduced to the Panathenaic procession: as early as the festival's inception;³² after Salamis, in order to glorify Athenian naval success and prowess;³³ or as late as 302 BCE under Demetrios Poliorcetes ("Besieger of Cities," ruled 294–288 BCE).³⁴ Iconographic and literary evidence are, however, late. The ship was not represented on the Parthenon frieze, perhaps because it did not process up the Acropolis (or because it was not a feature of the mid-fifth century Panathenaic procession).³⁵ Wachsmann (1967) marshals the limited iconographic evidence for the Panathenaic ship, including the damaged calendar frieze on which are still visible a ship's bow with a chisel-shaped ram, remnants of a mast and rigging, and two sets of wheels that would have conveyed the ship. Ships were central to Athenian national identity, an invention of the city's patron goddess, and a prominent feature in the most important festival that celebrated the *polis*' very existence. By the late-fourth century BCE, neither Athenian *polis*-identity nor its survival could be delinked from her harbors, through which the people were fed, and whereby the navy aimed to establish and maintain power over the Aegean under the divine authority of Athena, goddess of ships and navigation.

Hellenistic Thalassocracy

The ship on wheels may have been introduced to the Panathenaic festival during the Hellenistic era, a time when the Mediterranean world saw many changes, but the politicization of maritime endeavors endured. Larger ships of war were constructed, and they were called by the number of men pulling at each bank of oars (three men, for example, pulled the oar of a trireme).³⁶ By 399 BCE, Dionysius of Syracuse enhanced his fleet with tetrereis ("fours," probably a Carthaginian innovation: Pliny 7.207) and pentereis ("fives," his own invention: Diodorus Siculus 14.42.2). By 332 BCE, when Alexander besieged Tyre, both types were in service.³⁷ Dionysius' eponymous son commissioned hexereis ("sixes:" Aelian, *Various Histories* 6.12). Even larger ships followed. In 201 BCE, Philip V used a "ten" as his flagship (Polybius 16.3.3). As a display of his own power and the skill of his engineers, Ptolemy IV Philopater ("Father-loving," ruled 221–204 BCE),[38] commissioned a "forty" whose sea trial required 4,000 oarsmen.[39] Such large ships were impractical, and their construction was abandoned by the mid-third century BCE. Antony's largest ship at Actium was a "ten" and Octavian's flagship was a "six" (Plutarch, *Antony* 64.1; Dio 50.23.2). Instead, navies began to make use of swift, maneuverable *lemboi* (with a ram and as few as sixteen rowers) and other small craft.[40]

Fueling the advancement of ship technology, Alexander's successors contended for naval domination of the eastern Mediterranean (*didachoi*, Alexander's generals, including Ptolemy at Alexandria). In this way they established power (or the perception of power) over the seas, controlling markets, pilgrim sites, and the political development/autonomy of allied and enemy *poleis*. We focus here on Demetrios I of Macedon, "Poliorcetes," whose father, Antigonos Monophthalmos ("One-Eyed," ruled 306–301 BCE), had acquired "sevens" for his fleet. The son won a naval victory at Salamis in Cyprus in 306 BCE, vanquishing Ptolemaic forces with the aid of Athens who contributed 30 *quadriremes*.[41] Demetrios had also notoriously engaged in a year-long siege of Rhodes (305/4 BCE: Diodorus Siculus 20.81–100; Plutarch, *Demetrios* 21–22). Turning "siege craft into an art form,"[42] by 301 BCE Demetrios' fleet included "eights," "nines," "tens," "elevens," and a "thirteen." Within a decade, his navy also boasted a "fifteen" and "sixteen," engaging against ships of equal or greater strength.[43] With his fleet, Demetrios established a thalassocracy with control over much of the eastern Mediterranean (Cyprus, Corinth, Miletus, Ephesus, Tyre, and Sidon). He also aimed to restore his father's holdings in Asia (inherited from Alexander), and he financed an expeditionary fleet with coinage minted at Pella and Amphipolis. His coinage accentuates his thalassocratic ambitions.

Before the ascension of Alexander, portraits of living rulers were largely limited to the coinage of fourth-century Persian satraps (and perhaps, though inconclusively, also to Sicilian leaders). On the Persian model, Alexander included his own image on the obverse of his coinage. The practice of featuring images of contemporary leaders would be followed by the *didachoi* and their successors, enduring into the Roman era and beyond.[44] Even during civil war, each successful claimant would, in turn, quickly issue coinage showing his head as a statement of authority, however short-lived that might have been.[45] The reverses were intended to highlight significant events (e.g., land battles as indicated by horse and rider or quadriga) or to show divine favor with representations of deities (Zeus, Apollo, Athena, Herakles) or their emblems (eagles, thunderbolts).⁴⁶ Symbolism could be crowded. Alexander, for example, promoted his success on land and sea with coinage featuring a winged Nike on a ship's prow or mast together with Zeus' thunderbolt or eagle and a helmeted, Athena Promachos ("in the front-line of battle"), who granted victory.⁴⁷

Nautical imagery continued to be foregrounded on the coinage of many thalassocratic *poleis*. Such iconography includes Poseidon with ships,⁴⁸ and ships together with stars, evoking the Dioscuri who protected sailors on storm-tossed waters: see Chapter 8; *CWW*: Chapter 9).⁴⁹ Also spotlighting nautical imagery is the coinage of Hieron II of Syracuse (ruled 275–215 BCE), who maintained a large defensive fleet and assisted the Romans against the Carthaginians (Livy 21.49–51, 22.37, 23.21). On a typical example, a diademed Poseidon appears on the obverse, two dolphins on the reverse swim around a trident. The legend reads (H)IERONOS ("of Hieron").⁵⁰ In this way, Hieron emphasized that the deep authority enjoyed by his fleet to control the waters off Sicily was because of the divine favor of Poseidon, the god of the sea.

Demetrios, the Son of Poseidon

Thalassocracy was further highlighted by the association of the members of ruling families with marine deities, and these cults were promoted by Hellenistic ruling families (e.g., Arsinoë Euploia and others: *CWW*: Chapter 9). We focus on Demetrios Poliorcetes, whose thalassocratic coinage follows two paradigms. One type shows Demetrios diademed with bull's horns on the obverse;⁵¹ Poseidon Pelagaios ("of the sea") holds his trident and/or an *aplustre* (the ornamental finial on a ship's stern) on the reverse.⁵² The other type may be more common, and its iconography is more explicit. A winged Nike, playing a salpinx ("trumpet"), alights on the prow of a ship (Fig. 9.1); the reverse shows Poseidon brandishing a trident, a cloak sometimes (but not always) draped over his arm.⁵³ In Athenaeus (12.535e–536a), we have a description of the cloak, embellished with stars and the signs of the zodiac: "an image of the universe and the bright shining heaven (*ouranos*)."⁵⁴ The cloak may have strengthened Demetrios' claim that his mother was Aphrodite Ourania, the "heavenly" Aphrodite. Demetrios' usurpation of Poseidon with his trident (or the trident alone) underscores the king's military and political claim over the eastern Mediterranean with Poseidon's favor.⁵⁵ The winged Victory on the prow designates naval conquest and anticipates the Nike monument at Samothrace.⁵⁶

At Athens, Demetrios Poliorcetes was honored with an encomium that proclaimed him the son of Poseidon and Aphrodite.⁵⁷ Demetrios deliberately

Fig. 9.1 Silver *drachma* of Demetrios Poliorcetes, from Tarsus, ca. 298–295 BCE Newell 1937: #41.

linked himself with Poseidon (his father Antigonos had cultivated Aphrodite: Athenaeus 3.101e–f, 4.128b). Holton argues that the combination of Poseidon and Aphrodite Ourania declares Demetrios' conquest of both sea and sky.[58] Such may have been the intention of the Athenians who cultivated Aphrodite Ourania at the Piraeus with regular sacrifices,[59] but Aphrodite, born from the sea, was also a significant maritime deity who protected sailors at sea (cf., *CWW*: Chapter 9), as no doubt her cultivation at the Athenian harbor acknowledged. She was consequently an appropriate divine association for a thalassocratic ruler whose authority derived largely from naval successes.

The Nike of Samothrace

The Nike of Samothrace (whose mountain was an important landmark for navigators) is as fraught as it is beautiful. Discovered in 1863, the vibrantly fluid winged Nike in white Parian marble has held pride of place in the Louvre since 1884 (cleaned and restored 2013–14) (Fig. 9.2). The statue (2.75 meters tall [9 feet]) alighting on a prow of a storm-tossed ship, was carved from gray Rhodian marble excavated from the quarries of Lartos. Her elegant wings are

Fig. 9.2 Nike of Samothrace.

finished with an elaborately wrought, complex pattern of feathers that evokes the wing anatomy of raptors. Headless and armless, her right hand may have been raised, perhaps playing a salpinx or holding a victory wreath. Found nearby, a fragmentary statuette base is inscribed "Rhodios" (Rhodian). The ship seems to be the decked and armored "cataphract" quadrireme favored by the Rhodians.[60]

The Samothracian Nike is exceptionally depicted as in the throes of a storm. Her garments are tangled by the wind "crisscross[ing] ... like ocean breakers,"[61] and her "flowing panel of cloth flares ... like a rudder to her airborne flight."[62] Ancient galleys never engaged in battle during even moderate winds. Although Lartian marble was common for statue bases in the Dodecanese island group around Rhodes, the Samothracian ship is a unique example of Rhodian marble outside the Dodecanese.[63] There is, nonetheless, a strong, nautical link between the two islands. The cult of the Samothracian gods flourished in

Rhodian territory, where they were worshiped by guilds of *Samothraikiastai* ("Samothracian rowers").[64] One crew of *Samothraikiastai*, for example, honored their officer after the Great Gods saved their trireme off Libya during a storm at sea.[65] Rhodians are also well represented among the priests, ambassadors, and initiates to the Samothracian mysteries. The marble was thus deliberately imported against prevailing currents and winds more than 500 km away (310 miles/270 nautical miles). The sanctuary additionally featured other monuments of marble also imported for semiotic effect.[66]

The statue's date is uncertain.[67] Samothrace was dominated by Macedon until 168 BCE, and Rhodes had strong ties with the Ptolemies in Egypt whose dealings with Macedon were often strained and hostile.[68] Considering the Rhodian connection secure, Stewart concludes that the dramatic and "imposing" display may celebrate the Bithynian War of 156–154 BCE,[69] during which the Rhodians contributed 5 quadriremes (under the command of Prousias)[70] to their Pergamene ally, Attalus II Philadelphos, in advance of an invasion of Bithynia, after an unexpected storm in the Propontis had destroyed a significant portion of Prousias' fleet. Attalus was victorious, receiving Prousias' twenty cataphract ships together with 500 talents to be paid over twenty years (16,500 kg/3300 pounds). The Samothracian Nike may thus have been dedicated to the Great Gods at Samothrace in thanksgiving for a victory in a war whose outcome had been significantly shaped by a storm at sea.

Conclusion

For seafaring communities, ships were not merely planks of wood held together by treenails and rigged with masts and yards. Ships were districts of the *poleis* that fielded them, as in Homer's Catalogue of Ships. For Themistocles, as long as the Persians occupied Attic lands, the ships were the only country that the Athenians possessed. But even more resonant is the centrality of sea power and the initiatives to establish thalassocracies where leaders emphasized divine favor that granted and legitimized maritime conquest and consolidated maritime empires.

10

Rome: *Oceanus Domitus*

Introduction

The River Tiber, on whose banks Rome was founded, was fundamental to the city's self-definition. In the *Aeneid*, the Tiber declared his favor over the Trojan refugees and prophesied their victory over the Italian tribes (Vergil, *Aeneid* 8.31–78). The first casualties of Aeneas' Italian war, moreover, bear the names of local rivers, Almo and Galaesus.[1] On Aeneas' shield, the hegemony of the Roman Empire is delimited by major waterways: the Nile, Rhine, and Euphrates Rivers, and the Caspian Sea.[2] According to Rome's foundation legend, it was, in a sense, the River Tiber that saved the city's legendary founder, Romulus, and his twin brother, exposed on its flooding banks.[3] As the waters subsided, the basket carrying the babies eventually floated ashore, and the infants were rescued either by a shepherd or a she-wolf.[4]

We turn now to thalassocratic initiatives in the western Mediterranean. The Romans understood that control over the waters could ensure the success of military campaigns on land in terms of supplying troops and preventing enemy access to key ports. Alliances were sought for this very purpose. The Mamertines in Messana, after their defeat by Hieron II of Syracuse (ruled 270–215 BCE), sought protection from both Rome and Carthage (Polybius 1.9.7–8). Sicily, moreover, was prime nautical real estate (owning both to its wheat crop and its strategic position), over which both Rome and Carthage vied, hoping to assure their hegemony over the sea roads and trade routes of the western Mediterranean. Control of the sea, moreover, was so important that it inspired Rome, a city state whose military strength lay in the infantry, to reconsider its basic approach to war.

The Punic Wars

Increasing tensions between the two rising Mediterranean superpowers escalated into war in 264 BCE. The Carthaginians were formidable mariners, descended from the Phoenicians who had already explored the waters beyond the Strait of Herakles (*CWW*: Chapter 2) and whom Herodotus praised for their nautical skill in Xerxes' service (Chapter 5). But the aquaphobic Romans felt more secure in their successful citizen infantry. Nonetheless, the first Punic War was fought largely at sea, and in 260 BCE the Romans launched their first significant navy with 100 quinqueremes and 20 triremes (ancient ships were called according to the number of men at each oar: five men for a quinquereme). They had improved on the design of a captured Carthaginian quinquereme with the addition of a swiveling, spiked (grappling) gangway (*corvus*: "crow") that allowed Roman marines to board enemy ships. The Romans were a nation of infantry fighters, and they preferred to turn their naval battles into "land" battles by engaging in hand-to-hand combat on the decks of enemy ships. Under the admiralship of Gaius Duilius, the Romans secured their first naval victory against the Carthaginian fleet at Mylae (northern Sicily), employing their *corvi* to take full advantage of their skilled infantry (Polybius 1.23).[5] Duilius distributed the plunder to the people (*CIL* 6.103) in a coin issue that shows a prow on the reverse, eliciting Duilius' naval victory at Mylae, and on the obverse, Janus, protector of the Roman armory for whom Duilius built a temple at the *Forum Holitorium* (vegetable market) in thanksgiving.[6] The issue was probably the *aes grave* series of low-value Roman bronze coins. The fact that the imagery was imprinted on low-value coins shows that it was meant for wide distribution to folk of all economic standings. Duilius also commemorated his victory on a rostral column (decorated with ships' prows, "rostra") in the Roman forum, a clear declaration of his use of the sea to conquer a foreign enemy on the water.[7] In honor of his own victory at Naulochus (below), Augustus would later "refurbish" Duilius' monument—including recarving the inscription with archaic orthography—within the wider context of his own "restoration" of the archaic past (Suetonius, *Augustus* 31.5; *CIL* 6.1300; Goldschmidt 2017).

Victorious against Carthage, Rome consequently gained control of the western Mediterranean. In the wake of defeat, Carthage rebuilt its navy but not to its former strength, and this would prove a handicap, as navies were essential for transporting and supplying troops (e.g., Hasdrubal made "frequent and urgent requests" for men in 216 BCE: Livy 23.26). In 205 BCE, Romans intercepted Carthaginian supply ships off Sardinia (Livy 28.46.14). In the following year, while Scipio's fleet sailed southward without obstruction, the peoples of Carthage were in a panic, not knowing the size of the expeditionary force (Livy 29.3.9–10), and their speculation underscored the current impotence of their own navy. In the end, after the Punic defeat at Zama in North Africa, the Carthaginian navy was disbanded, the ships were conveyed to sea to be burned (Livy 30.43.11–12), and Carthage was allowed to maintain only ten ships for defending coastal waters (Polybius 15.18.3).

To Livy's Hannibal, the combined infantry and navy defined Carthage:

> then it was permitted (for us) to weep when our weapons were seized from us, our ships burned, and there was an interdiction from foreign wars. We perished from that wound.
>
> 30.44.7

The destruction of the navy meant the erosion of Carthaginian power in the western Mediterranean. Without a navy, Carthage could no longer maintain its vast empire or communicate easily with far-flung settlements. Without a navy, Carthage was powerless to instigate hostilities or defend itself from enemy attacks. Without a navy, Carthage could no longer protect mercantile interests in the western Mediterranean, and Carthage was thus weakened economically as well. The loss of power at sea was a political, economic, and moral defeat.

Hannibal may very well have expressed a Carthaginian sentiment, but his words apply equally to the Romans who gloried in war and exalted successful military leaders (e.g., Horatius Cocles, Scaevola, and Cincinnatus). The Roman fleet was featured in contemporary coinage. A very low-value bronze coin (*semis*) of Brundisium, for example, which had remained loyal to Rome in the Second Punic War, features Neptune (with trident and laurel wreath) and a weathered, wreath-wielding Victory; on the reverse, a well-muscled nude rider, astride a cavorting dolphin, holds a lyre; fluttering above, a winged Victory extends a wreath towards the youth's head.[8] Taken together, the iconography (the god of the sea, winged Victories, and dolphin) suggests divine approval for Roman conquest at sea.

Numismatic Iconography in the Second Century BCE

Throughout the second century, Rome continued to issue coinage with ships,[9] but it is difficult to connect the iconography with historical events. Coinage in the Roman Republic was usually designed and issued by the *tresviri monetales*,[10] who selected motifs that glorified their ancestors,[11] but coinage was also commissioned by quaestors (overseeing public revenue) and aediles (overseeing infrastructure and the grain supply).[12] An *as* (low-value bronze coin) issued by Gnaeus Domitius Ahenobarbus (consul 193 BCE and possibly a *vir monatalis* during the Second Punic War) shows a laureate Janus with a prow.[13] In 170 BCE, a delegation from Lampsacus (on the coast road south of the Hellespont) came to Rome seeking an alliance. As praetor (a high-ranking magistrate with extensive powers), Quintus Maenius, welcomed them, gifting each delegate with 20,000 *asses* (Livy 43.6.10), possibly from Maenius' Mercury/prow issue.[14] Coins convey meaning, as we have seen elsewhere, and the combination of Mercury with a ship sends a powerful message. As a god of travelers and rhetoricians, and the consummate divine orator and diplomat,[15] Mercury was an appropriate guarantor of diplomatic treaties. The coin issue either affirmed Roman maritime hegemony or codified the naval coalition between the Romans and Lampsacenes.

Other divine images with watery associations were also employed by Roman moneyers. Neptune with his trident is, for example, featured on the coinage of the Roman statesman, C. Plautius Hypsaeus.[16] The legend honors Gaius Plautius Decianus [Hypsaeus], who vanquished Privernum in 329 BCE, for which he received a triumph.[17] The sources give no indication of a naval engagement associated with the siege of Privernum, and Neptune's significance must be sought elsewhere. As Pompey's one-time ally, Hypsaeus may have participated in the campaign to eradicate the Cilician pirates, and his Neptune coinage may represent an attempt at self-aggrandizing his contributions to a victory at sea.

Pompeius Magnus (Pompey the Great)

We turn now to the late Republic when politicians sought glory and renown against the pirates of Rough Cilicia, with varying degrees of success. In 67 BCE, pacification of the eastern Mediterranean and conquest of the pirates fell to Pompey, who would later take full advantage of his achievement to this end. Nautical imagery is prominently noted on his civil war coinage (49–46 BCE), reminding the Romans of his successes at sea, successes which had secured peace in the eastern Mediterranean waters and allowed for the orderly conduct of trade (including safeguarding the perennially important grain imports that fed the city's masses). No doubt Pompey aimed to show through his coinage that he would again restore peace on the waters, which in turn would guarantee the supply of essential (and non-essential) goods into Rome. A *denarius* (the standard silver coin), issued together with Gnaeus Calpurnius Piso (when he was quaestor), has a laureate Numa Pompilius on the obverse, and a ship's prow on the reverse.[18] Rome's second king, Numa Pompilius had been responsible for instating the city's hallowed religious traditions. The message is clear. Pompey, who had vanquished Cilician pirates in 67–66 BCE, made the eastern Mediterranean waters safe for commerce, at least in the short term. With the approval of both the gods and their legendary agent, Numa, Pompey was now defending Rome's iconic institutions (religion, law, and modest lifestyle) with his fleet. Pompey was defeated, but his memory and compelling maritime imagery endured, cultivated by his youngest son.

Sextus Pompeius

By 44 BCE, Rome descended (once again) into civil war, fought on both land and sea. Pompey's youngest son, Sextus Pompeius, assumed leadership of his father's fleet. After the deaths of the Republican defenders, Brutus and Cassius, Sextus continued fighting for the senatorial cause, augmenting his name with the epithet *Pius*, "dutiful" (*Magnus Pompeius Magni filius Pius*, "Pompeius Magnus Pius son of Magnus") as a proclamation of filial loyalty, fully understanding the scope of his inherited struggle. To broadcast his ancestry (and his authority), he issued coins that glorified his father. The maritime imagery on his coinage also underscores his own successes at sea with an additional, equally potent message—that he fought under Neptune's protection.[19] Sextus furthermore "adopted" Neptune as his father (as had Demetrios Poliorcetes: Chapter 9; *CWW*: Chapter 9), "so great was the glory of his naval accomplishment" (Pliny 9.55); to Horace, Sextus was a "Neptunian leader" (*dux Neptunius*: Epode 9.7–8). A *denarius* minted at Massilia in 44/43 BCE shows Pompey's bust together with a trident, swimming dolphin, and the inscription NEPTUNUS. The reverse features a galley under both sail and oar, its lookout officer standing at the bow, and the moneyer's stamp (Q. Nasidius) below the ship.[20] Coinage strongly connects Sextus with the god of the sea, and proclaims Neptune's favor over Sextus' struggle and efforts at sea to restore the maritime legacy of his family.

Either glorifying Pompey's maritime successes or anticipating his own, Sextus issued several designs featuring Neptune with his trident on the obverse. On the reverse of one such issue, a naval trophy is embellished with an anchor, finials from a ship's prow and stern, and Scylla's dogs (the sea monster who is usually shown with six dog [shark] torsos emanating from her belly: *CWW*: Chapters 6 and 7) (Fig. 10.1).[21] Another type shows Pompey on the obverse, while the reverse features a nude figure resting a foot on a ship's prow (taken during Sextus' victories over Octavian), and holding an aplustre (the ornamental finial on a ship's stern). The pose is one of conquest. The figure is framed by the two Cataenei brothers who had carried their elderly, lame parents to safety during an eruption of Mt. Aetna (evoking *pius* Aeneas' escape from Troy?). Like the Cataenei who had saved fellow Roman citizens from disaster, Sextus gave refuge to proscriptees and runaway slaves.[22] The imagery underscores Sextus as a savior of the people, the one Roman who could or would restore peace and senatorial order by means of his victories at sea.

Fig. 10.1 *Denarius* of Sextus: Naval Trophy: Crawford 1974: #511/2b.

A third type shows a male figure atop a tower (probably a lighthouse). He has a foot on a ship's prow (to proclaim maritime victory) and a trident in one hand (showing Sextus as a courageous hunter at sea); an eagle, Jupiter's bird, appears in the foreground, showing that the king of the gods also favors Sextus. On the reverse, Scylla wields a ship's rudder like a club, an image that underscores the maritime defeat of an enemy (Fig. 10.2). Sextus clearly saw himself as the avenger of his father's murder, and he aimed to restore his family legacy by taking control of the seas. Despite the maneuverability of his ships and superior skill of his crew, Sextus was defeated in 36 BCE at Naulochus, Sicily, by Marcus Vipsanius Agrippa's forceful and inelegant use of grapnels.[23]

Octavian commemorated Sextus' defeat with equally forceful numismatic imagery: his foot on a globe (symbolizing hegemony over the *orbis terrarum*, "circle of the lands"), and an aplustre in one hand (to show that he has vanquished an enemy at sea) (Fig. 10.3).[24] Sextus' literary legacy would be mixed. He was

Fig. 10.2 *Denarius* of Sextus: Scylla smashing a ship's rudder: Crawford 1974: #511/4.

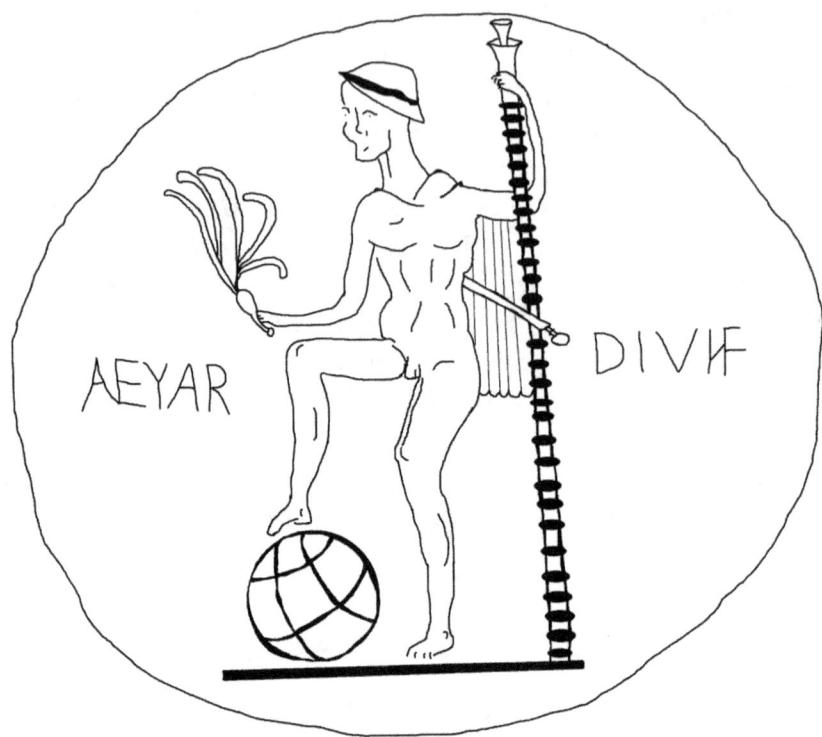

Fig. 10.3 *Denarius* of Octavian with globe and aplustre: Zanker 1990: 41 and figure 31a

characterized as a traitor (Horace, *Epode* 9.7–10), pirate (Lucan 6.419–24), and an ill-starred hero.[25] Although Sextus' execution was ordered by Antony in 35 BCE (Appian, *Civil Wars* 5.91–145), civil war continued as unresolvable tensions rose between the two former allies.[26]

The Battle of Actium

Although Marcus Antonius customarily accentuated his divine aspirations as a new Herakles or Dionysus, his pre-35 BCE fleet coinage is of nautical interest. Several types were struck, featuring Antony and his wife, Octavia, in various poses (sometimes with Octavia's brother, Octavian, who would soon come to be called Augustus),[27] with warships on the reverse. An issue, possibly from Corinth, has the blissful couple gazing into each other's eyes, while Neptune and Amphitrite drive a quadriga of hippocamps on the reverse: "the happy couple rides over the sea like Poseidon and Amphitrite."[28] After divorcing Octavia, Antony continued to issue coins that depicted ships, e.g., a war galley (on a coin of 31 BCE) framed by legends proclaiming Antony's augurship and his (defunct) status as triumvir (the triumvirate had expired in 33 BCE); the standards of Antony's seventh legion are featured on the reverse.[29]

The combined fleets of Antony and Cleopatra VII were defeated on September 2, 31 BCE at Actium (northwestern Greece) by Octavian's navy under the admiralship of Octavian's talented friend, Marcus Vipsanius Agrippa. Finally, "on both land and sea, he (Octavian) restored peace, long thrown into confusion" (Appian, *Civil Wars* 5.130). This victory (together with the conquest of Egypt) was an important component of Octavian's post-war propaganda. In honor of Apollo, his divine patron, and in celebration of the victory, Octavian rebuilt the hill-top temple of Apollo Actius ("of the Shores"), below which ship sheds housed the spoils of war: "where Augustus Caesar dedicated as first fruits ten ships, from one with a single bank of oars to one with ten banks."[30] He revived or instituted the Actian Games, observed first at Nikopolis and Rome in 27 BCE.[31] Along with traditional athletic competitions (foot races, wrestling, discus tosses, and musical competitions, etc.),[32] chariot and boat races may have also occurred. A boat race features prominently in the funeral games for Anchises in the *Aeneid* (widely interpreted as an allegory of Augustus' regime).[33] Across the gulf from the ancient city of Actium, Octavian built a "Nikopolis" (Victory City), where he erected a victory monument of bronze rams.[34] The rams (*rostra*) were fitted into sockets (33–35), perhaps representing the craft captured in battle.[35] In Rome, the senate decreed a rostral column (above) on top of which stood a golden statue of Octavian as *imperator* (victorious general) (Fig. 10.4). The monument was to be erected in the "agora" (forum: Appian, *Civil Wars* 5.130), and it would have been a vivid reminder that Augustus' *pax Romana* ("Roman peace") was won at sea. The naval victory consolidated Roman (Augustan) holdings in the eastern Mediterranean, and it brought the Mediterranean Sea (the body of water) under Roman authority.

Consequently, Octavian, who would assume the honorific title "Augustus" in 27 BCE,[36] glorified his Actian victory numismatically. Contemporary provincial coin issues were, furthermore, important declarations of loyalty to the new (and hopefully stable) regime. A *dupondius* (a "two-pounder" brass coin of low value) minted in Aurausio (Orange), ca. 30–28 BCE, recalls the earlier janiform type (above). With busts of Augustus and Agrippa back to back on the obverse, the reverse has a ship's prow decorated with an apotropaic eye.[37] A silver *denarius*, ca. 29 BCE, features a winged Victory alighting on the prow of a

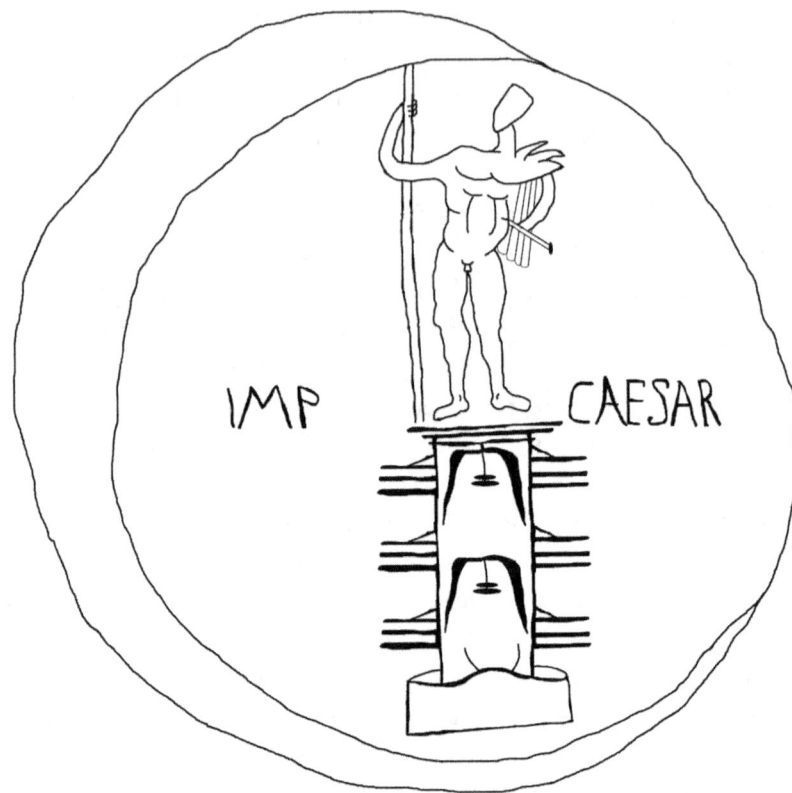

Fig. 10.4 *Denarius* of Octavian with rostral column: RIC 271; BMC 633; C 124.

ship, brandishing a wreath and a palm (of victory), in celebration of the naval victory at Actium.[38]

The naval victory was also eulogized in contemporary poetry both overtly and subtly.[39] Allusively anticipating Augustus' Actian festival, Aeneas and the Trojan refugees stopped at Actium to make sacrifices, celebrate games, and dedicate a shield to Apollo (*Aeneid* 3.278–293). Vergil, furthermore, patently placed the battle at the very center of Aeneas' magnificent shield (*Aeneid* 8.676–705), which, according to Putnam, shows its "importance for the poet's new cosmos."[40] In the *Aeneid's* opening scene, Vergil crafted a stunning storm that was quelled by Neptune (*Aeneid* 1.124–156). Angered that his provenance had been usurped by the king of the winds (at Juno's bidding), Neptune quickly calmed the swells, put the clouds to flight, and brought back the sun, opening up the vast sand bars with his trident, a powerful tool for hunting at sea and for causing devastation (earthquakes) or, as here, for mitigating the devastation caused by a minor god. Neptune is compared to a charismatic stateman who subdues a sudden riot among a crowd of low-born folk. The image is a metaphor for Augustus' appeasement of civil unrest and demonstrates his conquest of the seas as signifying national power.[41] Like Salamis for the Athenians, and the Battle of Trafalgar for Britain (1805),[42] the Battle of Actium was deeply resonant in Augustus' new Roman order.

Gaius Caligula

Augustus' vainglorious great-grandson, Gaius Caligula (ruled 37–41 CE), failed to secure naval (or other military) victories of his own, so he created them (e.g., Suetonius, *Gaius* 45). Like Hypsaeus, Caligula attempted to glorify himself by association with the maritime accomplishments of his grandfather, Agrippa. Issued during Caligula's reign and minted at Rome, an *as* shows Agrippa with a beaked rostral crown; on the reverse a cloaked Neptune holds a dolphin and a trident.[43] Agrippa was usually depicted on coinage with his rostral crown,[44] which, according to Dio, had been awarded for naval victories at Mylae and Naulochus in 36 BCE, an honor "given to nobody before or since" (49.14.3). Nor would Agrippa's role at Actium have been forgotten.[45]

Caligula, however, wanted to enhance his own glory (and vanity) by vanquishing the waters, and he made two attempts. First, he sought to outstrip Xerxes' seven-stade pontoon bridge that "yoked" the Hellespont in 480 BCE, recognized by Aeschylus (525/524–456/455 BCE) and Herodotus (ca. 484–ca. 425 BCE) as an act of flagrant *hubris*.[46] After construction was completed, a violent storm destroyed the bridge. Xerxes punished both the overseers (who were beheaded) and the very waters. Whipping the Hellespont 300 times, Xerxes' minions berated the Strait:

> O biter water, our master imposes this judgement on you because you wronged him who committed no injustice against you. And our king Xerxes will strut across you, whether you are willing or not. In accord with justice, no man will make sacrifices to you since you are a turbid and briny river.
>
> Herodotus 7.35.2

Xerxes was as much at war with the sea as he was with the city states of Hellas.

In 39 CE, Caligula aimed to surpass Xerxes and "inspire fear in Germany and Britain" with a pontoon bridge that extended 3.5 Roman miles (3.3 miles/5 km) between Baiae and Puteoli (Suetonius, *Gaius* 19), whether to impress Parthian hostages,[47] prove the discipline of his strike force,[48] or demonstrate his control over the army.⁴⁹ Gaius also sacrificed to Neptune before crossing the bridge, and, because the waters were so calm during his triumphal ride, he declared that even the sea god feared him, Gaius, the Roman emperor (Dio 59.17.4–11). For this feat, Caligula would also impersonate Neptune, "because he had yoked so great a measure of sea" (Dio 59.26.6). Caligula may have contemplated bridging the English Channel, and the Baiae bridge may have served as a test of the technology.⁵⁰ In 40 CE after his aborted plan to invade Britain, Caligula did declare victory over Ocean, ordering his men to gather the "spoils of Ocean" (*spolia Oceani*) owed to the Capitoline and Palatine. He furthermore planned a triumph of his "campaigns" in Germany, Gaul, and Britain. The parade was ultimately canceled (Suetonius, *Gaius* 46–49). Like Xerxes, Gaius was also at war with the ocean, which represented the successes of his ancestors in stark contrast with his own inactivity and failures. "Conquering" ocean, even if only in his imagination, would enhance his personal glory. Gaius' thalassocratic endeavors were not so much for Rome as for his own vanity.

The "Conquest" of Britain

Gaius' uncle Claudius (ruled 41–54 CE) succeeded where both his nephew and his ancestor the great Julius Caesar nearly a century before him had failed. In 43 CE, four legions together with *auxilia* under the command of Aulus Plautius embarked on an invasion of Britain. Claudius' military reputation was enhanced, Caligula was further discredited, Britain became a Roman province, and the emperor bestowed the honorific name "Britannicus" on his son.⁵¹ This campaign became the antiphon of Claudius' reign. At the triumph at Camulodunum (Colchester), Claudius displayed eleven elephants (eliciting the elephants spotlighted at Caesar's Roman triumph in 46 BCE). Throughout the empire—city and provinces—Claudian coinage promoted the British expedition with the legend *de Britannis* ("over Britain").⁵² Victory arches were erected in Cyzicus, Rome (where part of an inscription is extant), and likely also at Bononia (Boulogne: from where the expedition embarked, but no trace survives).⁵³ A spectacular relief from Aphrodisias in southwest Turkey advertises Claudius' subjugation of Britain, a type that was paralleled elsewhere.⁵⁴ His safe return from Britain was, moreover, recorded in a votive inscription at Rusellae, Etruria, which may have been cut into the base of a statue of Victory.⁵⁵ Claudius' successful invasion of Britain, and his actual triumph over the island (especially in light of Caesar's ultimate failure), underscores Roman success at sea and the extension of Roman national identity and authority to a realm beyond the Mediterranean.

Naumachiae

Roman naval prowess was celebrated in many ways, in coinage (intended for broad distribution), public monuments (that stood as constant reminders of naval success for residents as they went about their daily lives), in literature (for the enjoyment of the elite), and in *naumachia*, staged naval battles that were enjoyed by many residents from all walks of life. Julius Caesar, whose campaigns took him throughout the Mediterranean and beyond, to the Aegean, Alexandria, and Britain, celebrated his quadruple triumph in 46 BCE with, among other things, a staged naval battle in a basin, engineered for the event, in one of the Tiber's bends in the Campus Martius. Featuring Tyrian and Egyptian biremes, triremes, and quinqueremes, 1,000 men participated (prisoners of war and condemned criminals), attracting a large crowd of spectators, some of whom lost their lives in the melee of over-enthusiastic onlookers.[56]

Augustus also expressed Roman naval prowess with a *naumachia* in 2 BCE, which reenacted a naval battle between "the Persians and Athenians" (according to Dio in the third century CE) while celebrating the consecration of his temple to Mars Ultor ("Mars the Avenger:" which had been vowed in order to avenge the assassination of Julius). On the model of Caesar's *naumachia*, a large artificial lake—supplied by non-potable water from the newly constructed *Aqua Alsietina*—was constructed near Tiber Island, large enough to accommodate thirty triremes and biremes, together with smaller vessels.[57] Augustus listed the event among his gifts for the Roman people (*Res Gestae* 32). Nero took advantage of the lake for a party (Tacitus, *Annals* 14.15), and traces of the infrastructure were still visible in the third century CE. It is significant that Augustus' *naumachia* was staged to reenact the Battle of Salamis (between "the Persians and Athenians"), a battle that ensured Hellenic independence from the Persians in the fifth century BCE, a battle that put Athens on the path to thalassocracy. In this way, Augustus was directly connecting his own naval successes with those of Athens at the height of Attic glory and naval prowess. Those who fought in the naval battle at Salamis had saved the allied Hellenes from the despotism of Persia; Augustus, whose position and victory was owed to Agrippa's success at Actium, had saved the Roman people from the despotic tendencies of would-be kings.

Subsequent emperors also sponsored *naumachiae* to show their conquest of the natural watery world. Claudius' *naumachia* in 52 CE commemorated the draining of the Fucine Lake, a feat of engineering that reclaimed wetlands for agricultural use (see Chapter 3). Claudius' engineers had (not quite) succeeded in draining the lake. The event was celebrated by an almost Xerxes-like show of force over the water, with a naval battle featuring triremes, quadriremes, and 19,000 condemned criminals re-enacting a battle between "Sicilian and Rhodian" fleets, according to Dio (61.33.3; cf., Suetonius, *Claudius* 31.6; Tacitus, *Annals*, 12.56). It is unclear if this was an historical battle or a fabricated encounter between two renowned ancient sea powers.

Nero, who lacked military—much less naval—glory, staged another re-enactment of the battle of Salamis in 57 CE,[58] appropriating the glory of his ancestor Augustus, who had appropriated Athenian glory to celebrate his own maritime conquests. Nero's *naumachia* may have been intended to mark the emperor's unfinished canal project at Corinth (see Chapter 4), another attempt to conquer the watery realm through engineering.

Titus, moreover, included a *naumachia* in the lake as part of the celebration that opened the Flavian Amphitheater in 80 CE, also recalling Augustus' event eight decades earlier.[59] Titus' *naumachia* featured a reenactment of a naval engagement between the "Syracusans and Athenians," a battle which, historically, the Athenians lost (413 BCE: Thucydides 7.21, 36–45), but was won by the Attic reenactors. Thus, the outcomes were not foregone conclusions, and the spectators may have enjoyed the suspense of desperate men representing foreign navies and fighting for their lives (the Roman fleet is never featured in a *naumachia*, and therefore could not be defeated). Like Augustus, Titus' father, Vespasian, had restored peace to Rome after civil war (69–70 CE), and he commissioned the structure as a gift to the Roman people. Like Claudius, Titus' father had campaigned successfully in Britain (see below), and the *naumachia* would have reminded the spectators that Ocean had been (re)conquered and that Roman hegemony had been (re)extended beyond the Pillars of Hercules under the Flavian Dynasty.

Unlike chariot races and beast fights in the arena, where the infrastructure was in place, *naumachiae* required careful planning, specialized installations (that could not be left in place owing to the very real threat of the spread of water-borne pathogens), and they were expensive to produce. *Naumachiae* were imperial prerogatives, demonstrations of Roman maritime success, and expressions of Rome's conquest of the very waters of the Mediterranean.

Oceanus Domitus

To the Roman mind, Britain and Ocean were linked. Britain was both in the Ocean and beyond Ocean, a motif that also figured into British provincial iconography.[60] In 46 BCE, the procession of Caesar's quadruple triumph included the Rhine and Rhone Rivers and "captive Ocean rendered in gold" (Florus 2.13.88). Caesar had semiotically battled against Ocean (storms at sea and unfamiliar tidal activity that wreaked havoc on his ships: *BG* 4.29). Caesar's conquest, moreover, "raise(d) the scale of his achievement" in the popular imagination.[61] Anticipating potential expeditions under Augustus,[62] Horace described Britain as the "most distant (island) of the earth" (*Odes* 1.35.29–30) and Ocean as "abounding in monsters and roaring around Britain" (*Odes* 4.14.47–48: we recall the sea monsters that inhabited Germanicus' Ocean as described by Albinovanus Pedo: *CWW*: Chapter 6). Tibullus envisioned Ocean as surrounding the world (3.7.147–148). Despite decades of treaties and trade (e.g., Strabo 2.5.8, 4.5.3), Claudius' troops were on the brink of mutiny because they were being sent to campaign "outside the known world" (Dio 60.19).

Ocean was thus foregrounded in Claudius' victory celebrations.[63] A naval crown was displayed next to the civic crown at his Palatine house "as if a sign of vanquished Ocean" (*Domiti Oceani*: Suetonius, *Claudius* 17.3). In his argument in support of admitting Gauls to the Roman senate, Claudius proclaimed that he had extended the borders of empire and, furthermore, that the glory of advancing Roman authority "beyond Ocean" had fallen to him (*CIL* 13.1668 [*ILS* 212]). Ocean was also prominent in contemporary encomia. In Seneca, "Ocean, free for so long, obeyed (Claudius) and unwillingly received his ships" (*Octavia* 38–40). The world was bounded by Ocean, but not by Claudius' *imperium* (Anonymous, *Latin Anthology* 419, cf., 420–426 Reise).

The iconography on the temple pediment at Aquae Sulis (Bath), furthermore, was intended to emphasize the conquest of Ocean. Constructed in the decade after the turmoil of the revolts led by Caratacus (51–54 CE: Tacitus, *Annals* 12.31–40) and Boudicca (60/61 CE),[64] the temple complex was the ultimate project of Romanization and reconciliation. The magnificent "Gorgon's head" pediment features a wreathed sun figure with a Celtic moustache, flanked by winged Victories astride globes, Minerva's helmets, and conch-playing Tritons (Fig. 10.5).

A playing field for the vanity projects of emperors and governors, the island would remain remote and unknown, and its conquest would never be fully

Fig. 10.5 "Gorgon" Pediment at Aquae Sulis.

completed in the political or popular imagination. The Flavians would later appropriate the Julio-Claudian legacy, crediting Vespasian with the British conquest.[65] No doubt effective, if not instrumental in Claudius' British campaign, Vespasian had served as a legate (general) under Aulus Plautius in 43 CE. With the annexation of Britain came the subjugation of Ocean, as indicated in contemporary Flavian literature. In Silius Italicus (*Punica* 3.597–602), Jupiter extended Vespasian's hegemony to unknown Thule, on the remote edges of Ocean. Valerius Flaccus declared that Ocean, now "opened" (*pelagi aperti*), enhanced Vespasian's glory all the more. The efforts of Caesar and his descendants (especially Claudius) paled in comparison with Vespasian's far-reaching hold over the western waters: "Caledonian Ocean, which had earlier despised the Phrygian Julii, has conveyed your linen sails" (Valerius Flaccus, *Argonautica* 1.7–9). After Domitian's ascension, Britain and Ocean would be re-conquered, and Tacitus would credit his father-in-law, Agricola, with pacifying and Romanizing the Britons and with quelling Ocean (*victus Oceanus*: Tacitus, *Agricola* 25). Ocean would thus perennially remain a compelling symbol that affirmed divine approval for the Roman conquest of Britain.[66]

The ambitious and the power-hungry would continue to look towards Britain as an emperor maker, and Ocean would remain a prize to be conquered again and again, as stressed by the encomiasts. In his bid for power, relying heavily on naval forces, Carausius (ruled 286–293 CE) exploited the coastal areas of Britain and the continent. An unusual *denarius* of Carausius shows a nude Oceanus on the coin's reverse, with a trident on Ocean's right shoulder and a brace of crab claws protruding from his head.[67] The imagery recalls the crab iconography on Archaic Greek coins. Oceanus was likewise celebrated in contemporary mosaics as at Withington, Gloucestershire, St. Albans, Dorchester, and Cirencester.[68] Oceanus also seemed to have enjoyed a cult in Britain, as attested by the second-century dedication of an altar—narrow enough to fit unobtrusively on a bridge—erected by members of the sixth legion at Pons Aelius (Newcastle upon Tyne).[69] Underscoring maritime associations, other coins issued by Carausius show images of tridents, anchors, and dolphins.[70] *Quinquarii* of Allectus (ruled 293–296 CE), moreover, highlight the nautical foundation of his authority, with reverses showing anchor-wielding bearded figures reclining on ships.[71]

In Eusebius, Britain lies "within Ocean itself" (*Constantine* 263–339). Libanius heroized Constans, giving him hagiographic powers over the sea (*Oration* 18.82–23). As Constans prepared for an unusual winter crossing of the Channel in 242 CE, a storm arose. Not waiting it out, Constans instead "embarked a hundred men, and casting off he clove the Ocean, and straight-away everything became calm," thus affirming Constans' authority over Ocean. In Firmicus Maternus (*Error of Profane Religions* 28.6), where Ocean was still viewed as encircling the earth, Constans literally trampled the "swelling, raging waves of ocean," which "trembled under his oars," an ocean that was "still scarcely known to us" even in the fourth century CE. Jupiter's prophecy was thus fulfilled, and Rome did obtain, seriatim, an "empire without borders" (Vergil, *Aeneid* 1.279), which extended into and beyond the vast, boundless ocean.[72]

Conclusion

For the Romans, Ocean—water—was a realm to be conquered and controlled, facilitating private ambition and political expansion beyond the Italian archipelago. In the 60s, Pompey's victories were as much over the waters of the eastern Mediterranean as they were against the pirates of Rough Cilicia, ensuring that the sea roads would be safe for commerce and travel. Victories at sea would carry political resonance, especially Augustus' triumph at Actium (owing to Agrippa's admiralship), perhaps the most potent symbol of *pax Romana*. Roman conquest of Britain, on the "edge of empire" and within the hostile, monster-infested Ocean, would actualize Roman victory over Ocean, as Augustus' successors would continue to battle Ocean, a realm that remained "scarcely known to us," enhancing the danger and sheer glory of its repeated conquest in their efforts to control the political sphere and the physical world.

11

Conclusion

In this volume we have explored many of the technological and figurative modalities by which the ancients controlled and manipulated water for the ease and comfort of the many or for the ambitions and indulgences of the few. States legislated access to public and private waters. Human initiatives endeavored to improve the physical landscape, facilitating the distribution of water from distant sources, reclaiming fertile lands from otherwise uninhabitable fenlands, and improving and controlling access to ports and emporia for international trade. Water was both a commodity to be regulated (by legislation and infrastructure) and a tool for heavy industry, naval warfare, and maritime trade. Successful naval engagements justified thalassocratic initiatives and they were promoted as such widely across the Mediterranean Basin. Our focus has remained on the ancient literary and documentary evidence, as supported by material remains and geo-archaeological and hydro-archaeological data. Our aim has been to elucidate a part of the intellectual history of Classical antiquity and to explore how the Greeks and Romans understood and interacted with the natural world, especially its watery substrate.

It is now recognized that human activity had profound effects on the environment in antiquity (the consequences of deforestation are well documented in the primary sources). Lead content in glaciers in Greenland, for example, spiked in the second century BCE when Roman smelting became more efficient (and more environmentally destructive).[1] Much work has already been done on the far-reaching ramifications of anthropogenic activity in Classical antiquity, including Hughes' engaging and useful survey (2014), and collections of essays curated by Harris (2013) and Cordovana and Chiai (2016). The landscape and marinescape were changed by hydraulic infrastructure as waterways were diverted or depleted, and coastal areas were altered by the construction of artificial harbors. Investigations into the long-term consequences of anthropogenic activity on waterways and the Mediterranean hydrological cycle are ongoing, including on the question of climate change in late antiquity.[2]

Welcome would be a synoptic volume that surveys environmental change of the waters in the Greek and Roman world. Many questions have been asked (e.g., relative sea level change[3] and the effects of human activity on riparian systems[4]), but others have yet to be investigated.[5] What impact did human activity have on the marine environment in terms of marine fauna (size, populations, migrations)? To what degree might water chemistry have been affected and with what ramifications? What were the repercussions of invasive fish species on native marine fauna?

While our investigations have often begun with Homer, we end with Plutarch's "Whether Fire or Water is more Useful" (*Moralia* 955d–958), where we see explicitly that water is a commodity that is useful only when improved by fire, which Plutarch recognized as a metaphor for *techne* (skill, "technology"). Human reason, which distinguishes us from animals "who have no need of fire," bestows the ability to develop *techne* in order to manipulate the natural world. It is not possible to use water without heat (fire), he argued, and water's benefits are improved when heat is applied. "What is more profitable to life than art (τέχνης: *technes*)," which is developed and maintained by fire(τὸ πῦρ: *to pur* [958d])? In Plutarch, fire is explicitly an agent that conquers the natural world by removing the differences between day and night, enabling human work (958d). But on the whole, both fire and water are ambivalent, both are sustaining and destructive. In Plutarch, fire is at once an "all-consuming wild beast" (θηρίον παμφάγον: *therion pamhagon*), but also the catalyst that improves water and then, as the substrate of vision, the element that enables human beings to conform their souls to the heavens (thus we become more divine). In Plutarch, water is both the destructive element that extinguishes fire and also the very substance of life ("for the dead are deprived of moisture"), undiscovered by any deity (fire was discovered by Hephaestus, or it was found serendipitously and used superfluously), and never detrimental owing to its pervasive utility. Fire, moreover, and the technologies facilitated by it, are the gifts that Prometheus gave to humankind, improving our quality of life and enabling us to conquer the watery world with watercraft and the art of navigating.

Water is thus ambivalent, as recognized in the ancient initiatives to regulate and control it. It both nourishes and destroys. Legislation aimed to guarantee access to water and to protect against polluting it. Infrastructure conveyed water into communities for irrigation, drinking, or luxury applications. But installations also aimed to mitigate against the dangerous consequences of water by channeling gray and black water away from cities or by attempting to funnel floodwaters into the sea. Waterways were important conduits of commerce and communication, and of wealth and warfare, and they were vanquished by means of ships and the art of navigation. But water poses challenges that can be deadly, including virulent storms and complex seabed topography that was often poorly charted. Finally, waters were manipulated in the political arena, elevating some governments and individuals, devastating others who sought authority at sea.

Water, nonetheless, was the physical and symbolic frame of life in the Mediterranean Basin of classical antiquity. For Plato, those who dwelled along the Mediterranean shores were "living around the sea like ants or frogs around a marsh" (*Phaedo* 109b). Water was a source of life and of death. Ocean could be the "friendly boundary of the earth" (*Orphic Hymn* 83.7), but also a terrifying source of sea monsters and storms. Nonetheless, the waterways of the ancient Mediterranean Basin could also be the useful (and sometimes benevolently peaceful) arena that was to be conquered for human use:

> I call upon owl-eyed Tethys, bride of Ocean, dark-cloaked queen, rising in the quick surges, with sweet-smelling breezes treading around the earth. Her dark waves clash against the rocky strand, calmed in well-rolled soft swells, glorified by ships, nourisher of beasts, moistly-pathed, mother of Cypris (Venus), mother of gloomy clouds and of every source teeming with the springs of the nymphs. Hear me, O venerable lady, and may you approach well-disposed, sending a following breeze to straight-running ships.
>
> *Orphic Hymn* 21 (To Tethys)

Appendix

Major scientific, technical, historical, and medical writers and thinkers of Ancient Greece and Rome[1]

Thinker	Provenance	Language	Approximate dates of activity	Major scientific works or area(s) of scientific contributions
Claudius Aelianus (Aelian)	Praeneste	Greek	195–235 CE	*Historical Miscellany* *On the Nature of Animals (NA)*
Anaxagoras	Clazomene	Greek	480–428 BCE	Cosmology; Physics
Anaximander	Miletus	Greek	580–545 BCE	Cosmology; Physics
Anaximenes	Miletus	Greek	555–535 BCE	Cosmology; Physics
Archimedes	Syracuse and Alexandria	Greek	287–212 BCE	Mathematics, Engineering
Aristotle	Stagira and Athens	Greek	384–322 BCE	*Metaphysics* *Meteorology* *On Generation and Corruption (GC)* *On the Enquiry into Animals (HA)* *On the Generation of Animals (GA)* *On the Heavens* *On the Parts of Animals (PA)* *Physics* and others
Flavius Arrianus (Arrian)	Nicomedia	Greek	120–170 CE	*Periplus of the Black Sea* *Indica*

Appendix

Thinker	Provenance	Language	Approximate dates of activity	Major scientific works or area(s) of scientific contributions
Artemidorus	Ephesus	Greek	fl. ca. 100 CE	Geography
Athenaeus	Attaleia (or Tarsus)	Greek	Second century CE	*Deipnosophistae* ("Dinner Sophists")
Athenodorus	Tarsus	Greek	60–20 BCE	Wrote on tides
Basil	Caesarea	Greek	330–379 CE	Neoplatonism, Christian monk
Gaius Julius Caesar (Caesar)	Rome	Latin	100–44 BCE	*Commentaries on the Gallic Wars (BG)* / *Commentaries on the Civil Wars (BC)*
Marcus Porcius Cato (Cato the Elder)	Rome	Latin	Lived 234–149 BCE	*On Agriculture*
Aulus Cornelius Celsus	Rome	Latin	15–35 CE	*De Materia Medica* (*On Medical Matters*) (*On Medical Matters*)
Marcus Tullius Cicero (Cicero)	Rome	Latin	Lived 80–43 BCE	*Letters* / *On Divination* / *On Fate* / *On the Nature of the Gods* / *Tusculan Disputations*
Lucius Junius Moderatus Columella (Columella)	Rome	Latin	Lived 4–70 CE	*On Agriculture*
Democritus	Abdera	Greek	440–380 BCE	Cosmology, Mathematics, Physics
Diodorus Siculus	Sicily	Greek	80–20 BCE	*Bibliotecha* (*Universal History*)
Empedocles	Acragas	Greek	460–430 BCE	*On Nature* (fragmentary) / *Purifications* (fragmentary)
Epicurus	Samos and Athens	Greek	Lived 310–270 BCE	*Letter to Herodotus* (Physics) / *Letter to Pythocles* (Astronomy, Meteorology) / *On Nature* (fragmentary)

Thinker	Provenance	Language	Approximate dates of activity	Major scientific works or area(s) of scientific contributions
Eratosthenes	Cyrene and Alexandria	Greek	245–195 BCE	*Geographica* *Measurement of the Earth*
Euclid	Alexandria	Greek	300–260 BCE	*Elements*
Sextus Julius Frontinus (Frontinus)	Rome	Latin	Lived ca. 40–103/4 CE	*On the Aqueducts*
Galen	Pergamon and Rome	Greek	129–215 CE	Physician, Anatomy, Surgery
Heraclitus	Ephesus	Greek	510–490 BCE	Cosmology, Physics
Herodotus	Halicarnassus	Greek	circa 484–430/420 BCE	*Histories of the Persian Wars*
Heron	Alexandria	Greek	ca. 62 CE	Artillery Construction Automaton Construction *Catoptrics* *Mechanics* *Pneumatics*
Herophilus	Chalcedon	Greek	lived circa 330–260 BCE	Physician, Anatomy
Hesiod	Ascra	Greek	ca. 700 BCE	*Theogony* *Works and Days*
Hipparchus	Nicea	Greek	fl. ca.140–120 BCE	Astronomy, astrology, geography, mathematics
Hippocrates and his school		Greek		*Affections* *Airs, Waters, Places (AWP)* *Aphorisms* *Coan Prenotions* *Epidemics* *Internal Affections* *On Regimen in Acute Diseases* *Prorrhetic*

Thinker	Provenance	Language	Approximate dates of activity	Major scientific works or area(s) of scientific contributions
Homer	Unknown (Chios?)	Greek	750–700 BCE	*Iliad* *Odyssey*
Josephus	Jerusalem	Greek	37–100 CE	*Antiquities of the Jews* *Jewish War*
Juvenal	Rome	Latin	Second century CE	Satirist
Lucian	Samosata	Greek	120–180 CE	Satirist and rhetorician
Titus Lucretius Carus (Lucretius)	Rome	Latin	lived circa 99–55 BCE	*De Rerum Natura* (On the Nature of Things) Epicurean Physics
Manilius	Rome	Latin	10–30 CE	Astrology
Pomponius Mela	Baetica and Rome	Latin	fl. 30–60 CE	Geography
Oppian	Anazarbos in Cilicia	Greek	176–180 CE	*On Fishing* (Fishing) *On Hunting*
Publius Ovidius Naso (Ovid)	Rome	Latin	43 BCE–17/18 CE	*Fasti* *Heroides* *Metamorphoses* *Tristia ex Ponto*
Parmenides	Elea	Greek	490–450 BCE	Cosmology
Philo	Byzantium	Greek	fl. 240–200 BCE	Engineering and mechanics *Mechanical Collection* (Lever, Harbor Construction, Pneumatics, Artillery Construction, Siege Preparations, Siege Craft)
Plato	Athens	Greek	Lived 428/427–348/347 BCE	*Timaeus* Astronomy, Cosmology, Mathematics

Thinker	Provenance	Language	Approximate dates of activity	Major scientific works or area(s) of scientific contributions
Gaius Plinius Secundus (Pliny the Elder)	Novum Comum	Latin	Lived 23–79 CE	*Natural History*
Plutarch	Chaeronea	Greek	Lived 46–120 CE	*On the Cleverness of Animals* (*Cleverness*) *Table Talk* *Biographies*
Polybius	Megalopolis	Greek	180–120 BCE	*Histories*
Posidonius	Apamea	Greek	Lived 135–50 BCE	Cosmology, Geography, Hydrology, Mathematics, Meteorology
Protagoras	Abdera	Greek	487–412 BCE	
Claudius Ptolemaius (Ptolemy)	Alexandria	Greek	127–after 146 CE	Astronomy, Mathematics
Pythagoras	Samos, Croton	Greek	570–495 BCE	Cosmology, Mathematics, Physics
Pytheas	Massilia	Greek	fl. ca. 340–290 BCE	
Lucius Annaeus Seneca (Seneca the Younger)	Cordoba, Rome	Latin	Lived 4 BCE/1 CE–65 CE	*Natural Questions*
Seleukos	Seleukia (Tigris River)	Greek	165–135 BCE	Astronomy, mathematics
Socrates	Athens	Greek	469–399	Ethics
Strabo	Amasia	Greek	Lived 64 BCE–24 CE	Geography
Strato	Lampsakos	Greek	fl. 295–268 BCE	Natural philosophy, mechanics
Thales	Miletus	Greek	fl. 600–545	Cosmology, Physics
Theophrastus	Eresus and Athens	Greek	fl. 340–387/6 BCE	*On the Causes of Plants* (*CP*) *On the Enquiry* (*Historia*) *into Plants* (*HP*)

Thinker	Provenance	Language	Approximate dates of activity	Major scientific works or area(s) of scientific contributions
Marcus Terentius Varro (Varro)	Reate	Latin	Lived 116–27 BCE	On Farming
Vegetius Renatus (Vegetius)	Unknown	Latin	Circa 450 CE	On Military Matters
Publius Virgilius Maro (Vergil)	Mantua and Rome	Latin	Lived 70–19 BCE	Aeneid Eclogues Georgics
Marcus Vitruvius Pollio (Vitruvius)	Rome	Latin	Lived 85–20 BCE	On Architecture
Xenophon	Athens	Greek	430–355 BCE	Anabasis Economics Hellenika Symposium
Xenophanes	Colophon	Greek	Lived 570–478 BCE	Physics
Zeno	Elea	Greek	Lived 490–430 BCE	Physics
Zeno	Citium	Greek	333–262 BCE	Founder of the Stoic school of philosophy

Notes

Introduction

1 Cf., Aeschylus, *Prometheus Bound* 437–71; Irby-Massie 2008; *CWW*: Chapter 1.
2 Roller 2006.
3 Mason 2003; DeSantis 2017; Rees 2019; Bradford 2019.
4 Arnaud 2005; Kowalski 2012; Beresford 2013.
5 Morrison 1968; Casson 1995; McGrail 2001; Pitassi 2011.
6 Chapter 5 is adapted from Irby 2016b.

Chapter 1

1 https://www.un.org/waterforlifedecade/scarcity.shtml.
2 https://thewaterproject.org/water-crisis/water-in-crisis-india.
3 DeJong 2011: 15.
4 DeJong 2011: 49.
5 DeJong 2011: 53.
6 For a more technical overview of Greek and Roman water law: Bruun 2000a: 558–73; 2000b: 575–604; 2015. Cf., Rogers 2018: 10–19.
7 MacDowell 1978: 135.
8 *IC* 4.43 Bb; Bruun 2000a: 560.
9 Harrison 1999: 1.249–251. The drachma was a standard monetary unit in the ancient world, and it is difficult—if not impossible—to establish any accurate exchange rate since marketable goods have changed so much over the years.
10 *Digest* 10.1.13; Harrison 1999: 2.248.
11 Finley 1973: #2 (*IG* II² 2759), #116 (*IG* II² 2657), #159 (*IG* II² 2655).
12 Crouch 1993: 314.
13 Angelakis and Koutsoyiannis 2003: 999–1,007.
14 Angelakis and Koutsoyiannis 2003: 1,006.
15 *IG* xii (5) 569, trans. Arnaoutoglou 1998: #88.
16 See also Chiai 2017: 61–82, for a thoughtful assessment of the intersections between clean water and sacred law where environmental pollution is viewed as violence against the gods and a city's aesthetic appeal.
17 *IG* 14.645.

18 *SEG* xiii 521 (*OGIS* 483 = Arnaoutoglou 1998: #99).
19 Dead animals are naturally redolent. To arrest the decay of an animal hide, the pelt is dried and salted. Mulberry leaves soaked in urine and dung would be used to remove hair and flesh (Pliny 23.140). To increase malleability and prepare the skin for tanning, the hides were stretched and beaten, a process described by Homer as the Greeks and Trojans fought around Patroclus' corpse:

> Just as when a man might give the people the hide of a great bull in order to stretch it, dripping with fat. Then taking and standing around it, they stretch it in a circle. Straightaway the moisture comes out and the fat steeps in, while many men yank at it, and the entire hide is stretched out.
>
> *Iliad* 17.389–393

To complete the process, the cleaned, dried pelt was treated with wood and other vegetable tannins (compounds concentrated in the bark and skin) and dyes, including alum and salt (cf., Falcão and Araújo 2018: 1081).

20 *IG* i^3 257 (trans. Arnaoutoglou 1998: #87).
21 *SEG* xlii 785 (Arnaoutoglou 1998: #85).
22 IG xii (5) 107 (Arnaoutoglou 1998: #86).
23 Plato, *Laws* 8 845 d4–e9; see also *LSCG* 178 (ca. 400 BCE); Saunders 1990: 63–82.
24 Lytle 2012: 1–2.
25 Westcoat 2000: 91.
26 Bruun 2000b: 591–2.
27 Cicero, *To Quintus* #21 (3.1.3); see also Pliny the Younger, *Epistle* 3.19.51.2–4.
28 Cicero, *To Quintus* #21 (3.1.3); Bannon 2009: 172–5.
29 See Bannon 2009: 1–3.
30 Bannon 2009: 117.
31 Bannon 2009: 197.
32 Bannon 2009: 125–30; see also Syme 1986: 32, 35, 202, 212, 274.
33 Solazzi (1948: 184) argues that it is not clear whether the emperor was Augustus or Tiberius. Taurus, however, would have been at least forty when Tiberius ascended to power. Such a chronology is not impossible, but an Augustan date is more likely.
34 Bruun 2000b: 600.
35 Grimal 1961: 90–1.
36 North Africa: *CIL* 8.11338; Liguria: *CIL* 5.7783 (*ILS* 01128).
37 *CIL* 2.5439 (*ILS* 6087); de Kleijn 2001: 96.
38 See also *CIL* 10.4842 (*ILS* 5743).
39 Bruun 2012: 15–16.
40 Crawford et al., 1997: 1.5–6, 396.
41 *AE* 1978, 296; Bruun 2000b: 591.
42 Eck 1987: 79.
43 Bruun 2000a: 364.

44 Vitruvius 8.6.2; de Kleijn, 2001: 99.
45 18-11 BCE, *CIL* 10.4842 (*ILS* 5743); de Kleijn 2001: 100; Bruun 2012: 14.
46 Frontinus, *Aqueducts* 2.103, 105; *Digest* 30.39.5; Bruun 2012: 17.
47 Livy 39.44.4–5; Plutarch, *Cato the Elder* 19.1.
48 The plebeian Cato was a "new man," the first in his family to hold the consulship; the patrician majority endeavored to undermine Cato's position and authority.
49 de Kleijn 2001: 95n7.
50 Sesterces were the standard unit of Roman currency. Despite the difficulty of converting ancient currency to modern estimates, 100,000 would have been a hefty fine.
51 Crawford et al., 1997: 798.
52 One *pes* (plural *pedes*) is a Roman foot, equal to ca. 11.6 English inches (ca. 29.5 centimeters).
53 *I.Ephesus* 7.3217; Bruun 2000b: 593–4.
54 *CIL* 3.568 (*ILS* 5794); Bruun 2012: 22.
55 Cf., Wacke 2002: 11–12.
56 Geissler 1998: 225–6, 240–1.
57 Wacke 2002: 8–10.
58 Wacke 2002: 5–7.
59 *Digest* 43.12.1.3, 12.1.4. The diversion of the Durias River in the Alps deprived farmers in the lower valley of water for irrigation (Strabo 4.8).
60 Beltrán-Lloris 2006: 147–97; Bruun 2012: 27.
61 *Digest* 39.3.10.2, 43.12.2; cf., Bruun 2000b: 579.
62 E.g., Herodotus 4.126, 6.49; at 6.94; Balcer 1989: 130.

Chapter 2

1 See, for example, Hans Paerl, "Well Said: Hurricanes and water quality," The University of North Carolina at Chapel Hill (University Communications, September 19, 2018): https://www.unc.edu/discover/well-said-hurricanes-and-water-quality/.
2 Crouch 1993: 33–4.
3 DK 24B4 = Aëtius 5.30.1. See also Hippocratic *Diseases* 1.24.
4 Strabo 9.4.8. The Ozolians' unpleasant body odor may be attributable either to the waters or to the local goat-herding industry and goat-skin garments (Plutarch, *Greek Questions* 15; Pausanias 10.38). See Roller 2018: ad loc.; *CWW*: Chapter 3).
5 *AWP* 9; *Humours* 12.
6 Hippocratic *Aphorisms* 5.26; *Epidemics* 2.2.11.
7 Seneca, *NQ* 3.20.1. See also *CWW*: Chapters 2 and 3 for evidence from Aristotle and others.

8   Pliny 31.37; Hodge 2000b: 98. Taylor 2000: 30–3 for calcined accretions in the pipes.
9   E.g., Pliny 23.2, 23.9, 24.48, 25.169, 26.14, 27.101, 130, 29.31, 29.32.
10 Galen, *On Preserving Health* 1.11 Kühn.
11 *AWP* 7; Vitruvius 8.3.5; for industrial pollution see also Hughes 2014; Wacke 2002; Delile et al 2017; cf., Chapters 1 and 3.
12 Plato, *Critias* 111b–d, 117a, *Laws* 6.761 b–c.; Cato, *Agriculture* 1.7; Cicero, *To Quintus* #21 (3.1); Vitruvius 8.1.6–7; Pausanias 7.26.4; cf., Hughes, and Thurgood 1982.
13 Gildersleeve 1885, 129; see Race 1981.
14 Hodge 2000b: 98, with reference to André 1961: 163.
15 Galen, *On Preserving Health* 4.4 and 5.12 Kühn: Galen considered it unlikely for men with "warm dispositions" to have pursued such careers. These careers, however, were often not a matter of choice, as most athletes and ditch diggers would presumably have been slaves.
16 Pliny 22.102: *aquam bibentibus*. For silphium's digestive properties: Hippocratic *Regimen in Acute Diseases* 37; Theophrastus, *HP* 6.4.6; Columella 12.4. According to Strabo 15.2.10, silphium growing in bacteria aids in the digestion of raw meat. Silphium was widely used in antiquity for seasoning food and medicines (including gynecological). Like so many plants mentioned in Greco-Roman sources, it cannot be precisely identified. Growing along the coast of Cyrenaica (Herodotus 4.169; Theophrastus, *CP* 1.5.1; Strabo 17.3.22–23), the plant was crucial to the Cyrenaican economy where it was featured on coinage: Tameanko 1992: 26–28; Koerper and Kolls 1999: 133–143. In Pliny's day (19.39), the plant—worth its weight in silver—was rare in Cyrenaica, where sheep had grazed it to extinction: the emperor Nero had been gifted with the last stalk that had been found within living memory. Cf., Parejko 2003.
17 Galen, *On the Therapeutic Method* 5.13 Kühn; Modlin et al. 2004.
18 *AWP* 8; cf., Pliny 31.31.
19 Galen, *Good and bad Juices* 9.6 6.795 Kühn.
20 Galen, *On Preserving Health* 1.11 6.58 Kühn.
21 Oreibasios 5.3.6 = *CMG* 6.1.1. Cf., Strabo 5.3.8 where sewers wash refuse into the Tiber.
22 Pliny 31.32–33. Archimedes had discovered how to determine buoyancy in the third century BCE, but measuring specific gravity was beyond the technology available in the first century BCE.
23 Faust and Aly 1998: 2.
24 Faust and Aly 1998: 217.
25 Cf., McKenzie 2007: 77.
26 Plutarch, *Isis and Osiris* 383c–d. Scent was believed to enhance medical efficacy: Dioscorides, 1.48.1 and Pliny 21.163; Galen, *On Theriac for Piso* 16 14.281 Kühn; Butler 2014: 83–4; Totelin 2014.
27 Herodian, *Histories* 1.12. Cf., Totelin 2014.

28 Hughes 2014: 113.
29 Aristotle, *Meteorology* 2.3.359a1-6; Ahuja 2017: 441.
30 Pliny 31.70. For this same technique used by the Roman army: Rufus of Ephesus, pp. 298, 344 Daremberg = Oreibasios 5.3.29; *CMG* 6.1.1: 120.
31 Wilson 2008: 302.
32 *AWP* 8. Some translators use "filter" to translate the Greek here (ἀφέψεσθαι καὶ ἀποσήπεσθαι: *aphepsesthai kai aposepesthai*), but the meaning is unclear.
33 St. Basil of Caesarea, *Homily* 4: *Upon the Gathering Together of the Water*; cf., Haarhoff 2009.
34 Hales 1739; Haarhoff 2009: 237.
35 See Irby 2016d.
36 Plato, *Gorgias* 451e, *Laws* 744a, 747d-e; Aristotle, *Politics* 7.11.1330b3-6. Rykwert 1976: 41-4.
37 See Borca 2000: 56.
38 It is dangerous for men to encroach into the territory of wild animals and liminal deities. Actaeon, for example, was punished for trespassing into the sacred, ritually pure, celibate (but sexually charged) territory of the virginal Artemis/Diana (by gazing on the goddess and her nymphly entourage as they bathed); and Callisto, who had been raped by Zeus, was punished for violating Artemis' virginal proscriptions.
39 Athenaios of Attaleia in Oreibasios 9.12.6 = *CMG* 6.1.2.
40 Hippocratic *On Humors* 12; *Internal Affections* 45.
41 Strabo 5.1.7; Borca 2000: 57-60. For land reclamation: Chapter 3.
42 Vitruvius 1.4.11-12; Strazzulla 1989.
43 Vitruvius 1.4.12; Gabba 1983: 514-16; Borca 2000: 58.
44 Caesar, *BG* 4.11; *BC* 1.81, 3.66; *African War* 41, 51, 69, 76; *Spanish War* 41.
45 Caesar, *BG* 7.36, 8.40-41; *BC* 1.73, 1.84, 3.15, 3.17, 3.24, 3.49, 3.100.
46 Velleius Paterculus 2.84.1; Dio 50.13.4; Plutarch, *Antony* 63.2; Cf., Powell 2015: 86. From 27 BCE, Octavian would come to be addressed as "Augustus."
47 Davies 1970: 85; Scarborough 1981.
48 Davies 1970: 98.
49 See Frontinus, *Stratagems* 3.7; Mayor 2003: 99-118.
50 Vitruvius 8.6.10-11; Palladius 9.11; Columella 1.5.2; Pliny 31.57.
51 Vitruvius 8.6.1, 4-6; Frontinus, *Aqueducts* 25.2, 27.3, 29.1, 30.1, 39-63, 105.5, 106.3, 115.3, 118.4, 129.4-6; Hodge 2002: 307-15.
52 Hodge 1981; 2002: 2, 443n31, 308.
53 Vitruvius 8.3.5, 8.6.10-11.
54 Pliny 33.124, 34.176; Dioscorides 5.88.6. Soranus prescribed smearing white lead on the mouth of the uterus as a contraceptive (1.19.61).
55 Pliny 22.112, 23.80; Celsus 5.27.12b; Galen, *Antidotes* 14.144 Kühn.

Chapter 3

1. Wikander 2000a; Hodge 2002; Wilson 2008.
2. Luce 2006; Knauss, 2000: 99.
3. Crouch 1993: 23.
4. Bintliff 1975: 271.
5. Pliny 19.37, 26.30, 31.44–45.
6. Peltenburg et al. 2001; Garfinkel et al. 2006; Wilson 2008: 285.
7. Macgillivray et al. 2007.
8. Lang 1968: 6–7.
9. Keenan-Jones 2015: 196.
10. Thompson 1867.
11. Marlière 2002: 48–9, T35–T42.
12. Hodge 2000b 30; Wilson 2008: 286.
13. Hodge 2000b: 30.
14. Hodge 2000b: 31.
15. Hodge 2000a: 29–33.
16. Blair and Hall 2003.
17. Wilson 2008: 286.
18. Oleson 1991: 57–8.
19. Wilson 1998: 65–7.
20. Wilson 2001; Mays et al 2013.
21. Beaumont 2008.
22. Strabo 16.2.13; Pliny 5.128.
23. Wilson 2008: 288.
24. Klaffenbach 1954; Wilson 2008: 289.
25. Angelakis et al 2006, 2007.
26. Blegen and Rawson 1966: 332–6, 416.
27. Wilson 2008: 290.
28. Wilson 2008: 291; Baker 2019: 161–3.
29. Thompson, 1954.
30. Wilson 2008: 298.
31. Wilson 2008: 296.
32. Hodge 2000a: 47; cf., Bruun 1991: 113; de Kleijn 2001: 47–60.
33. Lassus 1983: 211.
34. Döring 2002; Wilson 2008: 298; Keenan-Jones 2013.
35. Gallardo 1975.
36. Wilson 2008: 299.
37. Grewe 1985; Hodge 2002: 171–97; Chanson 2000; Wilson 2008 300.
38. Wilson 2000b: 602.

39 Tölle-Kastenbein 1994: 18, 27; Jansen 2000: 108.
40 Wilson 2008: 303.
41 Frontinus, *Aqueducts* 1.7, 2.69, 75, 91.
42 Cicero, *To Quintus* #21 (3.1.3); cf., Bannon 2009: 173–4; Chapter 1.
43 Cicero, *To Quintus* #21 (3.1.1).
44 *Imp(eratore) Caesare Augusto XIII co(n)s(ule) desig(nato)/C(aius) Avillius C(ai) f(ilius) Caimus Patavinus/privatum*: CIL 5.6899. With thanks to Duane Roller for bringing this aqueduct bridge to my attention.
45 Ordinary consuls assumed their offices at the beginning of the Roman year, and the year would be named for them (e.g., 59 BCE was the year in which Julius Caesar and Marcus Bibulus were co-consuls). Under Augustus, the consulship was revised, allowing for the election of more than two consuls a year. Under the Julio-Claudians, the ordinary consuls vacated office mid-year, and they were replaced by suffect consuls. From 69 CE, it was common for the ordinary consuls to resign after four months, allowing for two sets of suffect consuls. The position of ordinary consul was the more prestigious.
46 *CIL* 11.3003 (*ILS* 5771); Bannon 2009: 73–4.
47 E.g., Pindar, *Pythian* 4.294; *Greek Anthology* 9.314–315, 327.
48 Homer, *Odyssey* 7.129–131; Glaser 2000: 436.
49 Glaser 2000: 416–17.
50 Christaki and Nastos 2018: 6.
51 Boston Museum of Fine Arts 61.195.
52 Glaser 2000, 416.
53 Longfellow, 2011.
54 Hellner 2006. The asphalt was imported from Mesopotamia or the Caspian, thus indicating long-distance trade with the Middle East.
55 Fahlbusch 1982: 78, 111–12; Glaser 2000: 429; Hellner 2006: 174.
56 Paga 2015: 356; Christaki and Nastos 2018: 6. Cf. Pausanias 1.14.1.
57 Glasser 2000: 424.
58 Glasser 2000: 425–7.
59 Kapossy 1969: 16, 27, 86–9; Glaser 2000: 433.
60 Eschebach 1982; Glaser 2000: 434.
61 Knell 1995: 73–81; Glaser 2000: 437–8.
62 Schweitzer 1938; Glaser 2000: 439; *CWW*: Chapter 9.
63 Bol 1984: 54; Glaser 2000: 440.
64 Longfellow 2011: 189–90; Rogers 2018: 65–6.
65 Rogers 2018: 64–7.
66 Hamilton and Tischbein 1791: 1.plate 58.
67 Hamilton and Tischbein 1791: 4.plate 30.
68 Panofka 1843: 9, 18.

69 Collitz and Bechtel 1884–1909: #4689.
70 Labor-intensive and expensive, hypocausts also heated elite homes, including Pliny the Younger's Laurentine Villa (*Epistle* 2.17.11), and commanders' quarters in northern provinces.
71 Wilson 2008: 308.
72 For the bath and temple complex at Aquae Sulis, see Cunliffe and Davenport 1985.
73 McMahon 2015: 22.
74 Antoniou and Angelakis 2015: 45–6.
75 Antoniou and Angelakis 2015: 52–3.
76 Antoniou and Angelakis 2015: 64–5.
77 Antoniou and Angelakis 2015: 61.
78 Orlandos 1940: Figure 52.
79 Antoniou 2007: 155; *CWW*: Chapter 5.
80 McMahon 2015: 22–3.
81 Antoniou and Angelakis 2015: 46–8.
82 Chiotis, E. and L.E. Chioti (2014).
83 For a detailed discussion of waste management in Rome: Havlíček and Morcinek 2016.
84 Wilson 2000b.
85 Wilson 2000a: 168–72.
86 Simmons 1979: 183–96.
87 Humphrey et al. 1998: 316.
88 Ctesibius, *Pneumatics* app.1. See also Chapter 2.
89 Hero of Alexandria, *Pneumatics* 1.28.
90 See *CIL* 6.1057–58, 2994, 31075 for *siphonatores* who handled the force pump fire extinguishers. By the seventh century CE (and probably earlier), force pumps were employed to clean vaulted ceilings: Isidore of Seville, *Origins* 20.6.9; Oleson 1984: 56.
91 Moschion (fl. 220–180 BCE) in Athenaeus 5.207, 208; Artemidorus 1.48.
92 See also Humphrey et al. 1998: 321; Casson 1995: 175–7.
93 Philo, *Pneumatics* 65.
94 See Wilson 2002: 16; Hierapolis: Ritti et al. 2007; Jerash: Seigne 2006; cf., Rogers 2018: 57.
95 Cf., Antipater of Thessaloniki, *Greek Anthology* 9.418; Roller 2018: 711; Wikander 2000b: 394–7.
96 Heron of Alexandria, *Pneumatics* 2.11.
97 Hughes 2014: 133, 135–6.
98 Mattingly 2006: 506.
99 Cf., Burnham 1997; Burnham and Burnham 2004: 4–6.

Chapter 4

1. Vergil replicated Homer's perfect natural harbor when Aeneas, after surviving a storm at sea, found refuge in a protected lagoon off Libya where "no chains hold tired ships, nor does an anchor bind them with a hooked bite" (*Aeneid* 1.168–169).
2. Shipley 2011.
3. Cf., Strabo 7.6.1; Arrian, *Periplus* 25.1.
4. Roller 2019b.
5. Oleson et al., 2004: 205–6.
6. Ancient ships had shallow draughts, and beaches remained a suitable alternative, especially for smaller vessels: Blackman 2008: 646.
7. Blackman 2008: 644.
8. Blackman 2008: 647.
9. Triantafillidis and Koutsoumba 2017: 169. It is highly unlikely that the man-made reef at Aegina formed part of the breakwater, as some have speculated. The gaps between the rock formations and distribution would have rendered them ineffectual: cf., *CWW*: Chapter 2.
10. Blackman 2008: 639.
11. Gibbins and Adams 2001.
12. Brandon et al. 2007.
13. The pier measured 372 meters long (1,220.5 feet), 16 meters wide (52.5 feet): Blackman 2008: 648; Piromallo 2004. The pier was rebuilt in 139 CE: *CIL* 10.1640, 1641.
14. Schreiner 2007; Meiggs and Lewis 26; In Cornelius Nepos (*Themistocles* 6–7), Themistocles refers to two naval defeats suffered by the Persians near Athens (including the defeat at Salamis); Diodorus 2.41 has Themistocles mention Athenian success against the Persians in an "unbroken succession of battles in naval war."
15. Lovén and Davis 2012.
16. Thucydides 1.121.3–4; DeSantis 2017: 36–7. At Rhodes, which would become a naval superpower, entering the docks illegally was a capital offence. At Carthage in the second century BCE, a double wall separated the military and commercial harbors; see Appian, *Punika* 96.
17. Lovén and Davis 2012.
18. DeSantis 2017: 40–3.
19. Meneghini 1985; Keay 2012b.
20. Quilici 1985–86; Maischberger 1997: 100–4.
21. Rice 2018: 209–10.
22. For the initial survey report: Keay et al. 2005. For results of the excavations: Keay and Paroli (2011), Keay 2012a.

23 Thylander B310. Contrary to much scholarly speculation, the harbor was likely completed under Claudius (confirmed by Dio 60.11), not Nero, who took no credit for it: "remarkable restraint for the notoriously vain emperor if he did indeed spend 10 years completing the massive project:" Tuck 2019: 441; cf., Suetonius, *Claudius* 20.
24 *RIC* 1.178: 162; BM R.6445.
25 Tuck 2019: 453–9.
26 The basin measured 357.77 meters (ca. 1200 Roman feet), with a maximum diameter of 715.54 meters (2345 feet), and a depth of 5 meters (16.5 feet).
27 Museo Torlonia #430; Testaguzza 1970: 171.
28 Blackman 2008: 651.
29 *RIC* 471; cf., *RIC* 632.
30 Zabehlicky 1995: 204–5; Wilson 2011: 51.
31 Fabre 2004–2005: 81; Morhange and Marriner 2008.
32 Morhange and Marriner 2008; Morhange et al. 2016.
33 Pomey 1995; Pomey and Rieth 2005: 47–53.
34 Blackman 2008: 662–3; Whitewright 2014.
35 Oleson 2000; Wilson 2000b.
36 On the evidence of an archaeological survey of the Wadi Tumilat, Necho may have been deepening an early canal (maintained during the New Kingdom: 1550–1070 BCE) in response to contemporary high siltation rates: Redmount 1995; Hassan 1997. Subsidence, however, was uneven, and the Delta's eastern side was subsiding more quickly than the western. For Darius' Persian inscription see Ch Wikander 2000: 325: "he arranged to excavate this canal which goes from Parsa from a river by the name of Pirava [the Nile] which flows in Egypt."
37 Cf. Ch Wikander 2000: 326–8.
38 Lewis 2000: 641; Roller 2018 ad loc.
39 Cf., Strabo 12.75–82; Suetonius, *Claudius* 20; Dio 60.113.
40 Cf., Lucan 8.184, 433; Statius, *Thebaid* 12.515–516.
41 For an English translation: Broadhurst 1952.
42 E.g., BM TOW.1182, TOW.1184, issued by Hadrian.
43 Behrens-Abouseif 2006.
44 We are not told what fuel was used for the fire, but it was likely papyrus or oil since wood is scarce in Egypt (Theophrastus, *HP* 4.8.3).
45 Chau Ju-Kua 1911: 146.
46 Empereur 1999: 36–43.
47 "Museums and Tourism: United Nations Educational, Scientific and Cultural Organization": http://www.unesco.org/new/en/culture/themes/underwater-cultural-heritage/museums-and-tourism/alexandria-museum-project/.

Chapter 5

1. Expanded from Irby 2016c.
2. Lusby et al. 2009/2010: 17–25; McGrail 2001: 315; Finney 1976.
3. Herodotus 2.43, 161, 153.
4. Wachsmann 1998: 297.
5. Aubet 1993: 140.
6. Herodotus 4.42; Kahanov 1999: 155–60; Roller 2006: 23–6.
7. *Codex Palatinus Graecus* 398; ninth century CE.
8. Aristotelian *On Marvelous Things Heard* 833a11–15; Mela 3.90; Pliny 6.200; Arrian, *Indika* 43.11.
9. Cf., Arrian, *Indika* 43.12; Polybius 3.4; Strabo 2.3.4. Pliny (2.67, 5.1), however, had confused Hanno's expedition with the earlier Phoenician venture, crediting Hanno with a circumnavigation of Africa without further comment on the feasibility of such an undertaking (Necho is nowhere cited in Pliny).
10. Strabo 1.1.6. Polaris is a post-antique designation for the contemporary "pole-star." In the Bronze Age the pole star was Thuban (Draco alpha). In the Greco-Roman era there was no pole star: Roseman 1994: 117–19.
11. Vegetius 4.41; Milner 1996: xxiii, 145n1.
12. Pitassi, 2011: 4–20.
13. Late-third-century-CE sarcophagus, Ny-Carlsberg Glyptothek, Inv.-No. 1299; Casson 1991: 196–7.
14. Plato, *Alcibiades* 1.125c; Ovid, *Metamorphoses* 13.365–367.
15. Philostratus, *Apollonius of Tyana* 3.35; cf., Plutarch, *Precepts of Statecraft* 812c; Casson 1994: 316–17.
16. Plato, *Republic* 346a; cf., *Gorgias* 511c–e, *Laws* 961e.
17. Homer, *Iliad* 23.316–17; cf., Plutarch, *Table Talk* 1.3.
18. Homer, *Odysseus* 11.10, 12.152.
19. Cyex had boastfully called his wife "Juno;" in answer she called her husband "Jupiter."
20. Cf., *Iliad* 18.483–489; *Odyssey* 5.271–275; Hannah 1993: 123–35.
21. Polybius 1.21.2; see also Propertius 3.21.12.
22. Cp., Pindar, *Isthmian* 4.71.
23. Casson 1995: 300–21.
24. Aeschylus, *Suppliant Maidens* 716.
25. Sophocles, *Assembly of the Greeks* 142. Wounded by Achilles, Telephus required a Hellenic cure, administered by Achilles: Hyginus, *Fabulae* 101; Apollodorus, *Library* 3.20.
26. Paris, Louvre E735: ABV 85, 2; Badd² 23; BAPD 300789; Morrison plate 11D.
27. Homer, *Iliad* 8.217; Ovid, *Metamorphoses* 13.4.
28. Medas 2004: 55–8.

29 In 212 BCE (Second Punic War), the Carthaginian general, Bomilcar, was able to evade Roman notice as he escaped from Syracuse into the sea during a violent storm that prevented the Romans from laying at anchor in the open water (Livy 25.25.11).
30 Pliny 9.100. Ovid compared the chariot of the Sun, driven by the inexperienced Phaethon, to an improperly ballasted boat that is rocked on stormy waves: *Metamorphoses* 2.184–185.
31 Thucydides 4.26.8; cf., Appian, *Spanish War* 6.91.
32 Cf., Holmes 1991: 272–4.
33 Frantantuono 2012: 121.
34 Pliny 35.139; Pausanias 10.19.1–2.
35 Skyllias is perhaps a watery doublet for Pheidippides' *hemerodromos* (day run) from Athens to Sparta, when he sought help against the Persians before the battle of Marathon (near Athens) in 490 BCE, a stunning Athenian victory despite the lack of Spartan help.
36 Dio Chrysostom, *Oration* 7.31–32.
37 E.g., Arrian, *Periplus* 3.2. The mythical king of Thessaly, Ceyx, left his beloved Alcyone under both sail and oar (Ovid, *Metamorphoses* 11.474–477).
38 Casson 1995: 281–96.
39 *Mechanical Problems* is attributed to Aristotle, but the authorship is disputed. It was probably the work of one of the students or followers of Aristotle's Peripatetic school.
40 Heath 1949: 237.
41 Ariadne had helped Theseus navigate the Minotaur's labyrinth. He promised marriage, but abandoned her as she slept.
42 Heath 1949: 239.
43 Heath 1949: 239.
44 Orpheus, *Argonautika* 276–277; Casson 1995: 224–8; Pitassi 2011: 56–8.
45 Lucian, *Navigium* 6. See also Chapter 6 for the *Isis*.
46 Beresford 2013: 231.
47 See Beresford 2013: 12–13, 32–6.
48 Oppian, *On Hunting* 120; Livy 35.44.5–6; Plutarch, *Pompey* 25.
49 Hesiod, *Works* 663–5; Casson 1995: 270.
50 Casson 1995: 270–3.
51 Beresford 2013: 47.
52 Beresford 2013: 3–4; cf. Chapter 6.
53 Aristotle *Meteorology* 2.4.359b27–65a12; Theophrastus, *On Winds*; Seneca, *NQ* 5; Vegetius 4.38. Coutant and Eichenlaub 1975; Kowalski 2012; Beresford 2013.
54 Coutant and Eichenlaub 1975: xviii.
55 Hannah 2008: 753–4; Hannah 2009: 63, 112.
56 Casson 1995: 283: Table 1, 284: Table 2.
57 Aristotle, *Meteorology* 2.4.359b27–65a12.

58 See Coutant and Eichenlaub 1975.
59 Varro's "Naval Books" are attested by Vegetius 4.41; see Theophrastus, *On Winds*; Pliny 2.117; Seneca, *NQ* 5.17.5; Vegetius 4.38.
60 Seneca, *NQ* 5.1.1, 5.6. For the Epicurean (atomic) poet Lucretius (1.295–298), winds are comprised of invisible particles of air, and their cumulative effects can produce rivals for "great rivers."
61 The Lips was first attested by Herodotus 2.25, as was the Apheliotes: 1.193.
62 Davis 2009: 94.
63 Pliny uses the word *mundus* (world), a reference, it would seem, to Seleukos' heliocentric theory. Cf. Irby 2019a: 80–1.
64 Phrixus had been conveyed to the Euxine on the back of the ram with the golden fleece; he later married one of the king's daughters.
65 The Chinese are known to have used magnetic devices for navigation by the first century CE. The Vikings would discover and implement this technology by the thirteenth century, more than a millennium later. Cf., *CWW*: Chapter 9.
66 Downey 2011.
67 Diogenes Laërtius 1.24.
68 Epirus, Messina: Casson 1995: 273; cf., *CWW*: Chapter 2.
69 King 2004: 16.
70 Roseman 1994: 42–3, 75–8, 107.
71 One of the aims of observing the transit of Venus on James Cook's first voyage (on the *Endeavor*: 1768–1771) was to determine the distance of the earth from the sun and, therefore, establish a procedure for estimating longitudes. During his second voyage on the *Resolution* (1772–1775), Cook had the use of a marine chronometer for calculating longitude with greater accuracy (Sobel 2007). Longitude gives positions east or west of some arbitrary prime meridian. The Greek prime meridian ran through Rhodes.
72 Marcian of Heraclea, *Epitome of Menippus' Periplus of the Inner Sea* 5, 1.567–568 Müller.
73 Arnaud 2005: 70–87.
74 E.g., Apollonius, *Argonautica* 1.580–608.
75 Cf., Sophocles fragment 776 Pearson. Such a shadow is feasible: Pearson 1917: 3.26–7.
76 *Gilgamesh* 11.3.117–130; *Genesis* 8.6–12; Apollonius, *Argonautica* 2.335–339; Pliny 6.83.
77 The Mediterranean waters into which the Nile debouches are alluvial, as noted by Herodotus (2.5.28). Olympiodoros (ad Aristotle *Meteorology* 1.13.351a) referred to sand brought up by weights (*Commentary on Aristotle's Meteorologica* 107.21–25). Aristotle eschewed mention of sediment samples. See further Casson 1995: 246, note 85; Oleson 2008b: 119–176, 126.

78 See Beresford 2013: 186–90.
79 Mark Twain, *Life on the Mississippi*, Chapter 9. Samuel Clemens took his pen name from the measurements used by the river boat pilots on the Mississippi: mark (measurement), twain (2 fathoms/12 feet). Nonetheless, Mississippi river pilots produced and used detailed pilot books, e.g., William W. Haydon's *Memoranda of the Bends, Places, Landings &c on Mississippi River Between St. Louis & New Orleans*, published between the mid-1850s and 1860.
80 Campbell 2012: 268.
81 *CIL* 13.1960, 1996, 2494 (*ILS* 9439).
82 *CIL* 13.2002 (*ILS* 7032).
83 Rome (*Inscriptiones Christianae Urbis Romae* 3.9117), Sulmo (*CIL* 9.3136), Etruria (*CIL* 11.6712,290), Numidia (CIL 8.19606; *Inscriptions latines d'Algérie* 2.1.1940; *Inscriptions latines d'Algérie* 2.3.8172), and Carthage (*Inscriptions latines d'Afrique* 412.57).
84 Philostratus, *Apollonius of Tyana* 2.18; Polybius 10.48; Seneca the Younger, *NQ* 6.7.
85 Herodotus 1.189, 1.193, 5.52; Arrian, *Anabasis* 6.14.5, 7.16.3.
86 E.g., the Oxus River: Polybius 10.48, Strabo 2.1.15.
87 Mela 1.51; cf., Strabo 1.3.7.
88 Hassan 1997: 67.
89 Aristotelian *On Marvelous Things Heard* 846b31–33; Claudian, *Gothic War* 335–9; Herodian 6.7.6. See also Aelian, *NA* 14.26 for the frozen Ister and ice fishing.
90 McGrail 2001: 17.
91 Homer, *Odyssey* 5.273–277. In Phaethon's wake, the Bear attempted, for the first time (*primum*), to plunge into the Ocean to escape the heat of the sun (Ovid, *Metamorphoses* 2.171–173).
92 Vegetius 4.39.
93 Casson 1995: 246; see Arnaud 2005: 46–50. Such treatises may not have survived or perhaps never existed: Dunsch 2012: 270–3.
94 The British Museum owns several Marshall Island navigational stick charts (late-nineteenth to early-twentieth century). Straight and curved palm-leaf sticks represent swells and currents, islands are indicated by shells: BM Oc1904,0621.34, Oc1941,01.4; Oc1944,02.931.
95 Wachsmann 1998: 295–9; Sauvage 2012.
96 Irby 2019a.
97 Apollonius, *Argonautica* 4.279–281; Irby 2016c: 867.
98 In Strabo, many roads are cited, including the Appian (6.3.7) and a road so narrow that travelers were affected by vertigo (4.6.6). Roads were also rendered on the Peutinger Map, a medieval copy of Agrippa's first-century-CE map of the Roman empire: see Talbert 2010.
99 Dunsch 2015.

Chapter 6

1. See Heslin 2011.
2. McGrail 2001: 17.
3. McGrail 2001: 129.
4. Pliny 16.201–202, 36.70; Pomey and Tchernia 1978: 246, 249; Swetnam-Burland 2010.
5. Crummy 1997:70–2.
6. Pliny 19.3.
7. Culham 2017.
8. From the first century BCE to the third century CE, about 2 km (1.5 miles) from the Roman forum, Mt. Testaccio was created as an ancient dumpsite. About 25 million amphorae, which had once contained olive oil mostly from Baetica (Haley 2003) but also from North Africa (Bonifay 2004), were smashed (21,188,800 cubic feet = 600,000 cubic meters). The shards were dusted with limestone, to prevent rancidification, and then deposited in terraces to prevent landslides. See Almeida 1984.
9. For ancient shipwrecks, see Parker 1992.
10. Rice 2016a: 111. For the effects of climate change (siltation, channel degradation, changes in seasonal patterns, and degradation of infrastructure) on riparian trade along the Rhine: Franconi 2016.
11. Rice 2016b: 181.
12. Russell 2013a.
13. Rice 2018: 207–8, 210. For Portus Vinarius: Peña 1999: 11–12; Holleran 2012: 79.
14. Russell 2013b: 79; Rice 2016b: 171, 189.
15. Ward 2010; Sherratt 2016; Sauvage 2017.
16. Pulak 2010: 871.
17. Pulak 2005, 2010; Tartaron 2013: 26.
18. Bass 2010.
19. Phleps 1999.
20. Cf., Caskey 1960; Wood 1985: 71–2, 166, 254.
21. Hellner 2006; cf. Chapter 3.
22. Arafat and Morgan 1994; Cerasuolo 2017; Bundrick 2019.
23. Grace 1979.
24. Bissa 2009: 169–91.
25. Romero 2016.
26. Casson 1989 for text, translation, and commentary. See also Marcotte 2017 for an analysis of Greco-Roman knowledge of the lands abutting the Indian Ocean, largely gleaned from trade contacts. See also Fabre 2004–2005: 36–43. For a more organic analysis of the economic, cultural, and political significance of the Erythraean Sea, see Cobb, ed. 2019.

27 For tariffs, which were strictly local: Günther 2016; Matthews 1984; *CIL* 6.1788 = 31931 (San Silvestro in Capite in response to Aurelian's wine ration: *SHA Aurelian* 3.48.4).
28 Produced in the Near East (e.g., Egypt [Zakik, Beni Salama, Bir Hooker, Marea Philoxenite, and Taposiris Magna] and Israel [Beth She'arim, Bet Eli'ezer, and Apollonia]), raw glass was then exported to workshops throughout the empire, including Cologne and Aquileia: Foy 2003; Lauwers 2012.
29 The best tortoise shell comes from Opone (*Erythraean Sea* 2), "finer than any other" (*Erythraean Sea* 13).
30 *Erythraean Sea* 54; cf., Strabo 16.4.19.
31 *Erythraean Sea* 4; cf., 6, 7, 10, 16, 17, 49, 56.
32 The Greek measurement of distance, the *stadion*, was 600 Greek feet, but there was no standard "foot:" at Olympia one *stadion* was 192.8 meters, the length of the stadium there, while the Athenian *stadion* measured 185 meters, and the Egyptian only 157.5 meters: see Irby 2012: 84–5.
33 Arnaud 2011: 61.
34 Arnaud 2011: 62; Rice 2016a: 104.
35 Polybius 3.22–25; cf., Arnaud 2011: 64.
36 Pliny 6.100; cf., Ptolemy, *Geography* 4.7.41. Hippalos may also not be a person but instead a metonym for whoever discovered the wind.
37 See Chapter 5 for a guild of local boatmen along the Rhone.
38 See Casson 1995: 285.
39 Casson 1995: 283–94, Tables 1–6. ORBIS: The Stanford Geospatial Network Model of the Roman World also offers a digital approximation of travel by sea and land: (http://orbis.stanford.edu/). See also Arnaud 2005: 97–148.
40 Parker 1992; Rice 2016b: 167.
41 Williams and Keay 2006.
42 Williams 2012.
43 Manning 2018: 221; cf., Strauss 2013.
44 Wilson 2009: 219–20; cf., Carlson 2011: 383–5.
45 Vegetius 4.39; *Codex Theodosianus* 13.9.3; cp. Hesiod, *Works* 66–9. See Rougé 1981: 15–16; Casson 1995: 270–1.
46 *Theodosian Codex* 13.9.3.
47 Beresford 2013: 23.
48 Caesar fastidiously recorded the movements of his troops from and to their winter quarters (*BG* 1.10.3, 54.2; 2.35.2; 3.29.3; 4.38.3; 5.1.1, 53.3; 6.3.3, 44.3; 7.1.7, 90.3).
49 See Fabre 2004–2005: 20–43 for a useful analysis of how sea-conditions and the "sailing season" differed within the various subdivisions of the eastern Mediterranean Sea.
50 Beresford 2013: 15.

51 Beresford 2013: 12–13.
52 Beresford 2013: 2.
53 *Demosthenic Corpus* 56.30; Beresford 2013: 17.
54 Beresford 2013: 17–18, 21.
55 Paul, Acts 27–28.
56 Beresford 2013: 36–40.
57 Heliodoros, *Aithiopica* 5.18; cf., Pliny 2.125; Beresford 2013: 19.
58 Peña 1998; Beresford 2013: 28–30.
59 Suetonius, *Claudius* 18–19. Containing provisions against celibacy and adultery, the *Lex Papia Poppaea* aimed to encourage marriage and the birth of legitimate citizen children. Such exemptions could be lucrative: see Field 1945.
60 Rickman 1980: 127.
61 Beresford 2013: 33.
62 Beresford 2013: 34.
63 Tacitus, *Annals* 15.18.2; Tuck 2019: 447.
64 Aelius Aristides, *Oration 26: Regarding Rome* 11.
65 *Demosthenic Corpus* 35.10; cf., *Demosthenic Corpus* 34; Beresford 2013: 45–51. Cf. Arnaud 2011: 69.
66 Pliny 6.104. Casson corrects this timeframe to twenty days: 1980: 32–3; 1984, 190–1.
67 The Somalians themselves, moreover, have suffered greatly by the piratical disruption of UN aid shipments: Axe 2009.
68 For a detailed history and analysis of ancient Mediterranean piracy: de Souza 1999. See also Casson 1991; Beresford 2013: 237–64.
69 Casson 1991: 177.
70 Thucydides 1.4; Meyer 2018:19.
71 E.g., *IG* 12.7 386; *IG* 12.8 53.
72 E.g., *IG* 2^2 844.1; *ICret.* 1; Brulé 1978: 1–24.
73 Lund 1999.
74 Pseudo-Aristotle, *Economics* 2.2.1352a16–17, 1352b15–19; Demetrios 56.6–8; *Syll.*[3] 354; Demetrios 56.17; Polybius 31.31.1–2.
75 Diodorus Siculus 20.81–88; Plutarch, *Demetrios* 21–22.
76 de Souza 1999: 46.
77 de Souza 1999: 45–6.
78 *I.Lindos* 88; *Syll.*[3] 581.50–8, 80–3; Gabrielsen 2012.
79 E.g., Agathokles (Diodorus Siculus 21.4); Philip V of Macedon (Diodorus Siculus 28.1.1; Polybius 5.95); Dikaiarchos the Aitolian (Polybius 13.3); cf., de Souza 1999: 56–60.
80 Appian, *Mithridates* 92; Dio 36.20–23; Plutarch, *Pompey* 24. We do not assume that all Cilicians engaged in piracy, or that all Mediterranean pirates were based in Rough Cilicia.

81 de Souza 1999: 100.
82 de Souza 1999: 117.
83 Plutarch, *Pompey* 24.6; de Souza 1999: 105–8; Chapter 5. Antonius Creticus' campaigns resulted in the *Lex de Provinciis Praetoriis* (Hassall, Crawford, and Reynolds 1974) which, in part, included provisions for safe maritime passage of Romans citizens, friends, and allies (de Souza 1999: 108–15).
84 de Souza and de Souza 2012.
85 Casson 1991: 209.
86 Launched in 1765, Horatio Nelson's flagship, the HMS *Victory*, had a beam of nearly 52 feet (15.8 meters), her weather deck was 186 feet (57 meters), her hold was 21.5 feet deep (6.5 meters), with a draught of nearly 29 feet (8.7 meters), and a capacity of 2,142 tons burthen. As a first-rate ship of the line, she was large for her time. In the east, in the early-fifteenth century, Zheng He traveled the Pacific between China and the shores of east Africa in mammoth nine-masted, four-decked ships.
87 *P. Teb.* 856; Casson 1995: 164; cf.: *Oxy.* 2415.
88 Pomey 1982; Delgado 1998: 252.; Rice 2016a: 101. The ship's cargo included wine from multiple production sites, as indicated by the amphora-stamps.
89 Phaneuf et al. 2001. The wreck was discovered in 1999.
90 Moschion in Athenaeus 5.206d–209b; *Brill's New Jacoby* #575.
91 Russell 2013a: 342.
92 Casson 1991: 209; Erdkamp 2005.
93 For Rome's system of "interlinked trade:" Virlouvet 2017.
94 Pavolini 1988: Figure 27.
95 Badisches Landesmuseum Karlsruhe, Germany; Rea 2001, Figure 23, third/fourth century CE.
96 Wilson 2018: 203 and Figure 12.13.
97 Suetonius, *Augustus* 41; *Digest* 5.1.52, 39.1.49, 39.1.87; Duncan-Jones 1974; cf., Morley 2007.
98 Statio 25 on the Piazzale delle Corporazioni at Ostia; Dunbabin 1999: 62–5.
99 Pavolini 1988: Figure 26; for maritime taxation, Purcell 2017.
100 Marlière 2014: 66–71; Marlière et al. 2017.
101 Strabo 4.1.14; Campbell 2012: 269–70.
102 Musée Lapidaire d'Avignon, Fondation Calvet.
103 Soon after his wedding, Herakles hired the centaur to ferry his new bride, Deianeira, across the Evenos River. But the centaur tried to rape the girl, and Herakles used his deadly arrows to dispatch Nessos. The centaur, however, advised Deianeira to collect some of his blood as a love charm. Years later, Deianeira unwittingly used the centaur blood to kill Herakles in a very painful way: Sophocles, *Women of Trachis*.

104 George 2013.
105 Cicero, *To Atticus* #104 (5.11.4); #105 (5.12), cf., #122 (6.8.4); Casson 1994: 151.
106 Tenney 1933–1940: 2.593–594, 715 #64, 66, 68.
107 Josephus, *Life* 3; also Tenney 1933–1940: 1.172.
108 Casson 1994: 153.
109 Strabo 17.1.45: "like sailors they [camel caravans] carried water as they travelled." Price gouging also occurred. The Samian captain, Dexikreon, had been warned by Aphrodite to bring potable water onboard for a sail to Cyprus. The wind ceased to blow, the passengers grew thirsty, and Dexikreon sold the water at a high price: Plutarch, *Greek Questions* 303c–d. Salassian locals sold their mountain water to Roman officials overseeing the mine works along the Douria River (Strabo 4.6.7). Where water was scarce in Anatolia, it was also sold (Strabo 12.6.1).
110 Blackman 2008: 653.
111 Casson 1994: 151.

Chapter 7

1 This chapter's title is a respectful nod to Marzano 2013 who treats this topic fulsomely.
2 Homer, *Iliad* 16.745–747; Aristotle, *HA* 8.20.603a.
3 *Erythraean Sea* 35, 59; cf., Athenaeus 3.93e; Arrian, *Indika* 8.11.
4 Pliny 9.153; Plutarch, *Cleverness of Animals* 981e.
5 Found in the north-eastern corner of Room 5 in the West House, Akrotiri (National Archaeological Museum, Athens).
6 Ganias et al 2017.
7 Marzano 2013: 17.
8 Marzano 2013: 3 et passim; Zucker 2017.
9 Agatharchades 5.32–33; Strabo 15.1.2; *Erythraean Sea* 2; Marzano 2013: 57–8.
10 Philostratus, *Images* 1.13; cp., Oppian, *Fishing* 3.637–644; Marzano 2013: 71–2.
11 Gianfrotta 1999; Thomas 2010.
12 Marzano 2013: 32–3, 52. See *CWW*: Chapter 6 for Claudius's Ostian orca.
13 Oppian, *Fishing* 5.131–151; Marzano 2013: 8; *CWW*: Chapter 6.
14 Pliny 9.29–32; Marzano 2013: 35.
15 Marzano 2013: 35.
16 Museum of Fine Arts, Boston, #1.8024.
17 El Alia, second/third century CE; Tunis, Bardo Museum: Koppermann Neg. D-DAI-Rom 63.3300.
18 Hadrumentum, third century CE, Sousse Museum Koppermann Neg. D-DAI-Rom 64.376; Marzano 2013: 21–3.
19 Cf. Schmidt 1976: 182–5; Davidson 1997: 13–20.

20 Archestratus, fr. 21; Davidson 1997: 4.
21 Davidson 1997: 5.
22 Davidson 1997: 4–11.
23 Synesius of Cyrene, *Epistle* 57; *SEG* 48.994.1–4.
24 Lytle 2016.
25 I. Parion 5 (*Die Inschriften von Parion* [IK 25, 1983]).
26 Strabo 7.6.2; cf., Aelian, *NA* 15.5; Oppian, *Fishing* 3.637–644; Philostratus, *Images* 1.13.
27 Aristotle, *HA* 8.13.598a20. For fish migration: Aelian, *NA* 9.46; Plutarch, *Cleverness of Animals* 979c–d. Plutarch remarked on the arithmetic prowess of the tuna who always travel in the formation of a perfect cube with six equal planes (979e–f).
28 Oppian, *Fishing* 3.620-622; see also Marzano 2013: 66–79.
29 Aristotelian *On Marvelous Things Heard* 136a.
30 Pliny 9.39; Seneca the Younger, *On Clemency* 1.18.2; Dio 52.23.2.
31 Potter 1971: 1.147–161; Africa 1995: 71.
32 Roberts 1988: 1.9; Marzano 2013: 201.
33 Aristotle, *HA* 8.2.592a; Athenaeus 7.300b–c.
34 Diodorus Siculus 11.25.4, 13.83; Athenaeus 12.541f.
35 Aristophanes, *Acharnians* 880; Dalby 2003: 36; Marzano 2013: 203.
36 Aristotle, *HA* 8.2.592a; cf., Pliny 9.171–172.
37 Marzano 2013: 207.
38 Bouffier 1999: 38; Marzano 2013: 206.
39 Bouffier 1999: 49; Marzano 2013: 206.
40 Etymologically, *piscinae* are installations for fish (*piscis*) but eventually came to also include purely decorative ponds (Aulus Gellius, *Attic Nights* 2.20.7).
41 E.g., Plautus, *Truculentus* 35; Varro, *Farming* 3.17; Columella 8.16–17. Owners of maritime villas utilized their property for private consumption and commercial profit: Marzano 2018b.
42 Varro, *Farming* 3.17.4; Columella 8.17.7.
43 Cf., Secord 2016: 220.
44 Marzano 2013: 216–17; Florido et al. 2011.
45 Pliny 32.321; Martial 3.85.
46 Aristotle, *HA* 5.15.547b12–13, 20; Pliny 32.59–60; Aelian, *NA* 9.6. See Llewellyn-Jones and Lewis 2018: 681.
47 Juvenal 4.139-142; cf., Pliny 32.21.
48 The oyster beds at Baiae are rendered on a delicate engraved glass flask from Piombino, perhaps a souvenir, now in New York Corning Museum, late-third to early-fourth century CE (after Günther 1897, Figure 1). Cf., Llewellyn-Jones and Lewis 2018: 681–2.
49 Pliny 9.168-169; cf., Horace, *Satire* 2.4.31–34; Martial 3.60, 5.37; Juvenal 8.85–86; Andrews 1948: 300.

50 Alcock 2001: 54–6.
51 BBC, "Roman villa unearthed 'by chance' in Wiltshire garden," April 17, 2016; Pruitt 2016.
52 Kraay 1976: #722–7.
53 Jameson 1913–1932: #39.
54 Kraay 1976: #724.
55 Kraay 1976: #725.
56 Horace, *Satire* 2.4.28; cf., Dalby 2003: 224.
57 Campbell 2012: 268; cf. Chapter 5.
58 *CIL* 5.7850 (*ILS* 3287).
59 *Corpus piscatorum et urinatorum*: *CIL* 6.1872 (*ILS* 7266), 6.29700, 29702.
60 *CIL* 6.9799.
61 *CIL* 6.1872.
62 I.Parion 5; Marzano 2013: 42–7.
63 *SEG* 14.638; Marzano 2013: 62.
64 Dioscorides 5.109.1; cf., Beck 2017, ad loc.
65 Marzano 2013: 125.
66 Bekker-Nielsen 2005; Lowe 2017.
67 Nonnis and Ricci 2007.
68 *P. Lond.* 3.1159.
69 *P. Got.* 3; Curtis 1991: 134n.114.
70 Marzano 2013: 26; Novello 2007. For garum and the economy of Galilee in the first century CE: Hakola 2017.
71 Dioscorides 2.32; Galen, *Mixtures and Properties of Simples*; Curtis 1991: 27–37; cf. *CWW*: Chapter 5.
72 Étienne and Mayet 1994: 48; see further Marzano 2013: 111.
73 Columella 12.55.4; Curtis 2008: 385; Marzano 2013: 121.
74 Morales Muñiz 1993.
75 González et al. 2004: 82.
76 Athenaeus book 3; Curtis 2008: 385–6.
77 Bottéro 1995: 253.
78 Curtis 1984b: 58.
79 Curtis 1979; Curtis 1984a.
80 Cassola and Gonzalez 2008: 21–3.
81 See Curtis 1991: 15–26, 191–2.
82 Cf., Lowe 2017: 309.
83 Part of Britain's attraction was as a source of pearls (valued as gifts), but they were small and discolored: Pliny 9.106. For *magaritarii*: *CIL* 6.641, 1925, 9544–49, 5972, 30973, 37803, 10.6492, 14.2655; *ILS* 7603; Marzano 2013: 170–1. Cf., Rice 2016a: 98–9 for other mercantile specializations: iron, pottery, grain, ivory, textiles.

84 Maeder 2008: 113.
85 The tale is recast by Commodus' arbiter of rhetoric, Julius Pollux. Writing in Greek in the second century CE, Pollux quoted Alexander of Macedon as attributing the discovery of the purple dye to Herakles' dog, whose snout was stained purple after eating the mollusks on the beaches in the Levant (*Onomasticon* 1.1.45–49). Ca. 1636, Peter Paul Rubens illustrated the Greek version in *Hercules' Dog Discovers Purple Dye* (Musée Bonnat, Bayonne). In yet another Greek tradition, Helen discovered the hue in Troy as she walked along the beach and saw the purple-stained snout of her dog, gifted to her by Herakles. She liked the color so much that she promised to marry the Trojan suitor who could produce a garment of the same color (cp., Homer, *Odyssey* 4.135 where Helen wove a tapestry of wool dyed a dark-color/purple [*iodnephēs*: ἰοδνεφής]). Into the nineteenth century, using lemon juice or sodium bicarbonate as a mordant, children along the same coast would dye rags with the fluids from mollusks as banners for their war games: Hitti 1962: 109–10; cf., Jensen 1963: 106; Hughes 2005: 339–41; Smith 2010: 599–611. For the Assyrian myth: Anonymous 1876.
86 Drews 1998: 48.
87 Cf., Marzano 2013: 143.
88 Ruscillo 2005: 103–4; Marzano 2013: 156–60. The shell heap at Sidon was 110 meters long (360 feet), 6.5–7.5 meters deep (21–25 feet).
89 Nieto 2006; *IG* I^3.377.
90 Lytle 2007: 258.
91 Pliny 6.201; Roller 2003: 115–17.
92 Meninx (Djerba, Tunisia: Strabo 17.3.18; Pliny 9.127), also Mogador (Pliny 6.201–203), Rhodes, Pontus, and Gaul (Vitruvius 7.13.1–3), Sidon in the Levant (Marzano 2013: 151), Lycia (Strabo 13.4.14), south-east Peloponnese (Plutarch, *Alexander* 36; Strabo 8.6.12; Pausanias 3.21.6), and Tarentum (Horace, *Epistle* 2.207–209, Pliny 9.137).
93 Suetonius, *Nero* 32.3; cf., Napoli 2004: 126–7.
94 SHA *Alexander Severus* 40.6; *CIL* 3.536.
95 Diocletian, *Price Edict* 24; Erim et al. 1970: 127. Under Augustus, a pound of the dye sold for 100 denarii: Pliny 9.137. We also note that the annual salary of a legionary soldier under Augustus was about 230 denarii.

Chapter 8

1 See, for example, Bradford 2019; Rees 2019; DeSantis 2017; Strauss 2004.
2 Irby 2016b: 829.
3 Marinatos 1973: 237.

4 Davis 1983: 3–14.
5 Koehl 1986: 407.
6 Koehl 1986: 417.
7 Koehl 1986: 413.
8 PM 1.519; Koehl 1986: 413, 416.
9 Evans 1935: 494, #429–30; Panagiotaki 1993: 67–8.
10 Weingarten 2010: 317.
11 Casson 1995: 32–35.
12 Casson 1995: #34–6.
13 Casson 1995: #37–45.
14 Casson 1995: #37–40, 42–4.
15 Casson 1995: #38.
16 Fifteen oars are depicted on Casson 1995: #47.
17 Casson 1995: #50, Tiryns, ca. 1300 BCE; #51, Crete, after 1400 BCE.
18 Casson 1995: #50.
19 Casson 1995: #34, 51.
20 Casson 1995: #34.
21 Evans 1935: 827 #805 = Casson 1995: #52.
22 Evans 1935: 952, #921; cf., Homer, *Odyssey* 12.222–259; *CWW*: Chapter 6.
23 See Platon 1984. Compare Evans 1935: #919, a gold-signet ring from Mochlos showing a boat also with a seahorse prow (Heraklion Archaeological Museum). A similar scene is depicted on a signet ring in the Ashmolean Museum (1500–1450 BCE). See Dimopoulou-Rethemiotaki 2004: 21, Figure 15.
24 Dimopoulou-Rethemiotaki 2004: 22–4; Heraklion Archaeological Museum, inv. #1700.
25 Homer, *Iliad* 2.494–759; Wilcock 1976: 22–3; Kullmann 2012. On the strength of the prominence of Boeotian and northern Greek contingents, Anderson 1995 argues that the author was a late-eighth century Boeotian poet who crafted the catalogue as a poetic record of his own travels.
26 The poet lists the Boeotians first, giving them the largest number of leaders and towns cited by name, and they provided the most ships: *Iliad* 2.494–510. Of the towns in Attica, only Athens is cited (*Iliad* 2.546–556), a significant Mycenaean stronghold, but led by the obscure Menestheus who made half a dozen or so cameo appearances (see also *Iliad* 4.327, 12.331, 12.373, 13.195, 13.690, 15.331). The text was likely edited in Athens in the sixth century: Wilcock 1976: 27.
27 Simpson 1968.
28 Fayer 2013.
29 Anderson 1995.
30 Allen 1921. Naval power in the Cyclades is attested as early as the early Cycladic Period (ca. 3000–2500 BCE) with depictions of many-oared ships on fan-like

utensils: Platon 1984: 65; Renfrew 1972: 357. By the mid-second millennium BCE, well-developed and well-organized harbors were installed throughout Crete: Platon 1984: 66.

31 Wilcock 1976: 23–5. Most of the contingents who supplied or manned ships at the battles of Artemisium and Salamis are represented in Homer's catalogue (cp. Herodotus 8.1, 43–47).

32 *Iliad* 15.415–419, 707–725. It is unclear if the ship attacked by Hector at 15.416 is identifiable with the one he attempted to destroy at 15.705: Wilcock 1976: ad loc.; Wilcock 1984: ad loc.

33 Exchange rates between ancient and modern money cannot be reasonably established.

34 Malkin 1998.

35 Tzetzes ad Lycophron, *Alexandra* 609, Servius, *ad Aen.* 8.9, 11.246; Strabo 6.3.9; Pliny 3.103, 104, 120. Hellenic settlers in Magna Graecia may have assimilated Diomedes with a local hero whose mythology was comparable: Malkin 1998: 235–6.

36 Strabo 6.1.3; cf., Malkin 1998: 215–26.

37 Strabo 5.3.5; Dionysius of Halicarnassus, *Roman Antiquities* 1.64.5; Livy 1.2.6; Ovid, *Metamorphoses* 14.608. Recent excavations at Lavinium suggest, inconclusively, a hero cult of Aeneas from the fourth century BCE. In both Livy and Dionysius of Halicarnassus, Aeneas was worshiped as Jupiter *Indiges* (indigenous?): Harrison 2015.

38 "Colonization" is a fraught term that in recent European contexts is intertwined with imperial aspirations, thus not according with Greek initiatives to relocate citizens in new settlements in response to the stresses of increasing population: Antonaccio 2017: 220.

39 Antonaccio 2017.

40 See Branigan 1981.

41 For Phoenician settlement of the western Mediterranean: Margarida 2009. For Phoenician incursions into the Aegean: Herodotus 1.1; Roller 2019a.

42 Hanno, *Periplus* 1. See also Chapter 5.

43 Apollo was widely accepted as a god of "colonization:" "Phoebus takes pleasure in the establishment of cities, and Phoebus himself contrives their foundation" (Callimachus, *Against Apollonius* 55–57). Some (though not all) of this activity was initiated from Apollo's shrine at Delphi where *poleis* would seek divine favor beforehand (Cicero, *On Divination* 1.3: Dodona and Ammon were also centers of "colonization").

44 Thucydides 1.13.2–3; 1.36.3. The other two superpowers were Athens and Corcyra.

45 Plutarch, *Greek Questions* 11; Timaios, *FGrHist* 566 F 80; cf., *PECS*, ad loc.; Boardman 1999: 225. See also Hall 2004; Porciani 2015.

46 Boardman 1999: 225; Kotsonas 2012.

47 Hahn 2001: 202–3; King 2004: 29–34. Although taking many forms, the connections between *apoikiai* and their *metropoleis* endured. On occasion, Delphi even encouraged subsequent immigration from mother city to *apoikia* (e.g., Cyrene): *SEG* 9.3, 30–33; Doukellis 2008: 115–16.
48 For a historical and regional overview of Greco-Roman coinage: Metcalf 2012.
49 Bubelis 2012.
50 Kemmers and Myrberg 2011.
51 See Llewellyn-Jones and Lewis 2018: 413; Barringer 2016: 117–29.
52 BM 1841; Jenkins 1971: #36 (sixth century BCE *stater*).
53 Kraay 1976: #276, #1001 (stater, ca. 460 BCE), with a bust of Athena on the reverse.
54 Jenkins 1971: #135; see Thucydides 6.4.5.
55 Kraay 1976: #800–15 (fifth to fourth centuries).
56 Phlanthos is occasionally assimilated with Taras, a son of Poseidon and a Tarentine nymph: Pausanias 10.10.8.
57 Cf., Strabo 6.3.2–3; Pausanias 10.10.6–8, 13.10. Ridgway 1970. See also Beaulieu 2016: 134–44.
58 Bowra 1963: 132. For Melicertes, see also Lucian, *Dialogues of the Sea Gods* 8.1; Pausanias 1.44.11; Philostratus, *Images* 2.16; *CWW*: Chapter 8.
59 See Herodotus 1.23–24; Hooker 1989: 142. See *CWW*: Chapter 8.
60 E.g., Kraay 1976: #664, 671.
61 Kraay 1976: #673 (Vlasto #260), #676 (Vlasto #79).
62 Kraay 1976: #665–666, 669; Vlasto #700: *nomos* (ca. 280 BCE); Vlasto #930 (mid-third century *nomos*).
63 Jenkins 1971: #492 (mid-fourth century *stater*).
64 Jenkins 1971: #131 (early-fifth century *stater*); Vlasto #501 (mid-fourth century *nomos*).
65 *Catalogue of the Greek Coins in the British Museum* #185 (mid-fourth century *nomos*).
66 Kraay 1976: #673–7.
67 Vlasto #773, 780, 959 (mid-fourth century *stater*). For the Dioscuri, *CWW*: Chapter 9.
68 Pausanias 7.3.10; see also Herodotus 1.146; Strabo 14.1.3; Nicolaus of Damascus, *FGrHist* 90 F 51; Draycott 2012.
69 E.g., Kraay 1976: #70 (ca. 600 BCE).
70 Jenkins 1971: #50 (ca. 520 BCE).
71 Jenkins 1971: #49 (ca. 500 BCE).
72 Herodotus 1.165–167; Pausanias 10.8.6. At Lade in 494 BCE, a Phocaean general was in command of the fleet that vanquished the Persians, but it is unclear if the *polis* outfitted any ships for that campaign (in the aftermath, Demetrios the Phocaean seized three enemy ships before escaping to Phoenicia: Herodotus 6.11, 17). In

Herodotus (8.43–48), no Phocaean contingent is included among the Greek ships assembled at Salamis. It was nonetheless between Cyme and Phocaea (boasting two good harbors in the Cymean Gulf) that Xerxes mustered his fleet before invading the Greek mainland in 480 BCE (Diodorus Siculus 11.2.3).

73 I.e., Gela: Kraay 1976: #828 (ca. 425 BCE).
74 Levine 2011.
75 Kraay 1976: #71.
76 di Natale 2014: #6a (450 BCE), #7a (450 BCE).
77 Kraay 1976: #955 (ca. 480 BCE).
78 di Natale 2014: #10b (ca. 600 BCE).
79 di Natale 2014: #11b (ca. 600 BCE).
80 di Natale 2014: #8a (fifth century BCE).
81 di Natale 2014: #9a (ca. 500 BCE).
82 di Natale 2014: #13b (fifth century BCE).
83 Kraay 1976: #952 (ca. 350 BCE).
84 di Natale 2014: #10a (ca. 500 BCE), #11a (ca. 550 BCE).
85 Jenkins 1971: #191= *Catalogue of the Greek Coins in the British Museum* #62 (late-fourth century *stater*).
86 di Natale 2014: #12b (ca. 500 BCE). After Poseidon raped the once beautiful, Medusa, in Athena's temple, Athena punished the girl by turning her into a hideous snaky-haired monster whose gaze, dead or alive, turned folk to stone: Hesiod, *Theogony* 287; Apollodorus, *Library* 2.4.3; Ovid, *Metamorphoses* 4.792.
87 Kraay 1976: #954 (ca. 500 BCE); #951 (395/4 BCE).
88 After the revolt of 445 BCE, the "well-cattled" island of Euboea came under the authority of Athens, which revoked their right to issue coins. After Euboea regained autonomy in 411, coins continued to show the bull, but the octopus became less common: e.g., Kraay 1976: #268–70, where cows are depicted on the obverse: #269 = BM 1884,0610.12. See further Llewellyn-Jones and Lewis 2018: 675.
89 Kraay 1976: #632, fifth century.
90 Kraay 1976: #817 (ca. 420 BCE).
91 Timocrean fragment 734; Hegemon fragment 1; Athenaeus 9.370c. See Apicius (*On Cooking* 9.5) for octopus cooked in broth with pepper and laser. Thiercy 2008.
92 Hippocratic *Nature of Women* 32.46, *On Barrenness* 5; See Athenaeus 9.370c for octopus as a remedy for nursing mothers.
93 Aelian, *NA* 9.25; see also Dalby 2003: 236–7; Llewellyn-Jones and Lewis 2018: 675.
94 Demosthenes, *Against Aristokrates* 211; Ephorus, *FGrHist* 70 F 176; cf., Pseudo-Skylax, *Periplus* 53. Artificial reefs at Aegina: Chapter 5.
95 Xenophon, *Hellenika* 6.2.1; cf., Aristotle, *Politics* 4.4.1291b24.
96 Baxter and Figueira 2012; for coins, see Kraay 1976: #113–17, 123, 126, 127, 137–8. See Lloyd Llewellyn-Jones and Lewis 2018: 563.

97 Pliny 6.109; Aelian described their flesh as "bitter:" *NA* 16.14.
98 Dalby 2003: 329. "Turtle-eaters" dwelt at Carmania on the Persian Gulf (Pliny 6.109).
99 Hippocratic *Affections of Women* 78.1, 78.151, 203.7; Dioscorides 2.78–79.
100 Pliny 9.35-36; Diodorus Siculus 3.21.3–4.
101 Diodorus Siculus 3.21.5; Strabo 16.4.13.
102 Dalby 2003: 105; Llewellyn-Jones and Lewis 2018: 679.
103 Aristotle, *HA* 4.4.529b20-530a31, 7.17.601a10; Pliny 9.97–99; Oppian, *Fishing* 1.283-304. For medicinal uses: Pliny 8.98, 19.28–29, and book 32. Pliny cited marine and freshwater crabs extensively as remedies for fevers, ulcers, phthisis, burns, hair loss, and for the stings of scorpions and other venomous animals: cf., Galen, *Mixtures and Properties of Simples* 11.24 = 12.343-344 Kühn; Llewellyn-Jones and Lewis 2018: 679.
104 Jenkins 1971: #141; cf., Kraay 1976: #18–19, 790-3, 795-8; Llewellyn-Jones and Lewis 2018: 680. See also Pliny 9.97–98 for the crab claw.
105 Calciati 1983–1987: #64 (the obverse of a *hexas*).
106 Sear #1022; cf., Sear #1023; Calciati: 1983–1987: #50 (*hemilitron*).
107 Calciati: 1983–1987: #40 (*hemilitron*).
108 Calciati: 1983–1987: #40, 43, 73, 80.
109 Kraay 1976: #861–2; cf., Thucydides 6.2.6.
110 Kraay 1976: #717 (ca. 330 BCE).
111 Deonna 1954: 55; Kraay 1976: Cyrenaica, #199.
112 Kraay 1976: #940 (ca. 350 BCE).
113 Plato, *Euthydemus* 297c; Pseudo-Hyginus, *Astronomika* 2. 23.
114 Remedies were repeated or augmented by Pliny 32.53–54, 56, 78, 88, 90, 100, 103, 107, 114, 117–118, 125, 130, 137.
115 Deonna 1954: 54.
116 *Sylloge nummorum graecorum* ANS 9.
117 BM 1971,1101.1.
118 Strabo 14.3.9.
119 Cicero, *2 Verres* 4.21; de Souza 1999: 129–30; Brandt 2012.
120 Kraay 1976: #991: ca. 520 BCE with a ship's prow on the obverse, incuse on the reverse. Kraay 1976: #992: ca. 460 BCE with a prow on the obverse with a stern on the reverse; the legend reads: Phas[elis].
Ship-prow coinage was issued during the Persian Wars (and shortly thereafter) also by Skione on the Chalcidean headland: Jenkins 1971: #93. Skione was among the city states that had supplied Xerxes with ships in 480 BCE: Herodotus 7.123, 8.127-129: they soon rebelled but were subsequently recaptured by Xerxes' general Artabazus.
121 Museum of Fine Arts Boston 63.820; bronze, 81 BCE or later. The reverse shows a helmeted Athena with aegis and thunderbolt.

Chapter 9

1. Thucydides 2.34.5; Pausanias 1.32.4.
2. Herodotus 7.144; Corvisier 2008: 132–3.
3. Herodotus 7.141.3–144; cf., Evans 1982; Robertson 1987.
4. Herodotus 8.1–21: they thus forced an attack of the pass by land.
5. Herodotus' numbers are inconsistent and probably exaggerated. The Persian fleet at Salamis may have included 600–800 craft: cf., Lazenby 1993: 174. Herodotus 7.89 recorded that the Persian fleet included 1,207 triremes, many of which were lost in a storm off Magnesia (7.188: "an overwhelming disaster"), several more were lost in a storm off the "Hollows" of Euboea (8.13), and thirty or so were taken in the Battle of Artemisium (8.11). According to Diodorus Siculus (11.3.7), 1,200 Persian ships mustered at Doriskos in 480 BCE. The Persian fleet included ships supplied and sailed by Egyptian, Ionian Greek, and Phoenician allies (7.89). The allied Greek force consisted of 300 or so war ships, 180 of which ("the most numerous and also the best") were supplied by the Athenians (8.42–46).
6. DeSantis 2017: 11–12. The significance of this maritime battle has been recognized in both scholarly and military circles: Custance 1970: 9–30; Ruffing 2006: 2–3.
7. Pindar, *Isthmian* 5.48–50; cf., Ruffing 2006: 7.
8. Thucydides 1.74.4; Ruffing 2006: 14–15; Kuch 1995; Parker 2007.
9. See Wallinga 2006.
10. Aeschylus' *Persians* was produced only eight years after the battle, and the wounds were fresh. The audience of Phrynicus' *Capture of Miletus* had burst into tears, and Phrynicus was fined for reminding the Athenians of those unhappy events (the city was sacked by the Persians in 494 BCE; play's production date is unknown). Aeschylus fictionalized the account, changing many details in order to soften the harsh realities of the Persian War, thus developing the story without opening fresh and deep wounds: "O most hateful name of Salamis!" (*Persians*, 284): Ferguson 1972: 37–8.
11. The allies had refused to fight under Athenian command, and so the Spartan Eurybiades technically had oversight of the fleet at Salamis, despite Themistocles' de facto leadership and role in laying the foundations for the Athenian navy (Herodotus 8.2, 42.2; Plutarch, *Themistocles* 4).
12. Aristotle, *Athenian Constitution* 23.5; cf., Thucydides 1.96.2.
13. Pomeroy et al. 2008: 229.
14. Corvisier 2008: 135; de Souza 2017.
15. The treasury was moved by Pericles, to protect resources from Persia, so he claimed, but his political rivals viewed the transfer as an effort to usurp the capital in order to underwrite an elaborate building program: Plutarch, *Pericles* 12.
16. E.g., *IG* I^3 259.

17 In 444/3, the *Hellenotamiai* paid 38,675 *drachmae* [?] for the Parthenon's construction: *IG* I³ 439.73.
18 Naxos was the first to withdraw but was besieged and forcefully dragged back into the "alliance." Thucydides 1.98.1–4.
19 See also Meiggs 1972; de Souza 2017: 415–18.
20 Isocrates, *Panegyric* 117f–118, 120; 7.80; 12.59; Demosthenes 15.29; 19.273–4; Diodorus Siculus 12.4.4–6, 12.26.2; Plutarch, *Cimon* 13.4–5. Theopompus implied that the "Peace of Callias" was a later myth intended to enhance the stature of Athens (f154). Thucydides omitted the treaty (but see Herodotus 7.151). Plutarch suggested that such a treaty was signed in 466 after an engagement at Eurymedon, or that it had never been signed at all (*Cimon* 13). Cf., Stockton 1959.
21 See further Mavrogiannis 2013-2014.
22 Thucydides 2.67.2; cf., Taylor 2010: 85–9.
23 Thucydides 5.97; cf., 5.109, 6.18.5.
24 Thucydides 6.10.5; Taylor 2010: 141–5.
25 At his trial, Socrates was also accused of being a "star-gazer" (*meteoroskopon*: Plato, *Apology* 18b; *Politics* 299b; cf., Aristophanes, *Clouds* 194–5). Plato compared the Demiourgos with a helmsman who takes charge when the cosmos is on the verge of foundering: *Statesman* 273E; cf., *Republic* 341d; cf., Brock 2013: 53–67. The metaphor is also common in Cicero's political writings: May 1980; Zarecki 2014.
26 Pausanias 5.9.1–2, 8.2.1; *CIG* 2374.
27 The peplos was woven by young, unmarried girls from elite families under priestly supervision (Norman 1983: 41–3); cf., Mansfield 1985: 2–18, 54–8. At Eleia, Hera received a new robe every four years (Pausanias 5.16.2), but in democratic Athens, Athena received a new *peplos* annually (Schol. Aristophanes, *Knights* 566).
28 John Tzetzes, ad Lycophron *Alexandra* loc cit.
29 Pausanias 1.5.3; Lycophron, *Alexandra* 359; cf., Lycophron, *Alexandra* 230; cf. *CWW*: Chapter 9.
30 For the Panathenaia: Xenophon, *Symposium* 4.17.4–4.18.1; Taylor 2012; Wachsmann 2012: 238.
31 Pausanias 1.29.1; Scholiast on Homer, *Iliad* 5.734.
32 Parke 1977: 39.
33 Mansfield 1985: 52–53, 68n62.
34 Norman 1983: 44.
35 Some have argued that the ship was omitted from the frieze because of its size and the challenges of proportionality in relation to human and animal figures: Wachsmann 2012, 238 and note 6.
36 Casson 1995: 99, 103–7.
37 Casson 1995: 97.
38 "Father-loving," perhaps an expression or projection of filial piety.

39 Athenaeus 5.203e–204b; Casson 1995: 98, 108–9.
40 Polybius 16.2.9, 4.8–12; Casson 1995: 125–35.
41 Demetrios had liberated Athens from Cassander and Ptolemy I Soter ("Savior") a year beforehand. In thanksgiving, the Athenians bestowed on Demetrios 1,200 panoplies, evoking the 1,200 Persian ships defeated by Athens and her allies at another, more famous, Battle of Salamis in 480 BCE: Herodotus 7.89.1 and Aeschylus, *Persians* 340–345 both recorded 1,207 ships. According to Diodorus Siculus, the ships "exceeded 1,200" (11.3.7; cf., Holton 2014: 373n10).
42 Wheatley 2012.
43 Casson 1995: 98.
44 Military leaders, kings, and emperors issued coins with their own busts (e.g., Alexander of Macedon, Cleopatra VII, Marc Antony, Augustus) or their womenfolk (Antoninus Pius' wife, Faustina; Severus Alexander's mother, Julia Mammaea). Marc Antony's third wife, Fulvia (he was her third husband), was the first woman featured on a Roman coin issued in the late 40s. See further Horr 1971–1972; Swingler 1976; Lorber 1985.
45 In 286 CE, the Menapian usurper, Carausius, declared himself emperor, issuing coinage on which he claimed to be a colleague of Diocletian and Maximianus. Legends on his coins read "Carausius and his brothers: (*Carausius et fratres sui*) while also emphasizing that the three emperors together were responsible for the current state of peace: *Pax Auggg(ustorum):* "the peace of the three Augusti," Carausius, Diocletian and Maximian (*RIC* 5.2 = Ireland 2008: #219). Note the triple 'gs' in Auggg, representing each of the three Augusti.
46 After 44 BCE, references to the Ides of March on Roman senatorial coinage underscored the Senate's role in vanquishing a tyrant; deified abstractions became particularly common on coinage of the mid-second century CE.
47 Carradice and Price 1988: #3128: gold *stater* from Salamis, ca. 332–323 BCE.
48 BM 2002,0101.1775: Philip V of Macedon (ruled 220–179 BCE).
49 ANS 1944.100.14155; Thonemann 2015: #171.
50 Copper alloy coin: BM 1994,0915.113.
51 The bulls' horns may elicit the Lysimachean portraits of Alexander with the horns of Ammon: Holton 2014: 378–9. Alexander had visited the oracle of Ammon at Siwah and there was declared a "son of Ammon."
52 E.g. Newell 1937: #121: silver tetradrachm from Amphipolis, ca. 289–288 BCE.
53 Newell 1937: #41: silver *drachma* from Tarsus, ca. 298–295 BCE, inscribed BASILEWS DHMHTRIOU ("of King Demetrios"). The Nike on a ship's prow was a widely minted and distributed type: Tarsus, Pella, Amphipolis, Salamis on Cyprus, Caria, Chalcis. Also Lycia (E.g., *Catalogue of the Greek Coins in the British Museum* #21 with Athena Promachos on the reverse, ca. 190–167 BCE) and Cilicia (Nicolet 1971).

54 Cf., Plato, *Symposium* 180e–180d; Aristotle, *Metaphysics* 12.8.1073b; Holton 2014: 380.
55 Holton 2014: 376–80.
56 Tarn 1905: 208; *ANS* 1964.79.6; Thonemann 2015: 159; see Casson 1995: 102; see *CWW*: Chapter 8.
57 Athenaeus 6.253b–f; Holton 2014.
58 Holton 2014: 382–5; For Aphrodite Ourania: Plato, *Symposium* 185b–c, 187d; *Philebus* 12b–c; *Laws* 8.840e.
59 Mikalson 1998: 103.
60 Stewart 2016: 404 and n23.
61 Stewart 2016: 400.
62 Carpenter 1960: 201.
63 Stewart 2016: 402.
64 Cf., Stewart 2016: 404.
65 *IG* 12.1.43; Rhodes, 100–50 BCE; cf., *CWW*: Chapter 9.
66 Stewart 2016: 403.
67 Some scholars have connected the Samothrace Nike with the naval victories of Demetrios or his son, Antigonos II Gonatas (ruled 246–229 BCE), including Gonatas' victory over the Ptolemaic fleet at Andros (246 BCE). The suggestion that it might have celebrated the Rhodian naval victories against the Seleukid Antiochus III at Side and Myonessos in 190 BCE is now rejected. Lehmann and Lehmann 1973: 188, Fig. 5; Hamiaux et al. 2015: 72–89, 164–79; Stewart 2016: 399–400.
68 *IG* 12.8.150. Aiming to preserve the Macedonian connection, Westcoat (2005: 167–71) entertains the possibility that the Nike can be linked with Arsinoë II, whose first husband was Lysimachus, king of Thrace and Macedon, himself honored with an inscribed altar at Samothrace for his successes against Cilician pirates. After Lysimachus' death in battle in 281 BCE, Arsinoë married her half-brother, Ptolemy Keraunos ("the thunderer"), but quickly fled from Cassandreia to Samothrace in a swift vessel (Polyaenus, *Stratagems* 8.57) such as "would have made a worthy dedication to the Great Gods," which was consecrated before she returned to Egypt in 279 BCE where she married her full brother, Ptolemy II Philadelphos.
69 Polybius 32.15.1; Diodorus Siculus 31.35.
70 Polybius 33.13.1.

Chapter 10

1 Running from the Alban Hills, the Almo flows into the Tiber; a small river, the Galaesus debouches into the Gulf of Tarentum at the instep of Italy's boot; Polybius 8.35; Vergil, *Aeneid* 7.531–537; Campbell 2012: 14.

2 Vergil, *Aeneid* 8.711–728; cf., Irby 2019a: 94–9.
3 Campbell 2012: 13.
4 Livy 1.3.10–11; Dionysius of Halicarnassus, *Roman Antiquities* 1.76.2; Plutarch, *Romulus* 3–4; Campbell 2012: 413n36.
5 As invaluable as the *corvus* was to Gaius Duilius during the First Punic War, there were design flaws. During a storm at Camarina in 255 BCE, the Romans lost about 300 ships, perhaps because the craft had been fitted with attached *corvi* that made them top-heavy and unbalanced. On a later design, the *corvus* would be detachable.
6 Crawford 1974: #35; Morello 2012; Tacitus, *Annals* 2.49. Duilius' naval victory was also recorded at Naples on a eulogistic inscription of uncertain date (*L'Année épigraphique*, 1927, p. 116). Janus' temple gates were closed during times of peace: Augustus, *Res Gestae* 13; Suetonius, *Augustus* 22; Dio 51.20, 53.27; cf., Vergil, *Aeneid* 7.607–610.
7 Other rostral monuments would be erected in the coming years. After a decisive naval victory in 338 BCE over the Volscians at Antium (a coastal city west of Rome), C. Maenius affixed rams from six of the confiscated ships to the speaker's platform: Livy 8.14.12; Pliny 34.20; Murray and Petsas 1989: 117–18. The speaker's platform thus acquired the name *Rostra* ("ships' beaks"). Ca. 100 BCE, Marc Antony's grandfather had celebrated a naval triumph over Cilician pirates and used his share of the plunder to adorn the new curved *rostra* with steps that had replaced Maenius' podium: Cicero, *On Oratory* 3.3.1; cf., Murray and Petsas 1989: 118; de Souza 1999: 102–8.
8 BM 1906,1103.2540.
9 With varying degrees of success, many have tried to employ the evidence of coins to reconstruct the design of ancient ships. As interesting and important as the evolution of ship design may be, it is beyond our scope. See Orna-Ornstein 1995.
10 *Tresviri monatales* are the three men, likely senatorial, who had oversight of casting and striking coins. According to Pomponius, they were appointed, probably annually, as in accord with other magistracies at Rome (*Digest of Roman Law* 1.2.30).
11 Crawford 1974: #433. As a *trevir monatalis* in 54 BCE, Lucius Iunius Brutus issued coins that glorified his own ancestor, the regicide who ushered in the Republic in 509 BCE. Many of Brutus' coins bear the legend *Libertas* ("freedom") (e.g., *Roman Republican Coinage* 433/1), a type motivated by concern for Pompey's growing power at Rome.
12 Hollander 1999, 2012.
13 Livy 35.10.10; Crawford 1974: #147/2. Evidence does not suggest that Ahenobarbus was politically or militarily active during the Carthaginian War.
14 Crawford 1974: #143/5.
15 Cf., *HH* #4 *to Hermes* 463–97.
16 E.g., BM R.8726.

17 Livy 8.19–21. Livy cited the victor of Privernum as C. Plautius without the cognomen. It is thus doubted if the victor of Privernum is the ancestor of Pompey's ally: see Crawford 1974: #444–5; Hölkeskamp 2004: 212–15; Oakley 2007: 485. It is also unclear if the Hypsaeus, whose senatorial troops were vanquished by Spartacus, is identical to Milo's political rival (Florus 2.7.8). Our Hypsaeus served as praetor ca. 55 BCE and, with the support of Clodius Pulcher, he stood for the consulship against Q. Metellus Scipio and Cicero's friend, T. Annius Milo (Livy, *Periochae* 107; Asconius 48c; Plutarch, *Younger Cato* 47.1), but was later tried for violence and corruption (Plutarch, *Pompey* 56).
18 Crawford 1974: #446.
19 Welch 2002: 17–20; Powell 2002: 121–7.
20 Crawford 1974: #483/2.
21 E.g., Crawford 1974: #511/2, *denarius*.
22 Anonymous, *Aetna* 625–34; Crawford 1974: #511/3, *denarius*; cf., Powell 2002: 124–5, plates 11–14; Zarrow 2003.
23 Suetonius, *Augustus* 16.1; Appian, *Civil Wars* 5.116–122.
24 *Denarius*: Zanker 1990: 41 and Figure 31a.
25 Ovid, *Tristia* 4.1.1–4; cf., Gowing 2002.
26 In 36 BCE, Lepidus had ceased to be a triumvir, and Antony—separated from Octavian's sister—"married" Cleopatra VII: Roller 2010: 79–84.
27 RPC #1463–5.
28 Zanker 1990: 61 and Figure 47.
29 See Southern 2010: #47 (silver *denarius*). At Rome, augurs observed natural signs that were thought to convey divine approval or disapproval. As there was no separation of church and state in Rome, consequently Roman state priesthoods were held by politicians.
The "triumvirate for organizing the republic," the alliance between Octavian, Antony, and Lepidus, was sanctioned by law (*Lex Titia*), allowing for two five-year terms. Under the law, the triumvirs had full authority, outranking the jurisdiction of all elected magistrates, including the consuls.
30 Strabo 7.7.6; cf., Suetonius, *Augustus* 18.2; Dio 51.1.2.
31 Thucydides 1.29; Strabo 7.7.6; Suetonius, *Augustus* 18.2; Gurval 1995: 65–7, 74–81. Murray and Petsas 1989: 10; Hubbard 1975: 117.
32 See Statius, *Silvae* 2.2.6–8.
33 Vergil, *Aeneid* 5.114–267. Vergil's boat race alludes to Octavian's naval campaigns against Sextus Pompey: Anderson and Dix 2013: 3–21.
34 Strabo 7.7.6; Philippus, *Palatine Anthology* 6.236; Plutarch, *Antony* 65.3; Suetonius, *Augustus* 96.2; Dio 51.1.3.
35 Murray and Petsas 1989.
36 Augustus, *Res Gestae* 34.2; Suetonius, *Augustus* 7.2.

37 BM 1901,0503.161.
38 BM R.6163.
39 E.g., Horace, *Odes* 1.37; Propertius 4.6; Manilius 1.914.
40 Putnam 1998: 136; See also Irby 2019a.
41 Austin 1971: 68; González Vázquez 1987; Cowan 2015: 105.
42 In 1808, William Worsley published *Trafalgaris Pugna*, in both Latin and English, eulogizing Horatio Nelson and mythologizing the naval battle off Cape Trafalgar.
43 BM 1921,1217.38.
44 E.g., *RIC* volume 1: Augustus #169, 172 (p. 77).
45 See also Zanker 1990: 223 and Figure 178 for a cuirassed statue of Caligula that features sea monsters evoking Naulochus and Actium.
46 Aeschylus, *Persians* 432–434; Herodotus 7.33–36. For a theoretical reconstruction of the engineering required to construct the bridge: Hammond and Roseman 1996.
47 Balsdon 1934: 50–4.
48 Speidel 1994: 21–2.
49 Kleijwegt 1994: 671.
50 Woods 2002. Caligula's Baiae bridge included running water, lodgings, and rest areas, perhaps superfluous for Xerxes' seven-stade bridge or Caligula's short infrastructure, but essential for troops who might be called upon to cross thirty miles over open water.
51 Such honorifics were a traditional method of eulogizing conquering heroes. Claudius' charismatic older brother had been called "Germanicus" in honor of the German conquests of their father, Drusus; Scipio had assumed the honorific agnomen "Africanus" after the defeat of Hannibal at Zama in 202 BCE.
52 e.g., *RIC* volume 1: Claudius #8 (p. 125). The issue recalled the *de Germanis* coins that broadcast his father Drusus' successful campaigns in the Rhineland
53 *Britannia* 22 (1991) 12.
54 Erim 1982.
55 Braund 1996: 103; Saladino 1980.
56 Suetonius, *Caesar*, 39.4; Dio 43.23.4; Appian, *Civil Wars* 2.102; Plutarch, *Caesar* 55. The senate feared the risk of disease from stagnant waters in the basin and voted to fill it in (Dio 45.17.8). Caesar may also have intended the site for a temple to Mars (Suetonius, *Caesar* 44.1).
57 Velleius Paterculus 2.100.2; Pliny, 16.190, 200; Frontinus, *Aqueducts* 1.11; Tacitus, *Annals* 12.56; Suetonius, *Augustus* 43.1; Dio 55.10.7–8.
58 Pliny 19.24; Tacitus, *Annals* 13.31; Suetonius, *Nero* 13.1; Dio 61.9.5.
59 Suetonius, *Titus* 7.3; Martial, *On the Spectacles* 28; Dio 66.25.3–4.
60 Braund 1996: 12–15.
61 Mattingly 2006: 67.
62 Dio 49.38.2, 53.22.5, 53.25.2.

63 Braund 1996: 106.
64 Tacitus, *Annals* 14.29–39; Dio 62.1–12.
65 Tacitus, *Agricola* 13.3; Suetonius, *Vespasian* 4.1.
66 Braund 1996: 10–24.
67 CM 1998.4–1.1. Williams' (1999) emphasis of Ocean as symbolic of "peaceful government" and "general contentment" may be overstated.
68 Williams 1999: 311; see also "Oceanus" in *Lexicon Iconographicum Mythologiae Classicae* (*LIMC*) 8.1.907–914.
69 *RIB* 1320; Braund 1996: 12–13; Irby-Massie 1999: 21; *RIB* 1319–20. See *CIL* 4.8634 (Pompeii): *Oceano*; *CIL* 8.23184 (Thelepte): *Oceano* ("to Ocean"); *CIL* 13.8811 (Vechten, Germania Inferior): *Neptuno, Oceano, Rheno* ("to Neptune, Ocean, the Rhine" for the health of Elagabalus); *L'Année épigraphique* 2008 #641 (Lusitania): *Soli, Lunae, Oceano* ("to the Sun, Moon, and Ocean").
70 *RIC* volume 5.2: Carausius #8, 86, 213–14, 552–3, 709, 754, 764–5.
71 *RIC* volume 5.2: Allectus #59 (Neptune on a galley holding an anchor). Also #55, 128–31 (*Virtus Augusti* [manliness of Augustus], on a galley), #56 (a bird alights on the mast), #58 (Victory on a galley), #75–81 (*Latitia Augusti* [Joy of Augustus], with anchors or rudders), #124–7 (*Laetitia*, Augusti on a galley).
72 E.g., Hesiod, *Theogony* 678; Catullus 31.3; Ovid, *Fishing* 84, *Fasti* 4.419.

Chapter 11

1 Hong et al. 1994.
2 E.g., Manning 2013; Harper 2017.
3 Schmiedt 1972; Slim et al. 2004.
4 Franconi 2017.
5 Wilson 2013: 275.

Appendix

1 For further information about individual authors, the interested reader is invited to consult *Encyclopedia of Ancient Natural Scientists: The Greek Tradition and Its Many Heirs* edited by P.T. Keyser and G.L. Irby-Massie, Routledge (2008); *Oxford Classical Dictionary*; *Internet Encyclopedia of Philosophy*: https://iep.utm.edu/; or *Stanford Encyclopedia of Philosophy*: https://plato.stanford.edu/

Bibliography

Africa, TW (1995) "Adam Smith, the Wicked Knight, and the use of Anecdotes," *Greece and Rome* 42 (1), pp 70–5

Ahuja, S (ed.) (2017) *Chemistry and Water: The Science Behind Sustaining the World's Most Crucial Resource*, Amsterdam

Alcock, JP (2001) *Food in Roman Britain*, Stroud

Allen, TW (1921) *The Homeric Catalogue of Ships*, Oxford

Almeida, ER (1984) *Il Monte Testaccio*, Rome

Anderson, CA and TK Dix (2013) "Vergil at the Races: The Contest of Ships in Book 5 of the 'Aeneid,'" *Vergilius* 59, pp 3–21

Anderson, JK (1995) "The Geometric catalogue of ships," in Jane P Carter and, Sarah P Morris (eds.) *The Ages of Homer: A Tribute to Emily Townsend Vermeule*, Austin, pp 181–91

André, J. (1961) *L'alimentation et la cuisine à Rome*, Paris

Andrews, AC (1948) "Oysters as a Food in Greece and Rome," *CJ* 43, pp 299–303

Angel, JL (1966) "Porotic Hyperostosis, Anemias, Malarias, and Marshes in the Prehistoric Eastern Mediterranean," *Science* 153 (3,737), pp 760–3

Angelakis, AN and D Koutsoyiannis (2003) "Urban Water Engineering and Management in Ancient Greece," in A Stewart and T Howell (eds.) *Encyclopedia of Water Science*, London, pp 999–1,007

Angelakis, AN, YM Savvakis, and G Charalampakis (2007) "Aqueducts during the Minoan Era," *Water Supply* 7, pp 95–101

Angelakis, AN, YM Savvakis, and G Charalampakis (2006) "Minoan Aqueducts: A Pioneering Technology," IWA 1st International Symposium on Water and Wastewater Technologies in Ancient Civilizations, Iraklio, Greece, October 28–30, 2006

Anonymous (1876) "Tyrian purple," *The Ladies' repository* 4, pp 424–7

Antonaccio, C (2017) "Greek colonization, connectivity, and the Middle Sea," in de Souza and Arnaud, pp 214–23

Antoniou, GP (2007) "Lavatories in Ancient Greece," *Water Supply* 7, pp 155–64

Antoniou, GP (2010) "Ancient Greek Lavatories: Operation with Reused Water," in LW Mays (ed.) *Ancient Water Technologies*, Dordrecht, pp 67–86

Antoniou, GP and AN Angelakis (2015) "Latrines and Wastewater Sanitation Technologies in Ancient Greece," in PD Mitchell (ed.) *Sanitation, Latrines and Intestinal Parasites in Past Populations*, London and New York, pp 41–68

Arafat, Karim W and Morgan, Catherine A (1994) Athens, Etruria and the Heuneburg: mutual misconceptions in the study of Greek-barbarian relations, in I Morris (ed.) *Classical Greece: Ancient Histories and Modern Archaeologies*, Cambridge, pp 108–34

Arnaoutoglou, I (1998) *Ancient Greek Laws: A Sourcebook*, London

Arnaud, P (2005) *Les routes de la navigation antique: Itinéraires en Méditerranée*, Paris

Arnaud, P (2011) "Ancient Sailing-Routes and Trade Patterns: the Impact of Human Factors," in Robinson and Wilson, pp 61–80

Arnaud, P (2016) "Entre mer et rivière: les ports fluvio-maritimes de Méditerranée ancienne. Modèles et solutions," in C Sanchez and M-P Jézégou, (eds) *Les ports dans l'espace méditerranéen antique. Narbonne et lessystèmes portuaires fluvio-lagunaires. Actes du colloque international tenu à Montpellier du 22 au 24 Mai 2014, Montpellier-Lattes*, Revue Archéologique de Narbonnaise, Supplément 44

Aubet, ME (1993) *The Phoenicians and the West*, Cambridge

Austin, RG (ed.) (1971) *P. Vergili Maronis: Aeneidos Liber Primus*, Oxford

Axe, D (2009) "10 Things You Didn't Know About Somali Pirates," *The Wall Street Journal* April 27, https://www.wsj.com/articles/10-things-you-didn-t-know-about-somali-pirates-11601915298

Bagnall, RS, K Brodersen, CB Champion, A Erskine, and SR Huebner (eds.) (2012) *The Encyclopedia of Ancient History*, Malden, MA

Baker, Jill L (2019) *Technology of the Ancient Near East: From the Neolithic to the Early Roman Period*, London

Balcer, JM (1989) "The Persian Wars Against Greece: A Reassessment," *Historia* 38 (2), pp 127–43

Balsdon, JPVD (1934) *The Emperor Gaius*, Oxford

Bannon, C (2009) *Gardens and Neighbors: Private Water Rights in Roman Italy*, Ann Arbor

Barringer, JM (2016) "The Shefton dolphin-rider," in J Boardman, A Parkin, and S Waite (eds.) *On the Fascination of Objects: Greek and Etruscan Art in the Shefton Collection*, Oxford, pp 117–29

Basch, L (1989) "Les graffiti de Délos," in H Tzalas (ed.) *Tropis I: 1st International Symposium on Ship Construction in Antiquity, Piraeus, 30 August – 1 September 1985: Proceedings*, Athens, pp 17–24

Bass, GF (2010) "Cape Gelidonya Shipwreck," in E Cline (ed.) *The Oxford Handbook of the Bronze Age Aegean (ca. 3000–1000 BC)*, Oxford, pp 797–803

Baxter, K and KB Figueira (2012) "Aigina," in Bagnall et al. 2012

Beaulieu, M-C (2016) *The Sea in the Greek Imagination*, Philadelphia

Beaumont, P (2008) "Water Supply at Housesteads Roman Fort, Hadrian's Wall: the case for Rainfall Harvesting," *Britannia* 39, pp 59–84

Beck, L (2017) *Dioscorides: De Materia Medica*, Third edition, Hildesheim

Behrens-Abouseif, D (2006) "The Islamic History of the Lighthouse of Alexandria," *Muqarnas* 23, pp 1–14

Bekker-Nielsen, T (2005) "The Technology and Productivity of Ancient Sea Fishing," in T Bekker–Nielsen (ed.) *Ancient Fishing and Fish Processing in the Black Sea Region*, Aarhus, pp 83–96

Beltrán Lloris, F (2006) "An Irrigation Decree from Roman Spain: The *Lex Rivi Hiberiensis*," *JRS* 96, pp 147–97

Beresford, J (2013) *The Ancient Sailing Season*, Leiden

Bintliff, J (1975) "Mediterranean alluviation: New evidence from archaeology," *Proceedings of the Prehistoric Society* 41, pp 78–84

Bissa, MAE (2009) *Governmental Intervention in Foreign Trade in Archaic and Classical Greece*, Boston

Blackman, DJ (2008) "Sea Transport, Part 2: Harbors," in Oleson 2008a, pp 638–70

Blair, I and J Hall (2003) *Working Water: Roman Technology in Action*, London

Blegen, CW and M Rawson (1966) *The Palace of Nestor at Pylos in Western Messenia, Vol. 1: The Buildings and Their Contents*, Princeton

Boardman, J (1999) *The Greeks Overseas: Their Early Colonies and Trade*, 4th ed., London

Bol, R (1984) *Das Statuenprogramm des Herodes-Atticus-Nymphäuns*, Mainz

Bonifay, M (2004) *Études sur la céramique romaine tardive d'Afrique*, Oxford

Borca, F (2000) "Towns and Marshes in the Ancient World," in Hope and Marshall, pp 56–62

Bottéro, J (1995) "The Most Ancient Recipes of All," in J Wilkins, D Harvey, and M Dobson (eds.) *Foods in Antiquity*, Exeter, pp 248–55

Bouffier, SC (1999) "La Pisciculture dans le monde grec: État de la question," *MEFRA* 111, pp 37–50

Bowra, CM (1963) "Arion and the Dolphin," *Museum Helveticum* 20, pp 121–34

Bradford, AS (2019) *Cruel and Ancient Sea: A History of Naval Warfare in the Ancient World*, London

Brandon, C, R Hohlfelder, and JP Oleson (2007) "Constructing the Harbour of Caesarea Palaestina, Israel: New Evidence from the ROMACONS Field Campaign of October 2005," *International Journal of Nautical Archaeology* 36 (2), pp 409–15

Brandt, H (2012) "Phaselis," in Bagnall et al. 2012

Branigan, K (1981) "Minoan Colonialism," *Annual of the British School at Athens* 76, pp 23–33

Braund, D (1996) *Ruling Roman Britain: Kings, Queens, Governors and Emperors from Julius Caesar to Agricola*, London

Broadhurst, RJC (1952) *The Travels of Ibn Jubayr: Being the chronicle of a mediaeval Spanish Moor concerning his journey to the Egypt of Saladin, the holy cities of Arabia, Baghdad the city of the Caliphs, the Latin kingdom of Jerusalem, and the Norman kingdom of Sicily*, London

Brock, R (2013) *Greek Political Imagery from Homer to Aristotle*, London

Brulé, P (1978) *La piraterie crétoise hellénistique*, Paris

Bruun, C (1991) *The Water Supply of Ancient Rome: A Study of Roman Imperial Administration*, Helsinki

Bruun, C (2000a) "The Greek World," in Wikander, pp 558–73

Bruun, C (2000b) "The Roman World: Roman Water Legislation," in Wikander, pp 575–604

Bruun, C (2012) "Roman emperors and legislation on public water use in the Roman Empire: clarifications and problems," *Water History* 4, pp 11–33

Bruun, C (2015) "Water Use and Productivity in Roman Agriculture: Selling, Sharing, Servitudes," in P Erdkamp, K Verboven, and A Zuiderhoek (eds.) *Ownership and Exploitation of Land and Natural Resources in the Roman World*, Oxford, pp 132–4

Bubelis, WS (2012) "Coinage, Greek," in Bagnall et al. 2012

Bundrick, S D (2019) *Athens, Etruria, and the Many Lives of Greek Figured Pottery*, Madison, WI

Burnett, A, M Amandry, C Howgego, and J Mairat (1992–) *Roman Provincial Coinage Online*, London: https://rpc.ashmus.ox.ac.uk/

Burnham, BC (1997) "Roman Mining at Dolaucothi: The Implications of the 1991-3 Excavations near the Carreg Pumsaint," *Britannia* 28, pp 325–36

Burnham, BC and H Burnham (2004) *Dolaucothi-Pumsaint: Survey and Excavations at a Roman Gold-Mining Complex, 1987–1999*, Oxford

Butler, S (2014) "Making Scents of Poetry," in M Bradley (ed.) *Smell and the Ancient Senses*, London, pp 74–89

Cagnat, R (ed.) (1901–1927) *Inscriptiones graecae ad res romanas pertinentes*, Paris

Calciati, R (1983–1987) *Corpus Nummorum Siculorum. The Bronze Coinage*, Milan

Campbell, B (2012) *Rivers and the Power of Ancient Rome. Studies in the History of Greece and Rome*, Chapel Hill

Carlson, DN (2011) "The Seafarers and Shipwrecks of Ancient Greece and Rome," in A Catsambis, B Ford, and DL Hamilton (eds.) *The Oxford Handbook of Maritime Archaeology*, Oxford, pp 379–405

Carpenter, R (1960) *Greek Sculpture: A Critical Review*, Chicago

Carradice, I and M Price (1988) *Coinage in the Greek World*, London

Casasola, DB (2010) "Fishing Tackle in Hispania," in T Bekker Nielsen and DB Casasola (eds.) *Ancient Nets and Fishing Gear: Proceedings of the International Workshop on 'Nets and Fishing Gear in Classical Antiquity: A First Approach.' Cadiz, November 15–17, 2007*, Cadiz and Arhaus, pp 83–138

Caskey, JL (1960) "The Early Helladic Period in the Argolid," *Hesperia* 29 (3), pp 285–303

Cassola, DB and AA Gonzalez (2008) "Baelo Claudia y sus industrias helieuticas: sintesis des las ultimas actuaciones arqueologicas (2000–2004)," in J Napoli (ed.) *Ressources Et Activités Maritimes Des Peuples De L'antiquite: Actes Du Colloque International De Boulogne-Sur-Mer, 12, 13, Et 14 Mai 2005*, Boulogne-sur-Mer, pp 9–30

Casson, L (1980) "Rome's Trade with the East: The Sea Voyage to Africa and India," *Transactions of the American Philological Association* 110, pp 21–36

Casson, L (1984) "The Sea Route to India: *Periplus Maris Erythaei* 57," *CQ* 34 (2), pp 473–9

Casson, L (1989) *The Periplus Maris Erythaei: Text with Introduction, Translation, and Commentary*, Princeton

Casson, L (1991) *The Ancient Mariners: Seafarers and Sea Fighters of the Mediterranean in Ancient Times*, 2nd ed., Princeton
Casson, L (1994) *Travel in the Ancient World*, Baltimore
Casson, L (1995) *Ships and Seamanship in the Ancient World*, Baltimore
Cerasuolo, O (2017) "Greek Geometric incised coarse ware, Euboea, and its connections to central Italy," in Ž Tankosić, F Mavridis, and M Kosma (eds.) *An Island Between Two Worlds: The Archaeology of Euboea from Prehistoric to Byzantine Times: Proceedings of International Conference, Eretria, 12–14 July 2013*, Athens, pp 235–52
Chanson, H (2000) "Hydraulics of Roman Aqueducts: Steep Chutes, Cascades, and Dropshafts," *AJA* 104, pp 47–72
Chau Ju-Kua. (1911) *His work on the Chinese and Arab trade in the twelfth and thirteenth centuries, entitled Chu-fan-chï*. Translated from the Chinese and Annotated by Friedrich Hirth and WW Rockhill, St Petersburg
Chiai, GF (2017) "Rivers and Waters Protection in the Ancient World: How Religion can Protect the Environment," in Cordovana and Chiai 2017, pp 61–82
Chiotis, E and LE Chioti (2014) "Drainage and Sewerage Systems at Ancient Athens, Greece," in AN Angelakis and JB Rose (eds.) *Evolution of Sanitation and Wastewater Technologies through the Centuries*, London, pp 315–32
Christaki, M and PT Nastos (2018) "Springs, Fountains of Athens," *Hellenic Geographical Society* 1075
Cobb, MA (ed.) (2019) *The Indian Ocean Trade in Antiquity: Political, Cultural and Economic Impacts*, London and New York
Collitz, H and F Bechtel (1884–1909) *Sammlung der griechischen Dialekt-inschriften*, Göttingen
Connolly, P and H Dodge (1998) *The Ancient City*, Oxford
Conwell, D (2008) *Connecting a City to the Sea: The History of the Athenian Long Walls*, Leiden
Cordovana, OD and GF Chiai (eds.) (2017) *Pollution and the Environment in Ancient Life and Thought*, Stuttgart
Corvisier, JN (2008) *Les Grecs et La Mer*, Paris
Coutant, V and VL Eichenlaub (1975) *Theophrastus: De Ventis*, Notre Dame
Cowan, R (2015) "On the Weak King according to Vergil: Aeolus, Latinus and Political Allegoresis in the Aeneid," *Vergilius* 61, pp 97–124
Crawford, HEW, R Killick, and J Moon (eds.) (1997) *The Dilmun Temple at Saar: Bahrain and Its Archaeological Inheritance*, London
Crawford, M (1974) *Roman Republican Coinage*, Cambridge
Crouch, DP (1993) *Water Management in Ancient Greek Cities*, Oxford
Crummy, P (1997) *City of Victory: The story of Colchester: Britain's first Roman town*, Colchester Archaeological Trust
Culham, P (2017) "The Roman Empire and the Seas," in de Souza and Arnaud, pp 283–93
Cunliffe, B and P Davenport (1985) *The Temple of Sulis Minerva at Bath*, Oxford

Curtis, RI (1979) "The Garum Shop of Pompeii (I.12.2)," *Cronache Pompeiane* 5, pp 5–23
Curtis, RI (1984a) "A Personalized Floor Mosaic from Pompeii," *AJA* 88, pp 557–66
Curtis, RI (1984b) "The Salted Fish Industry of Pompeii," *Archaeology* 37, pp 58–75
Curtis, RI (1991) *Garum and Salsamenta: Production and Commerce in Materia Médica*, Leiden
Curtis, RI (2008) "Food Processing and Preparation," in JP Oleson (ed.) *Oxford Handbook of Engineering and Technology in the Classical World*, Oxford, pp 369–92
Custance, R (1970) *War at Sea: Modern Theory and Ancient Practice*, London
Dalby, A (2003) *Food in the Ancient World from A to Z*, London
Davidson, JN (1997) *Courtesans and Fishcakes: The Consuming Passions of Classical Athens*, Chicago
Davies, RW (1970) "The Roman Military Medical Service," *Saalburg-Jahrbuch* 27, pp 84–104
Davis, DL (2009) *Commercial Navigation in the Greek and Roman World*, Ph.D. dissertation, Austin
Davis, EN (1983) "The Iconography of the Ship Fresco from Thera," in WG Moon (ed.) *Ancient Greek Art and Iconography*, Madison, WI, pp 3–14
de Kleijn, G (2001) *The Water Supply of Ancient Rome: City Area, Water, and Population*, Amsterdam
de Souza, P (1999) *Piracy in the Graeco-Roman World*, Cambridge
de Souza, P (2017) "The Athenian Maritime Empire of the Fifth Century BC," in de Souza and Arnaud, pp 412–25
de Souza, P and P Arnaud (eds.) (2017) *The Sea in History: The Ancient World*, Woodbridge
de Souza, P and P de Souza (2012) "Piracy," in Bagnall et al. 2012
DeJong, DH (2011) *Forced to Abandon our Fields: The 1914 Clay Southworth Gila River Pima Interviews*, Salt Lake City
Delgado, JP (1998) *Encyclopedia of Underwater and Maritime Archaeology*, New Haven, CT
Delile, H, D Keenan-Jones, J Blichert-Toft, J-P Goiran, F Arnaud-Godet, and F Albarède (2017) "Rome's urban history inferred from Pb-contaminated waters trapped in its ancient harbor basins," *Proceedings of the National Academy of Sciences of the United States of America* www.pnas.org/cgi/doi/10.1073/pnas.1706334114
Deonna, W (1954) "The Crab and the Butterfly: A Study in Animal Symbolism," *Journal of the Warburg and Courtauld Institutes* 17, pp 47–86
DeSantis, MG (2017) *A Naval History of the Peloponnesian War: Ships, Men and Money in the War at Sea, 431–404 BC*, South Yorkshire
di Natale, A (2014) "The ancient distribution of Bluefin Tuna Fishery: How Coins can Improve our Knowledge," *Collective Volume of Scientific Papers: International Commission for the Conservation of Atlantic Tunas* 70, pp 2,828–44
Dimopoulou-Rethemiotaki, N (2004) *The Ring of Minos and Gold Minoan Rings: The Epiphany Cycle*, Athens
Döring, M (2002) "Wasser für den 'Sinus Baianus.' Römische Ingenieur- und Wasserbauten der Phlegraeischen Felder," *Antike Welt* 33, pp 305–19

Doukellis, PN (2008) "Between Greek Colony and Mother- City: Some Reflections," in M Rozen (ed.) *Homelands and Diasporas: Greeks, Jews and their Migrations*, London, pp 110–12

Downey, WS (2011) "Orientations of Minoan Buildings on Crete may indicate the First Recorded Use of the Magnetic Compass," *Mediterranean Archaeology and Archaeometry* 11, pp 9–20

Draycott, CM (2012) "Phocaea," in Bagnall et al. 2012

Drews, R (1998) "Canaanites and Philistines," *Journal for the Study of the Old Testament* 81, pp 39–61

Dunbabin, KMD (1999) *Mosaics of the Greek and Roman World*, Cambridge

Duncan-Jones, R (1974) *The Economy of the Roman Empire*, Cambridge

Dunsch, B (2012) "*Arte rates reguntur*: Nautical Handbooks in Antiquity?" *Studies in History and Philosophy of Science* 43 (2), pp 270–83

Dunsch, B (2015) "Why do we Violate Strange Seas and Sacred Waters? The Sea as Bridge and Boundary in Greek and Roman Poetry," in M Grzechnik and H Hurskainen (eds.) *Beyond the Sea: Reviewing the Manifold Dimensions of Water as Barrier and Bridge*, Cologne, pp 14–42

Dürrbach, F (ed.) (1923–37) *Inscriptions de Délos*, Paris

Eck, W (1987) "Die Wasserversorgung im Römiscen Reich: Sozio-politische Bedingungen, Recht und Administration," in *Die Wasserversorgung antiker Städte. Pergamon. Recht/Verwaltung. Brunnen/Nymphäen. Bauelemente* (Geschichte der Wasserversorgung, II), Mainz, pp 49–101

Empereur, J (1999) "Diving on a Sunken City," *Archaeology* 52 (2), pp 36–43

Erdkamp, P (2005) *The Grain Market in the Roman Empire*, Cambridge

Erim, KT (1982) "A New Relief Showing Claudius and Britannia from Aphrodisias," *Britannia* 13, pp 277–81

Erim, KT, J Reynolds, JP Wild, and MH Balance (1970) "The Copy of Diocletian's Edict on Maximum Prices from Aphrodisias in Caria," *JRS* 60, pp 120–41

Eschebach, H (1982) "Katalog der pompejanischen Laufbrunnen und irhe Reliefs," *Antike Welt* 13, pp 21–6

Étienne, R and F Mayet (1994) "A propos de l'amphore Dressel 1C de Belo," *Mélanges de la Casa de Veláquez* 30, pp 131–8

Evans, AJ (1935) *The Palace of Minos. Volume IV Part II: Camp-stool Fresco, long-robed priests and beneficent genii; Chryselephantine Boy-God and ritual hair-offering; Intaglio Types, M.M. III – L. M. II, late hoards of sealings, deposits of inscribed tablets and the palace stores; Linear Script B and its mainland extension, Closing Palatial Phase; Room of Throne and final catastrophe*, London.

Evans, JAS (1982) "The Oracle of the 'Wooden Wall,'" *CJ* 78, pp 24–9

Fabre, D (2004–2005) *Seafaring in Ancient Egypt*, London

Fahlbusch, H (1982) *Vergleich antiker griechischer und römischer Wasserversorgungsanlagen*, Leichtweiss-Institut für Wasserbau der Technischen Universität Braunschweig

Falcão, L and MEM Araújo (2018) "Vegetable Tannins Used in the Manufacture of Historic Leathers," *Molecules* 23 (5), pp 1,081

Faust, SD and O Aly (1998) *Chemistry of Water Treatment*, 2nd ed., Ann Arbor

Fayer, VV (2013) "The Art of Memory and Composition of 'The catalogue of the ships,'" in NN Kazansky (ed.) *Indo-European Linguistics and Classical Philology. 17: Proceedings of the 17th conference in memory of Professor Joseph M. Tronsky: June 24–26, 2013*, Sankt-Peterburg, pp 895–906

Ferguson, J (1972) *A Companion to Greek Tragedy*, Austin

Field, JA (1945) "The Purpose of the *Lex Iulia et Papia Poppaea*," *CJ* 40, pp 398–416

Finley, MI (1973) *Studies in Land and Credit in Ancient Athens, 500–200 B.C.; Horos-inscriptions*, New York

Finney, BR (ed.) (1976) *Pacific Navigation and Voyaging*, Wellington

Florido, ER Auriemma, S Faivre, IR Rossi, F Antonioli, S Furlani, and G Spada. (2011) "Istrian and Dalmatian fishtanks as sea-level markers," *Quaternary International* 232, pp 105–13

Foy, D (2003) "Quid de l'Occident?" in D Foy (ed.) *Cœur de verre. Production et diffusion du verre antique*, Gollion, pp 34–5

Franconi, T (2016) "Climatic Influences on Riverine Transport on the Roman Rhine," in C Schäfer (ed.) *Connecting the Ancient World: Mediterranean Shipping, Maritime Networks and their Impact*, Princeton, pp 27–44.

Franconi, T (2018) "The Import and Distribution of Eastern Amphorae within the Rhine Provinces," *Journal of Roman Pottery Studies* 17, pp 1–10

Franconi, T (ed.) (2017) *Fluvial Landscapes in the Roman World*, JRA Supplement 104

Frantantuono, L (2012) *Madness Triumphant: A Reading of Lucan's Pharsalia*, Lanham

Gabba, E (1983) "La rifondazione di Salapia," *Athenaeum* 61, pp 514–16

Gabrielsen, V (2012) "Rhodes," in Bagnall et al. 2012

Gallardo, AR (1975) *Supervivencia de una obra hidráulica: el acueducto de Segovia*, Segovia

Ganias, Kostas, Ch Mezarli, and E Voulstiadou (2017) "Aristotle as an Ichthyologist: Exploring Aegean Fish Diversity 2,400 Years Ago," *Fish and Fisheries, Oxford, England* 18, pp 1,038–55

Garfinkel, Y, A Vered, and O Bar-Yosef (2006) "The Domestication of Water: The Neolithic Well at Sha'ar Hagolan, Jordan Valley, Israel," *Antiquity* 309, pp 686–96

Geissler, K (1998) *Die Öffentliche Wasserversorgung Im Römischen Recht*, Berlin

George, R (2013) *Ninety Percent of Everything: Inside Shipping, the Invisible Industry That Puts Clothes on Your Back, Gas in Your Car, and Food on Your Plate*, New York

Gianfrotta, PA (1999) "Arceologia subacquea e testimonianze di pesca," *Mélanges d'Archéologie et d'Histoire de l'école Française de Rome, Antiquité* 111, pp 9–36

Gibbins, D, and J Adams (2001) "Shipwrecks and Maritime Archaeology," *World Archaeology* 32, pp 279–91

Gildersleeve, BL (1885) *Pindar: The Olympian and Pythian Odes*, New York

Glaser, F (2000) "Fountains and Nymphaea," in Wikander, pp 413–51

Goiran, J-P, F Salomon, C Vittori, H Delile, J Christiansen, C Oberlin, G Boetto, P Arnaud, I Mazzini, L Sadori, G Poccardi, and A Pellegrino (2017) "High chrono-stratigraphical resolution of the harbor sequence of Ostia: palaeo-depth of the basin, ship draught, and dredging," *JRA* Supplement 104, pp 67–83

Goldschmidt, N (2017) "Textual Monuments: Reconstructing Carthage in Augustan Literary Culture," *CPh* 112 (3), pp 368–83

González Vázquez, J (1987) "La imagen-comparación de Neptuno (Aen. 1, 148–156)," *Latomus* 46, pp 363–68

González, A A, D B Cassola, and A T Silva (eds.) (2004) *Garum y Salazones en el Circulo del Estrecho*, Armilla, Granada

Gowing, A M (2002) "Pirates, Witches, and Slaves: The Imperial Afterlife of Sextus Pompeius," in Powell and Welch, pp 187–211

Grace, V R (1979) *Amphoras and the Ancient Wine Trade*, American School of Classical Studies at Athens

Grewe, K (1985) *Planung und Trassierung romischer Wasserleitungen*, Wiesbaden

Grimal, P (établi, traduit, et commenté) (1961) *Frontin, Les Aqueducs de la Ville de Rome*, Paris

Günther, RT (1897) "The Oyster Culture of the Ancient Romans," *Journal of the Marine Biological Association of the United Kingdom* 4, pp 360–5

Günther, S (2016) *Taxation in the Greco-Roman World: The Roman Principate*. Oxford Handbooks online: https://www.oxfordhandbooks.com/view/10.1093/oxfordhb/9780199935390.001.0001/oxfordhb-9780199935390-e-38?rskey=RVJJ5o&result=1

Gurval, R A (1995) *Actium and Augustus: The Politics and Emotions of Civil War*, Ann Arbor, MI

Haarhoff, J (2009) "The Distillation of Seawater on Ships in the 17th and 18th Centuries," *Heat Transfer Engineering* 30, pp 237–50

Hahn, R (2001) *Anaximander and the Architects: The Contributions of Egyptian and Greek Architectural Technologies to the Origins of Greek Philosophy*, Albany

Hakola, R (2017) "The Production and Trade of Fish as Source of Economic Growth in the First Century CE Galilee," *Novum Testamentum* 59 (2), pp 111–30

Hale, John R (2009) *Lords of the Sea: The Epic Story of the Athenian Navy and the Birth of Democracy*, Viking

Hales, S (1739) *Philosophical Experiments: Showing How Sea-Water May Be Made Fresh and Wholesome: And How Fresh-Water May Be Preserved Sweet*, London

Haley, E W (2003) *Baetica Felix*, Austin

Hall, J (2004) "How 'Greek' Were the Early Western Greeks?" in K Lomas (ed.) (2004), *Greek Identity in the Western Mediterranean: Papers in Honour of Brian Shefton*, Leiden-Boston, pp 35–54.

Hamiaux, ML Laugier, and J-L Martinez (2015) *The Winged Victory of Samothrace: Rediscovering a Masterpiece*, Paris

Hamilton, W and JHW Tischbein (1791) *Collection of Engravings from Ancient Vases Mostly of Pure Greek Workmanship Discovered in Sepulchres in the Kingdom of the Two Sicilies but Chiefly in the Neighbourhood of Naples During the Course of the Years MDCCLXXXIX and MDCCLXXXX Now in the Possession of Sir Wm. Hamilton, with Remarks on Each Vase by the Collector*, Naples

Hammond, NGL and LJ Roseman (1996) "The Construction of Xerxes' Bridge over the Hellespont," *JHS* 116, pp 88–107

Hannah, R (1993) "The Stars of Iopas and Palinurus," *AJP* 114, pp 123–36

Hannah, R (2008) "Timekeeping," in Oleson 2008a, pp 740–58

Hannah, R (2009) *Time in Antiquity*, London and New York

Harper, K (2017) *The Fate of Rome: Climate, Diseases, and the End of an Empire*, Princeton

Harris, WV (ed.) (2013) *The Ancient Mediterranean Environment between Science and History*, Leiden

Harrison, AR (1999) *The Law of Athens*, 2 vols, 2nd ed., Indianapolis

Harrison, SJ (2015) "Aeneas," *Oxford Classical Dictionary*, 4th ed., Oxford

Hassall, M, M Crawford, and JM Reynolds (1974) "Rome and the Eastern Provinces at the End of the Second Century BC," *JRS* 64, pp 195–220

Hassan, F (1997) "The Dynamics of a Riverine Civilization: A Geoarchaeological Perspective on the Nile Valley, Egypt," *World Archaeology* 29 (1) pp 51–74

Havlíček, F and M Morcinek (2016) "Waste and Pollution in the Ancient Roman Empire," *Journal of Landscape Ecology* 9 (3): https://www.researchgate.net/publication/312266467_Waste_and_Pollution_in_the_Ancient_Roman_Empire

Heath, T (1949) *Mathematics in Aristotle*, Oxford

Hellner, N (2006) "The Krene in Megara. The Analysis of a Thin Black Layer on the Floor Plaster," Πρακτικά: *2nd Annual International Conference on Ancient Greek Technology*, Athens, pp 172–8

Heslin, K (2011) "Dolia Shipwrecks and the Wine Trade in the Roman Mediterranean," in Robinson and Wilson 2011, pp 157–68

Hesnard, A (2004) "Vitruve, *De Architettura*, V, 12 et le port romain de Marseille," in A Gallina Zevi and R Turchetti (eds.) *Le strutture dei porti e degli approdi antichi, II Seminario ANSER, 16-17 Aprile 2004, Roma-Ostia Antica*, Soveria Mannelli, pp 175–204

Hine, HM (2010) *Lucius Annaeus Seneca: Natural Questions*, Chicago

Hitti, PK (1962) *Lebanon in History from the Earliest Times to the Present*, 2nd ed., London

Hodge, AT (1981) "Vitruvius, Lead Pipes and Lead Poisoning," *AJA* 85, pp 486–91

Hodge, AT (1990) "A Roman Factory," *Scientific American* 263, pp 106–13

Hodge, AT (2000a) "Aqueducts," in Wikander, pp 39–65

Hodge, AT (2000b) "Wells," in Wikander, pp 29–33

Hodge, AT (2002) *Roman Aqueducts and Water Supply*, London

Hohlfelder, RL (ed.) (2008) *The Maritime World of Ancient Rome: Proceedings of 'The Maritime World of Ancient Rome' Conference held at the American Academy in Rome 27-29 March 2003*, Ann Arbor

Hölkeskamp, K-J (2004) *Rekonstruktionen einer Republik. Die politische Kultur des antiken Rom und die Forschung der letzten Jahrzehnte*, München

Hollander, DB (1999) "The Management of the Mint in the Late Roman Republic," *Ancient History Bulletin* 13, pp 14–27

Hollander, DB (2012) "Tresviri Monetales," in Bagnall et al. 2012

Holleran, C (2012) *Shopping in Ancient Rome: The Retail Trade in the Late Republic and the Principate*, Oxford

Holmes, N (1991) "Notes on Lucan," *CQ* 41, pp 272–4

Holton, JR (2014) "Demetrios Poliorketes, Son of Poseidon and Aphrodite: Cosmic and Memorial Significance in the Athenian Ithyphallic Hymn," *Mnemosyne*, Series 4, 67, pp 370–90

Hong, S, J-P Candelone, C Patterson, and C Boutron (1994) "Greenland Ice Evidence of Hemispheric Lead Pollution Two Millennia Ago by Greek and Roman Civilizations," *Science* 265, pp 1,841–3

Hooker, JT (1989) "Arion and the Dolphin," *Greece & Rome* 36, pp 141–6

Hope, VM and E Marshall (eds.) (2000) *Death and Disease in the Ancient City*, London

Horden, P and N Purcell (2000) *The Corrupting Sea: A Study of Mediterranean History*, Malden, MA

Horr, WD (1971–1972) "Political Aspects of Roman Coin Reverses," *Journal of the Society of Ancient Numismatics* 3: 42 and 52

Houston, G (1987) "Lucian's Navigium and the Dimensions of the Isis," *AJP* 108, pp 444–50

Hubbard, M (1975) *Propertius*, London

Hughes, B (2005) *Helen of Troy: The Story Behind the Most Beautiful Woman in the World*, New York

Hughes, JD (2014) *Environmental Problems of the Greeks and Romans: Ecology in the Ancient Mediterranean*, Baltimore

Hughes, JD and J Thurgood (1982) "Deforestation, Erosion and Forest Management in Ancient Greece and Rome," *Journal of Forest History* 26, pp 60–75

Humphrey, JW, JP Oleson, and AN Sherwood (1998) *Greek and Roman Technology: A Sourcebook*, London

Iossif, P, Fr. De Callataÿ, and R Veymiers (eds.) (2018) *ΤΥΠΟΙ: Greek and Roman Coins Seen through their Images*, Liège

Irby, GL (2012) "Mapping the World: Greek Initiatives from Homer to Eratosthenes," in RJA Talbert (ed.) *Ancient Perspectives: Maps and Their Place in Mesopotamia, Egypt, Greece, and Rome*, Chicago, pp 81–108

Irby, GL (ed.) (2016a) *The Blackwell Companion to Ancient Science, Medicine, and Technology*, Malden, MA

Irby, GL (2016b) "Greek and Roman Cartography," in Irby 2016a, pp 819–35

Irby, GL (2016c) "Navigation and the Art of Sailing," in Irby 2016a, pp 854–69

Irby, GL (2016d) "Climate and Courage," in R Futo Kennedy and M Jones-Lewis (eds.) *The Routledge Handbook of Identity and the Environment in the Classical and Medieval Worlds*, London and New York, pp 247–65

Irby, GL (2019a) "The Politics of Cartography: Foundlings, Founders, Swashbucklers, and Epic Shields," in DW Roller (ed) *New Directions in Ancient Geography: Proceedings of the Association of Ancient Historians* 14, pp 80–102

Irby, GL (2019b) "Tracing the *orbis terrarum* from Tingentera," in DW Roller (ed) *New Directions in Ancient Geography: Proceedings of the Association of Ancient Historians* 14, pp 103–34

Irby, GL (forthcoming) "Knowledges of the Sea in Classical Antiquity," in M-C Beaulieu (ed) *A Cultural History of the Sea in Antiquity*, London

Irby-Massie, GL (1999) *Military Religion in Roman Britain*, Leiden

Irby-Massie, GL (2008) "*Prometheus Bound* and Contemporary Trends in Greek Natural Philosophy," *Greek Roman and Byzantine Studies* 48, pp 133–57

Ireland, S (2008) *Roman Britain: A Sourcebook,* 3rd ed., London

Jameson, R (1913–1932) *Collection R. Jameson. Monnaies grecques antiques*, Paris

Jansen, CM (2000) "Urban Water Transport," in Wikander, pp 103–25

Jenkins, GK (1971) *Ancient Greek Coins*, New York

Jensen, L (1963) "Royal Purple of Tyre," *Journal of Near Eastern Studies* 22, pp 104–18

Kahanov, Y (1999) "Ma'agan-Michael ship (Israel)," in P Pomey and E Rieth (eds.) *Construction navale maritime et fluviale Archaeonautica* 14, pp 155–60

Kapossy, B (1969) *Brunnenfiguren der hellenistichen und römischen Zeit*, Zürich

Keay, S (2012b) "The port system of Imperial Rome," in Keay 2012a, pp 33–67

Keay, S (ed.) (2012a) *Rome, Portus and the Mediterranean: Recent Archaeological Research*, Archaeological monographs of the British School at Rome 21, London

Keay, S, and L Paroli (eds.). (2011) *Portus and its Hinterland: Recent Archaeological Research*, Archaeological Monographs of the British School at Rome 18, London

Keay, S, M Millett, and K Strutt (2014) "The canal system and Tiber delta at Portus. Assessing the nature of man-made waterways and their relationship with the natural environment," *Journal of Water History* 6, pp 11–30

Keay, S, M Millett, L Paroli and K Strutt (eds.) (2005) *Portus: An Archaeological Survey of the Port of Imperial Rome*, Archaeological Monographs of the British School at Rome 15, London

Keenan-Jones, D (2013) "Large-scale Water Management Projects in Roman Central-Southern Italy," in Harris 2013, pp 233–56

Keenan-Jones, D (2015) "Somma-Vesuvian Ground Movements and the Water Supply of Pompeii and the Bay of Naples," *American Journal of Archaeology* 119, pp 191–215

Kemmers, F and N Myrberg (2011) "Rethinking numismatics: The Archaeology of Coins," *Archaeological Dialogues* 18, pp 87–108

King, C (2004) *The Black Sea: A History*, Oxford

Klaffenbach, G (1954) *Die Astynomeninschrift von Pergamon*, Berlin

Kleijwegt, (1994) "Caligula's 'Triumph' at Baiae," *Mnemosyne* 47, pp 652–71

Knauss, J (2000) *Spiithelladische Wasserbauten*, Munich

Knell, H (1995) *Die Nike Von Samothrake: Typus, Form, Bedeutung Und Wirkungsgeschichte Eines Rhodischen Sieges-Anathems Im Kabirenheiligtum Von Samothrake*, Darmstadt

Koehl, RB (1986) "A Marinescape Floor from the Palace at Knossos," *American Journal of Archaeology* 90, pp 407–17

Koerper, H and AL Kolls (1999) "The Silphium Motif Adorning Ancient Libyan Coinage: Marketing a Medicinal Plant," *Economic Botany* 53, pp 133–43

Koloski-Ostrow, AO (2015) *The Archaeology of Sanitation in Roman Italy: Toilets, Sewers, and Water Systems*, Chapel Hill

Kotsonas, A (2012) "Corcyra," in Bagnall et al. 2012

Kowalski, J-M (2012) *Navigation et géographie dans l'antiquité gréco-romaine: La terre vue de la mer*, Paris

Kraay, CM (1976) *Archaic and Classical Greek Coins*, London

Kuch, H (1995) "La tradizione poetica sulla battaglia di Salamina," in AM Hakkert (ed.) *Atti del convegno internazionale « Intertestualità: il "dialogo" fra testi nelle letterature classiche »: Cagliari, 24–26 Novembre 1994*, Amsterdam, pp 145–55

Kullmann, W (2012) "The relative chronology of the Homeric catalogue of ships and of the lists of heroes and cities within the catalogue," in Ø Andersen and DTT Haug (eds.) *Relative Chronology in Early Greek Epic Poetry*, Cambridge, pp 210–23

Lang, M (1968) *Waterworks in the Athenian Agora*, Princeton

Lassus, J (1983) "L'eau courante à Antioch," in J-P Boucher (ed.) *Journées d'études sur les aqueducts romains: Tagung über römische Wasserversorgungsanlagen, Lyons, 26–28 Mai, 1977*, Paris, pp 207–29

Lauwers, V (2012) "Glass, Greece and Rome," in Bagnall et al. 2012

Lazenby, JF (1993) *The Defence of Greece 490–479 BC*, Oxford

Lehmann, W and K Lehmann (1973) *Samothracian Reflections*, Princeton

Leidwanger, J (2013) "Between local and long-distance: A Roman shipwreck at Fig Tree Bay off SE Cyprus," *JRA* 26 pp 191–208

Leveau, P, M Provansal, H Bruneton, J-M Palet-Martinez, P Poupet, and K Walsh (2002) "La Crise Environnementale De La Fin De L'antiquité Et Du Haut Moyen Âge: Définition D'un Modèle Et Retour Aux Milieux Réels," in H Richard and A Vignot (eds.) *Équilibres et ruptures dans les écosystèmes durant les 20 derniers millénaires en Europe de l'Ouest, Actes du colloque international de Besançon, Septembre 2000*, Besançon, pp 291–303

Levine, DB (2011) "Tuna in Ancient Greece and Modern Tuna Population Decline," https://sites.uark.edu/dlevine/daniel-b-levine-tuna-lecture-copywrite-2011/

Lewis, M (1999) "Vitruvius and Greek Aqueducts," *Papers of the British School at Rome* 67, pp 145–72

Lewis, M (2000) "The Hellenistic Period," in Wikander, pp 631–48

Liddle, A (2003) *Arrian: Periplus Ponti Euxini*, London

Llewellyn-Jones, L and S Lewis (2018) *The Culture of Animals in Antiquity: A Sourcebook with Commentaries*, London

Longfellow, B (2011) *Roman Imperialism and Civic Patronage: Form, Meaning, and Ideology in Monumental Fountain Complexes*, Cambridge

Lorber, CC (1985) "Greek Imperial Coins and Roman Propaganda. Some issues from the Sole Reign of Caracalla: I," *Journal of the Society of Ancient Numismatics* 16, pp 45–5

Lorenz, WF, and EJ Wolfram (2014) "The Wells of Pompeii," *Groundwater* 52, pp 808–81

Lovén, B and D Davis (2012) "Piraeus," in Bagnall et al. 2012

Lowe, B (2017) "The Consumption of Salted Fish in the Roman Empire," in de Souza and Arnaud, pp 307–18

Luck, G (2006) *Arcana Mundi: Magic and the Occult in the Greek and Roman Worlds: A Collection of Ancient Texts,* 2nd ed., Baltimore

Lund, J (1999) "Rhodian Amphorae in Rhodes and Alexandria as indicators of trade," in V Gabrielsen (ed.) *Hellenistic Rhodes: Politics, Culture and Society*, Aarhus, pp 187–204

Lusby, S, R Hannah, and P Knight (2009/2010) "Navigation and Discovery in the Polynesian Oceanic Empire," *Hydrographic Journal* 131/132, pp 17–25

Lytle, E (2007) "The Delian Purple and the *lex portus Asiae*," *Phoenix* 61, pp 247–69

Lytle, E (2012) "Ἡ θάλασσα κοινή: Fishermen, the Sea, and the Limits of Ancient Greek Regulatory Reach," *Classical Antiquity* 31, pp 1–55

Lytle, E (2016) "One Fish, Two Fish, Bonito, Bluefish: Ancient Greek ἀμία and γομφάριον," *Mnemosyne* Series 4 (69), pp 249–61

MacDowell, DM (1978) *The Law in Classical Athens*, Ithaca, NY

Macgillivray, JA, LH Sackett, JM Driessen, C Doherty, D Evely, EM Hatzaki, D Mylona, DS Reese, A Sarpaki, G Shipley, SM Thorne, S Wall-Crowther, and J Weingarten (2007) *Palaikastro: Two Late Minoan Wells*, The British School at Athens. Supplementary Volumes no. 43

Maeder, F (2008) "Sea-Silk in aquincum: First Production Proof in Antiquity," in C Alfaro and L Karali (eds.) *Purpureae Vestes II. Vestidos, textiles y tintes. Estudios sobre la producción de bienes de consumo en Antigüedad*, Valencia, pp 109–18

Maischberger, M (1997) *Marmor in Rom. Anlieferung, Lager- und Werkplatz in der Kaiserzeit*, Weisbaden

Maiuri, A (1958) "Navalia pompeiana," *Rendiconti della Accademia di Archaeologia di Napoli* 33, pp 7–34

Malissard, A (1994) *Les romains et l'eau: Fontaines, salles de bains, themes, égouts, aqueduts*, Paris

Malkin, I (1998) *The Returns of Odysseus: Colonization and Ethnicity*, Berkeley

Manning, JG (2018) *The Open Sea: The Economic Life of the Ancient Mediterranean World from the Iron Age to the Rise of Rome*, Princeton

Manning, SW (2013) "The Roman World and Climate: Context, Relevance of Climate Change, and Some Issues," in Harris, pp 103–70

Mansfield, JM (1985) *The Robe of Athena and the Panathenaic πέπλος*, Berkeley
Marcotte, D (2017) "L'Océan Indien dans l'Antiquité: Science, Commerce et Géopolitique," in de Souza and Arnaud, pp 511–22
Margarida, AA (2009) "Phoenician Colonization on the Atlantic Coast of the Iberian Peninsula," in M Dietler and C López-Ruiz (eds.) *Colonial Encounters in Ancient Iberia*, Chicago, pp 113–30
Marinatos, S (1973) "A Fresco of Historical Nature from Thera," *Praktika tes Akademias Athenon* 48, pp 231–37
Marlière, S (2014) "La batellerie fluviale gallo-romaine: Le chaland Arles-Rhône 3," *Dossiers d'Archéologie* 364, pp 66–71
Marlière, S, P Poveda, and N Ranchin (2017) "The Arles-Rhône 3 project, Arles, France. From the excavation and raising of a Gallo-Roman barge to its documentation and 3D-modelling," in J Gawronski, A Van Holk, and J Schokkenbroek (eds.) *Ships and Maritime Landscapes: Proceedings of the Thirteenth International Symposium on Boat and Ship Archaeology, Amsterdam 2012* (2011–2012), Eelde, pp 383–9
Marlière, É (2002) *L'outre et le tonneau dans l'Occident romain*, Montagnac
Marzano, A (2013) *Harvesting the Sea: The Exploitation of Marine Resources in the Roman Mediterranean*, Oxford
Marzano, A (2018a) *The Roman Villa in the Mediterranean Basin: Late Republic to Late Antiquity*, Cambridge
Marzano, A (2018b) "Maritime Villas and the Resources of the Sea," in Marzano 2018a, pp 125–40
Mason, DJP (2003) *Roman Britain and the Roman Navy*, Stroud, Gloucestershire
Matthews, J (1984) "The Tax Law of Palmyra: Evidence for Economic History in a City of the Roman East," *JRS* 74, pp 157–80
Mattingly, DJ (2006) *An Imperial Possession: Britain in the Roman Empire, 54 BC–AD 409*, London
Maiuri, A (1958) "Navalia pompeiana," *Rendiconti della Accademia di Archaeologia di Napoli* 33, pp 7–34
Mavrogiannis, T (2013–2014) "Consequences of the Battle at Salamis in Eastern Mediterranean and Cyprus from 478 to 449 B.C. The « Hellenikos polemos »," *Ostraka* 22–23, pp 129–68
May, JM (1980) "The Image of the Ship of State in Cicero's *Pro Sestio*," *Maia* 32, pp 259–64
Mayor, A (2003) *Greek Fire, Poison Arrows, and Scorpion Bombs: Biological and Chemical Warfare in the Ancient World*, New York
Mayor, A (2018) *Gods and Robots: Myths, Machines and Ancient Dreams of Technology*, Princeton
Mays, L, GP Antoniou, and AN Angelakis (2013) "History of Water Cisterns: Legacies and Lessons," *Water* 5, 1916–1940
McGrail, S (2001) *Boats of the World: From the Stone Age to Medieval Times*, Oxford
McKenzie, J (2007) *The Architecture of Alexandria and Egypt, C. 300 B.C. to A.D. 700*, New Haven, CT

McMahon, A (2015) "Waste Management in Early Urban Southern Mesopotamia," in P.D. Mitchell (ed.) *Sanitation, Latrines and Intestinal Parasites in Past Populations*, London and New York, pp 19–40

Medas, S (2004) *De rebus nauticis: l'arte della navigazione nel mondo antico*, Rome

Meiggs, R (1972) *The Athenian Empire*, Oxford

Meijer, F van (1992) *Trade, Transport, and Society in the Ancient World: A Sourcebook*, London

Meneghini, R (1985) "Attivita e installazioni portuali lungo il Tevere. La riva dell' Emporium," in R Bussi and V Vandelli (eds.) *Misurare la terra: centuriazione e coloni nel mondo romano: citta, agricoltura, commercio: materiali da Roma e dal suburbia*, Modena, pp 162–72

Merker, I (1975) "A Greek Tariff Inscription in Jerusalem," *Israel Exploration Journal* 25, pp 238–44

Metcalf, WE (ed.) (2012) *The Oxford Handbook of Greek and Roman Coinage*, Oxford

Mikalson, JD (1998) *Religion in Hellenistic Athens*, Berkeley

Milner, NP (1996) *Vegetius: Epitome of Military Science*, Liverpool

Modlin, IM, SF Moss, M Kidd, and KD Lye (2004) "Gastroesophageal Reflux Disease: Then and Now," *Journal of Clinical Gastroenterology* 38, pp 390–402

Morales Muñiz, A (1993) "Where are the Tunas? Ancient Iberian Fishing Industries from an Archaeozoological Perspective," in A Claxton (ed.) *Skeletons in her Cupboard: Festschrift for Juliet Cluton-Brock*, Oxford, pp 135–41

Morello, A (2012) "PRORAE The First Punic War, Gaius Duilius, the first *Aes Grave* Prow Bronzes of the Roman Republic, and the CORVUS or boarding bridge of the Romans. A web-essay:" https://andrewmccabe.ancients.info/Corvus.html

Morhange, C and N Marriner (2008) "Mind the (stratigraphic) gap: Roman dredging in ancient Mediterranean harbours," *Bollettino di Archeologia online*

Morhange, C, N Marriner, and N Carayon (2016) "The eco-history of ancient Mediterranean harbours," in T Bekker-Nielsen and R Gertwagen (eds.) *The Inland Seas: Towards an Ecohistory of the Mediterranean and the Black Sea*, Stuttgart, pp 85–106

Morley, N (2007) *Trade in Classical Antiquity*, Cambridge

Morrison, SJ (1968) *Greek-Oared Ships: 900–322 BC*, Cambridge

Murray, WM and PM Petsas (1989) *Octavian's Campsite Memorial for the Actian War*, Philadelphia

Napoli, J (2004) "Art purpuraire et législation a l'époque romaine," in C Alfaro, JP Wild, and B Costa (eds.) *Purpureae Vestes: Actas del I Symposium internacional sobre textiles y tintes del Mediterraneo en epoca romana (Ibiza 8–10 de Novembre, 2002)*, Valencia, pp 123–36

Neuerburg, N (1965) *L'architettura delle fontane e dei ninfei nell'Italia antica*, Napoli

Newell, Edward T (1937) *Royal Greek Portrait Coins*, Atlanta

Nicolet, H (1971) "Monnaies de bronze de Cilicie (Séleucie du Kalykadnos)," *Revue Numismatique* 13, pp 26–37

Nieto, FJF. (2006) "Titularidad y cesión de los derechos de la pesca marítima en la antigua Grecia," in H-A Rupprecht (ed.) *Symposion 2003: Vorträge Zur Griechischen Und Hellenistischen Rechtsgeschichte (Rauischholzhausen, 30. September – 3. Oktober 2003)*, Wien, pp 207–32

Nonnis, D and C Ricci. (2007) "Supplying the Roman army. La *legio II Augusta* in *Britannica*: il contributo dei materiali iscritti," in E Papi, and M Bonifay (eds.) *Supplying Rome and the Empire: The Proceedings of an international Seminar held at Siena-Certosa di Pontignano on May 2–4, 2004, on Rome, the Provinces, Production and Distribution, Journal of Roman Archaeology*, Portsmouth, RI, pp 193–207

Norman, NJ (1983) "The Panathenaic ship," *Archaeological News*, 12, pp 41–6

Novello, M (2007) *Scelte tematiche e committenza nelle abitazioni dell'Africa Proconsolare*, Pisa

Oakley, SP (2007) *A Commentary on Livy: Books VI–X: Vol. 4: Book X*, Oxford

Oleson, JP (1984) *Greek and Roman Mechanical Water-Lifting Devices: The History of a Technology*, Toronto

Oleson, JP (1991) "Aqueducts, cisterns, and the strategy of water supply at Nabataean and Roman Auara (Jordan)," in AT Hodge (ed.) *Future Currents in Aqueduct Studies*, Leeds, pp 45–62

Oleson, JP (2000) "Irrigation," in Wikander, pp 183–215

Oleson, JP (ed.) (2008a) *The Oxford Handbook of Engineering and Technology in the Classical World*, Oxford

Oleson, JP (2008b) "Testing the Waters: The Role of Sounding Weights in Ancient Mediterranean Navigation," in Hohlfelder, pp 119–76

Oleson, JP, C. Brandon, SM Cramer, R Cucitore, E Gotti, and RL Hohlfelder (2004) "The ROMACONS Project: A contribution to the historical and engineering analysis of hydraulic concrete in Roman maritime structure," *International Journal of Nautical Archaeology* 33, pp 199–229

Orlandos, A (1940) "The role of the Roman building located northern of horologe of *Andronikos Kiristos*," Paper presented at the Athens Academy, Athens, Greece 1940 (in Greek)

Orna-Ornstein, J (1995) "Ships on Roman coins," *Oxford Journal of Archaeology* 14, pp 179–200

Paga, J (2015) "The Southeast Fountain House in the Athenian Agora: A Reappraisal of its Date and Historical Context," *Hesperia* 84, pp 355–87

Panagiotaki, M (1993) "The Temple Repositories of Knossos: New Information from the Unpublished Notes of Sir Arthur Evans," *The Annual of the British School at Athens* 88, pp 49–91

Panofka, Theodor (ed.) (1843) *Bilder Antiken Lebens*, Berlin

Parejko, K (2003) "Pliny the Elder's Silphium: First Recorded Species Extinction," *Conservation Biology* 17, pp 925–7

Parke, HW (1977) *Festivals of the Athenians*, London

Parker, AJ (1992) *Ancient Shipwrecks of the Mediterranean and the Roman Provinces*, Oxford

Parker, V (2007) "Herodotus' use of Aeschylus' « Persae » as a source for the battle of Salamis," *Symbolae Osloenses* 82, pp 2–29

Pavolini, C (1988) *Ostia Antica. Raymond Chevallier, Ostie Antique, Ville Et Port*, Paris

Pearson, AC (1917) *The Fragments of Sophocles*, 3 vols., Cambridge

Peltenburg, E, P Croft, A Jackson, C McCartney, and MA Murray (2001) "Well-established Colonists: *Mylouthkia* I and the Cypro-Pre-Pottery Neolithic B," in S Swiney (ed.) *The Earliest Prehistory of Cyprus: From Colonization to Exploitation*, Cyprus American Archaeological Research Institute Monograph Series 2: American Schools of Oriental Research, Boston, pp 61–93

Peña, JT (1998) "The Mobilization of State Olive Oil in Roman Africa: The Evidence of late fourth century Ostraca from Carthage," *JRA supplementary series* 28, pp 117–238

Peña, JT (1999) *The Urban Economy During the Early Dominate: Pottery Evidence from the Palatine Hill*, Oxford

Phaneuf, BA, TK Dettweiler, and T Bethge (2001) "Special Report: Deepest Wreck," *Archaeology Magazine* 54, https://archive.archaeology.org/0103/etc/wreck.html

Phelps, WW (1999) *The Point Iria wreck: Interconnections in the Mediterranean, ca. 1200 BC: Proceedings of the international conference, Island of Spetses, 19 September 1998*, Athens: Hellenic Institute of Marine Archaeology

Piromallo, M (2004) "Puteoli, porto di Roma," in AG Zevi and R Turchetti (eds.) *Le strutture dei porti e degli approdi antichi, II Seminario ANSER, 16–17 Aprile 2004, Roma-Ostia Antica*, Soveria Mannelli, pp 267–78

Pitassi, M (2011) *Roman Warships*, Rochester, NY

Platon, N (1984) "The Minoan Thalassocracy and the Golden Ring of Minos," in R Hägg and N Marinatos (eds.) *The Minoan Thalassocracy. Myth and Reality. Proceedings of the Third International Symposium at the Swedish Institute in Athens, 31 May–5 June, 1982*, Stockholm, pp 65–9

Pomeroy, SB, SM Burstein, W Donlan, and JT Roberts (2008) *Ancient Greece: A Political, Social, and Cultural History*, Oxford

Pomey, P (1982) "Le Navire Romain de la Madrague de Giens," *Comptes rendus des séances de l'Académie des Inscriptions et Belles-Lettres* 126, pp 133–54

Pomey, P (1995) "Les épaves grecques et romaines de la place Jules Verne à Marseille," *Comptes Rendus de l'Académie des Inscriptions et Belles Lettres* 139, pp 459–84

Pomey, P (ed.) (1997) *La Navigation dans L'Antiquite*, Aix-en-Provence

Pomey, P and A Tchernia (1978) "Le tonnage maximum des navires de commerce romains," *Archaeonautica* 2, pp 233–52

Pomey, P and E Rieth (2005) *L'archéologie navale*, Paris

Porciani, L (2015) "Early Greek Colonies and Greek Cultural Identity: Megara Hyblaia and the Phaeacians," *Dialogues d'Histoire Ancienne* 41 (2) pp 9–18

Potter, IC (1971) *The Biology of Lampreys*, 2 vols., New York

Powell, A (2002) "'An Island Amid the Flame': The Strategy and Imagery of Sextus Pompeius," in Powell and Welch, pp 103–33

Powell, A and K Welch (eds.) (2002) *Sextus Pompeius*, London

Powell, L (2015) *Marcus Agrippa: Right-Hand Man of Caesar Augustus*, Barnsley

Pruitt, S (2016) "Huge Roman Villa Discovered Underneath a Garden in Britain," April 21: https://www.history.com/news/huge-roman-villa-discovered-underneath-a-garden-in-britain

Pulak, C (2005) "Discovering a Royal Ship from the Age of King Tut: Uluburun, Turkey," in GF Bass (ed.) *Beneath the Seven Seas: Adventures with the Institute of Nautical Archaeology*, New York, pp 34–47

Pulak, C (2010) "Uluburun Shipwreck," in E Cline (ed.) *The Oxford Handbook of the Bronze Age Aegean (ca. 3000–1000 BC)*, Oxford, pp 862–76

Purcell, N (2017) "Taxing the Sea," in de Souza and Arnaud, pp 319–34

Putnam, MCJ (1998) *Vergil's Epic Designs: Ekphrasis in the Aeneid*, New Haven, CT

Quilici, L (1985–86) "Il Tevere e l'aniene come vie d'acqua a monte di Roma in eta imperial," *Archeologia Laziale* 7, pp 198–217

Race, WH (1981) "Pindar's 'Best is water': Best of What?" *Greek, Roman and Byzantine Studies* 22, pp 119–24

Ravel, OE (1947) *Descriptive Catalogue of the Collection of Tarentine Coins formed by M.P. Vlasto*, London, reprinted

Rea, R (2001) "Gli animali per la *venatio*: cattura, transporto, custodia," in A La Regina (ed.) *Sangue e arena*, Rome, pp 245–77

Redmount, C (1995) "The Wadi Tumilat and the 'Canal of the Pharaohs,'" *Journal of Near Eastern Studies* 54 (2), pp 127–35

Rees, O (2019) *Great Naval Battles of the Ancient Greek World*, South Yorkshire

Reese, DS (1980–1979) "Industrial exploitation of murex shells. Purple-dye and lime production at Sidi Khrebish, Benghazi (Berenice)," *Society for Libyan Studies* 11, pp 79–9

Reinholdt, C (2009) *Das Brunnenhaus der Arsinoe in Messene: Nutzarchitektur, Reprasentationsbaukunst und Hydrotechnologie im Rahmen hellenistisch romischer Wasserversorgung*, Wien

Renfrew, C (1972) *The Emergence of Civilisation: The Cyclades and the Aegean in the Third Millennium BC*, London

Ricciardi, M and VSM Scrinari (1996) *La civiltà dell'acqua in Ostia Antica*, Rome

Rice, C (2016a) "Mercantile Specialization and Trading Communities: Economic Strategies in Roman Maritime Trade," in A Wilson and M Flohr, *Urban Craftsmen and Traders in the Roman World*, Oxford, pp 97–111

Rice, C (2016b) "Shipwreck cargoes in the western Mediterranean and the organization of Roman maritime rrade," *JRA* 29, pp 165–92

Rice, C (2018) "Rivers, Roads, and Ports," in C Holleran and A Claridge (eds) *A Companion to the City of Rome*, Malden, pp 199–217

Rickman, GE (1980) *The Corn Supply of Ancient Rome*, Oxford

Ridgway, BS (1970) "Dolphins and Dolphin-Riders," *Archaeology* 23, pp 86–95

Ritti, T, K Grewe, and P Kessner (2007) "A Relief of a Water-powered Stone Saw Mill on a Sarcophagus at Hierapolis and its Implications," *JRA* 20, pp 138–63

Roberts, BK (1988) "The Re-Discovery of Fishponds," in M Aston (ed.) *Medieval Fish, Fisheries, and Fishponds in England*, Oxford, pp 1.9–16

Robertson, N (1987) "The True Meaning of the Wooden Wall," *CPh* 82 (1), pp 1–20

Robinson, D (1906) "Ancient Sinope: First Part," *AJP* 27, pp 125–53

Robinson, D and AI Wilson (2011) *Maritime Archaeology and Ancient Trade in the Mediterranean*, Oxford

Robinson, E (1944) "Greek Coins found in the Cyrenaica," *The Numismatic Chronicle and Journal of the Royal Numismatic Society*, pp 105–13

Rogers, DK (2018) *Water Culture in Roman Society*, Leiden

Roller, DW (2003) *The World of Juba II and Kleopatra Selene: Royal Scholarship on Rome's African Frontier*, London

Roller, DW (2006) *Through the Pillars of Herakles: Greco-Roman Exploration of the Atlantic*, London

Roller, DW (2010a) *Cleopatra: A Biography*, Oxford

Roller, DW (2014) *The Geography of Strabo: An English Translation, with Introduction and Notes*, Cambridge

Roller, DW (2018) *A Historical and Topographical Guide to the Geography of Strabo*, Cambridge

Roller, DW (2019a) "Phoenician Exploration," in C Lopez-Ruiz and B Doak (eds.) *Oxford Handbook on the Phoenicians*, Oxford, pp 645–53

Roller, DW (2019b) "Timosthenes of Rhodes," in DW Roller (ed.) *New Directions in Ancient Geography: Proceedings of the Association of Ancient Historians* 14, pp 56–79

Romeo, N (2016) "How Archaeologists Discovered 23 Shipwrecks in 22 Days," *National Geographic*: https://www.nationalgeographic.com/news/2016/07/greece-shipwrecks-discovery-fourni-ancient-diving-archaeology/

Roseman, CH (1994) *Pytheas of Massalia: On the Ocean*, Chicago

Rougé, J (1981) *Ships and Fleets of the Ancient Mediterranean*, trans. S. Frazer, Middletown, CT

Ruffing, K (2006) "Salamis: die grösste Seeschlacht der alten Welt," *Grazer Beiträge* 25, pp 1–3

Ruscillo, D (2005) "Reconstructing Murex Royal Purple and Biblical Blue in the Aegean," in DE Bar-Yosef Mayer (ed.) *Archaeomalacology: Molluscs in Former Environments of Human Behaviour, Proceedings of the 9th ICAZ Conference, Durham 2002*, Oxford, pp 99–106

Russell, B (2013a) "Roman and Late Antique shipwrecks with stone cargoes: A new inventory," *JRA* 26, pp 331–61

Russell, B (2013b) *The Economics of the Roman Stone Trade*, Oxford

Rykwert, J (1976) *The Idea of a Town: The Anthropology of Urban Form in Rome, Italy and the Ancient World*, Princeton

Saladino, V (1980) "Iscrizioni Latine di Roselle (II)," *Zeitschrift Für Papyrologie Und Epigraphik* 39, pp 215–36
Saunders, TJ (1990) "Plato and the Athenian Law of Theft," in P Cartledge, P Millett, and SC Todd (eds.) *Nomos: Essays in Athenian Law, Politics, and Society*, Cambridge, pp 63–82
Sauvage, C (2012) *Routes maritimes et systèmes d'échanges internationaux au Bronze récent an Méditerranée orientale*, Lyon
Sauvage, C (2017) "The development of maritime exchange in the Bronze Age Eastern Mediterranean," in de Souza and Arnaud, pp 151–64
Scarborough, J (1981) "Roman medicine and Public Health," *Proceedings of the Fifth International Symposium on the Comparative History of Medicine East and West 1980*, Kyoto, pp 33–74
Schmidt, M (1976) *Die Erklärungen zum Weltbild Homers und zur Kultur der Herenzeit in den bT-Scholien zur Ilas*, Munich
Schmiedt, G (ed.) (1972) *Il livello antico del Mar Tirreno. Testimonianze dei resti archeologici*, Firenze
Schreiner, JH.(2007) "The battle of Phaleron in 490 BC," *Symbolae Osloenses* 82, pp 30–4
Schweitzer, B (1938) *Ein Nymphäum des frühen Hellenismus*, Leipzig
Seaby, HA, DR Sear, and CE King (1967–1987) *Roman Silver Coins*, 2nd ed. 5 vols., London
Secord, J (2016) "Overcoming Environmental Determinism: Introduced Species, Hybrid Plants and Animals, and Transformed Lands in the Hellenistic and Roman Worlds," in R Futo Kennedy and M Jones-Lewis (eds.) *Routledge Handbook of Identity and the Environment in the Classical and Medieval Worlds*, London and New York, pp 210–29
Seigne, J (2006) "Water-Powered Stone Saws in Late Antiquity. First Step on the Way to Industrialisation?" in G Wiplinger (ed.) *Cura Aquarum in Ephesus, BABesch* Supplement 12, Leuven, pp 371–8
Shelmerdine, CW (ed.) (2008) *The Cambridge Companion to the Aegean Bronze Age*, Cambridge
Sherratt, S (2016) "From 'institutional' to 'private': Traders, route sand commerce from the Late Bronze Age to the Iron Age," in García J (ed.) *Dynamics of Production in the Ancient Near East*, Oxford: digital
Shipley, G (2011) *Pseudo-Skylax's Periplous: The Circumnavigation of the Inhabited World. Text, Translation and Commentary*, Exeter
Silberman, A (texte établi, trad. et annoté) (1988) *Pomponius Mela: Chorographie*, Paris
Simmons, BB (1979) "The Lincolnshire Car Dyke: Navigation or Drainage?" *Britannia* 10, pp 183–96
Simpson, RH (1968) "The Homeric Catalogue of Ships and its Dramatic Context in the Iliad," *Studi Micenei ed Egeo-Anatolici* 6, pp 39–44
Slim, H, P Trousset, R Paskoff, and A Oueslati (2004) *Le littoral de la Tunisie. Étude géoarchéologique et historique*, Paris

Smith, AC (2011) *Polis and Personification in Classical Athenian Art*, Leiden

Smith, AM (2010) "From Murex shells to purple Cloth," *Journal for Semitics* 19 (2), pp 599–611

Sobel, D (2007) *Longitude: The True Story of a Lone Genius Who Solved the Greatest Scientific Problem of His Time*, London

Solazzi, S (1948) *Specie ed estinzione delle servitù prediali*, Naples

Southern, P (2010) *Mark Antony: A Life*, Stroud

Speidel, MP (1994) *Riding for Caesar: The Roman Emperors' Horse Guard*, London

Stewart, A (2016) "The Nike of Samothrace: Another View," *AJA* 120, pp 399–410

Stockton, D (1959) "The Peace of Callias," *Historia: Zeitschrift Für Alte Geschichte* 8, pp 61–79

Strauss, B (2004) *The Battle of Salamis: The Vaval Encounter that Saved Greece—and Western Civilization*, New York

Strauss, J (2013) *Shipwrecks Database*. Version 1.0: http://oxrep.classics.ox.ac.uk/databases/shipwrecks_database/

Strazzulla, MJ (1989) "In paludibus moenia constituta: problemi urbanistici di Aquileia in età repubblicana alla luce della documentazione archeologica e delle fonti scritte," *Antichità Altoadriatiche* 35, pp 187–225

Swetnam-Burland, M (2010) "'Aegyptus Redacta': The Egyptian Obelisk in the Augustan Campus Martius," *The Art Bulletin* 92 (3), pp 135–53

Swingler, DH (1976) "Roman Coinage. Its Use as an Implement of Propaganda," *Journal of the Society of Ancient Numismatics* 7, pp 48–51

Syme, R (1986) *The Augustan Aristocracy*, Oxford

Talbert, RJA (2010) *Rome's World: The Peutinger Map Reconsidered*, Cambridge

Tameanko, M (1992) "The Silphium Plant: Wonder Drug of the Ancient World Depicted on Coins," *Celator* 6, pp 26–8

Tarn, WW (1905) "The Greek Warship," *JHS* 25, pp 137–56, 204–24

Tartaron, T (2013) *Maritime Networks in the Mycenaean World*, Cambridge

Taylor, MC (2010) *Thucydides, Pericles, and the Idea of Athens in the Peloponnesian War*, Cambridge

Taylor, MC (2012) "Panathenaia," in Bagnall et al. 2012

Taylor, RM (2000) *Public Needs and Private Pleasures: Water Distribution, the Tiber River and the Urban Development of Ancient Rome*, Rome

Tchernia, A, P Pomey, and A Hesnard (1978) *L'Épave romaine de la Madrague de Giens*, Paris

Tenney, Frank (ed.) (1933–40) *An Economic Survey of Ancient Rome*, Baltimore

Testaguzza, O (1970) *Portus: Illustrazione Dei Porti Di Claudio E Traiano E Della Città di Porto a Fiumicino*, Roma

Thiercy, P (2008) "Bête comme ses huit pieds," in D Auger and J Peigny (eds.) *Phileuripidès: Mélanges F. Jouan*, Nanterre, pp 93–101

Thomas, R (2010) "Fishing Equipment from Myos Hormos and Fishing Techniques on the Red Sea in the Roman Period," in T Bekker Nielsen and D Bernal Casasola (eds.)

Ancient Nets and Fishing Gear: Proceedings of the International Workshop on "Nets and Fishing Gear in Classical Antiquity: A First Approach." Cadiz, November 15–17, 2007, Cadiz and Arhaus, pp 139–59

Thompson, F (1867) "Roman Well at Biddenham, 1858," *Notes of the Bedfordshire Architectural and Archaeological Society*, 1 (8), Bedfordshire, UK

Thompson, HA (1954) "Excavations in the Athenian Agora: 1953," *Hesperia* 23, pp 31–67

Thonemann, P (2015) *The Hellenistic World: Using Coins as Sources*, Cambridge

Tölle-Kastenbein, R (1990) *Antike Wasserkultur*, Munich

Tölle-Kastenbein, R (1994) *Das archaische Wasserleitungsnetz für Athen und seine späteren Bauphasen*, Mainz

Totelin, L (2014) "Smell as Sign and Cure in Ancient Medicine," in M Bradley (ed.) *Smell and the Ancient Senses*, London, pp 17–29

Triandafyllou, V and I Tsakiri (2018) "Ancient shipwrecks found in Greek waters tell tale of trade routes," Reuters: https://www.reuters.com/article/us-greece-ancient-shipwrecks/ancient%E2%80%93shipwrecks-found-in-greek-waters-tell-tale-of-trade-routes-idUSKCN1ML1MJ

Triantafillidis, I and D Koutsoumba (2017) "The harbour landscape of Aegina (Greece)," in Jerzy Gawronski, André Van Holk, and Joost Schokkenbroek (eds.) *Ships and Maritime Landscapes: Proceedings of the Thirteenth International Symposium on Boat and Ship Archaeology, Amsterdam 2012* (2011–2012), Eelde, pp 165–70

Tuck, S (2019). "Was the *tempestas* of AD 62 at Ostia Actually a Tsunami?," *CJ* 114 (4), pp 439–62

Virlouvet, C (2017) "La Mer et L'Approvisionnement de la Ville de Rome," in de Souza and Arnaud, pp 268–82

Wachsmann, S (1998) *Seagoing Ships and Seamanship in the Bronze Age Levant*, College Station, TX

Wachsmann, S (2012) "Panathenaic Ships: The Iconographic Evidence," *Hesperia* 81, pp 237–66

Wacke, A (2002) "Protection of the Environment in Roman Law?" *Roman Legal Tradition* 1, pp 1–24

Wallinga, HT (2006) "Aeschylus and the last act of Salamis," in A Pierre, MH Lardinois, MGM Van der Poel, and VJC Hunink (eds.) *Land of Dreams: Greek and Latin Studies in Honour of A. H. M. Kessels*, Leiden, pp 97–105

Ward, C (2010) "Seafaring in the Bronze Age Aegean: Evidence and Speculation," in D Pullen (ed.) *Political Economies of the Aegean Bronze Age*, Oxford, pp 149–60

Webb, PA (2017) *The Tower of the Winds in Athens. Greeks, Romans, Christians, and Muslims: Two Millennia of Continual Use*, Philadelphia

Weingarten, J (2010) "Minoan Seals and Sealings," in Eric H Cline (ed.) *The Oxford Handbook of The Bronze Age Aegean (ca. 3000–1000 BC)*, Oxford and New York, pp 317–28

Welch, K (2002) "Both Sides of the Coin: Sextus Pompeius and the so-called *Pompeiani*," in Powell and Welch, pp 1–30

Westcoat, B (2005) "Buildings for Votive Ships on Delos and Samothrace," in JJ Coulton, M Yeroulanou, and M Stamatopoulou (eds.) *Architecture and Archaeology in the Cyclades: Papers in Honour of J.J. Coulton*. British Archaeological Reports International Series, 1,455, Oxford, pp 153–72

Westcoat, J (2000) "Landscape of Roman Water Law," *Environmental Design: Journal of the Islamic Environmental Design Research Centre*, pp 88–99

Wheatley, P (2012) "Demetrios Poliorketes," in Bagnall et al. 2012

Whitewright, J (2014) *Roman Mediterranean Shipping*: http://moocs.southampton.ac.uk/portus/2014/06/03/roman-mediterranean-shipping/

Wikander, Ch (2000) "Canals," in Wikander, pp 321–30

Wikander, Ö (2000a) *Handbook of Ancient Water Technology*, Leiden

Wikander, Ö (2000b) "The Water-Mill," in Wikander, pp 371–400

Willcock, MM (1976) *A Companion to the Iliad*, Chicago

Wilcock, MM (1984) *Homer: Iliad XIII–XXIV*, Cambridge

Williams, DF (2012) "Trade, Roman," in Bagnall et al. 2012

Williams, DF and SJ Keay (2006) *Roman Amphorae: A Digital Resource* [online]: https://archaeologydataservice.ac.uk/archives/view/amphora_ahrb_2005/index.cfm

Williams, JHC (1999) "Septimius Severus and Sol, Carausius and Oceanus: Two new Roman Acquisitions at the British Museum," *Numismatic Chronicle* 159, pp 307–13

Wilson, A (2001) "Urban Water Storage, Distribution, and Usage in Roman North Africa," in AO Koloski-Ostrow (ed.) *Water Use and Hydraulics in the Roman City*, Dubuque, IA, pp 83–125

Wilson, AI (2000a) "Drainage and Sanitation," in Wikander, pp 152–79

Wilson, AI (2000b) "Land Drainage," in Wikander, pp 303–17

Wilson, AI (2002) "Machines, Power, and the Ancient Economy," *JRS* 92, pp 1–32

Wilson, AI (2008) "Hydraulic Engineering and Water Supply," in Oleson, pp 285–318

Wilson, AI (2009) "Approaches to Quantifying Roman Trade," in AK Bowman and A Wilson (eds.) *Quantifying the Roman Economy: Methods and Problems*, Oxford, pp 213–49

Wilson, AI (2011) "Developments in Mediterranean Shipping and Maritime Trade from the Hellenistic Period to AD 100," in Robinson and Wilson, pp 33–59

Wilson, A (2013) "The Mediterranean Environment in Ancient History: Perspectives and Prospects," in Harris, pp 259–76

Wilson, RJA (1998) "Piazza Armerina," in T Akiyama (ed.) *The Dictionary of Art: 24: Pandolfini to Pitti*, Oxford, pp 65–7

Wilson, RJA (2018) "Roman Villas in Sicily," in Marzano, pp 195–219

Wood, M (1985) *In Search of the Trojan War*, New York

Woods, D (2002) "Did Caligula plan to bridge the English Channel?" *Ancient World* 33, pp 157–69

Zabehlicky, H (1995) "Preliminary View of the Ephesian Harbor," in H Koester (ed.) *Ephesos: Metropolis of Asia*, Harvard Theological Studies 41, Cambridge, pp 201–15

Zanker, P (1990) *The Power of Images in the Age of Augustus*, trans. Alan Shapiro, Ann Arbor

Zarecki, J (2014) *Cicero's Ideal Statesman in Theory and Practice*, London

Zarrow, E (2003) "Sicily and the Coinage of Octavian and Sextus Pompey: Aeneas or the Catanean Brothers?" *The Numismatic Chronicle* 163, pp 123–35

Zeng, L, SB Jacobsen, DD Sasselov, and A Vanderburg (2018) "Survival function analysis of planet size distribution with Gaia Data Release 2 updates," *Monthly Notices of the Royal Astronomical Society* 479, pp 5,567–76

Zucker, A (2017) "Les Techniques de Pêche dans l'Antiquité," in de Souza and Arnaud, pp 294–306

Index of Places Cited

Poleis and City States

Abdera 169, 210, 212
Abydos 78
Ace 136
Acragas 124, 139, 171, 209
Actium 33, 101, 182, 194, 195, 196, 197, 199, 203, 247
Adulis 117, 118, 119, 125
Aegina 41, 69, 77, 170, 222, 239
Agrigentum 144
Aia 109
Akrotiri 139, 158, 232
Alexandria 26, 28, 61, 62, 68, 70, 79, 80, 81, 102, 111, 117, 120, 121, 124, 131, 132, 133, 136, 137, 149, 171, 182, 199, 208, 210, 212, 222
Amastris 71
Amphipolis 182, 243
Andania 54
Antioch 46, 53, 133
Antipolis 150
Antium 245
Aphrodisias 198
Apollonia 171, 229
Aquileia 32, 229
Arados 41
Arelate (Arles) 134, 63, 134
Argos 163, 166
Arnazabus 149
Arpinum 13
Arsinoë 117
Artemisium 93, 175, 237, 241
Asturia 65
Athenai 91, 101
Athens 2, 3, 8, 9, 10, 11, 20, 33, 40, 43, 48, 50, 51, 57, 58, 67, 69, 71, 72, 90, 97, 111, 114, 122, 124, 131, 136, 164, 166, 170, 175, 176, 177, 178, 179, 180, 182, 183, 199, 208, 209, 211, 212, 213, 222, 225, 232, 236, 237, 239, 242, 243
Aulis 78, 100, 162
Aurausio (Orange) 195

Avalites 118, 119, 126
Axomites 118
Azarium 104

Baelo Claudia 150
Baesuri 169
Baiae 145, 147, 197, 198, 233, 247
Barake 119
Barbegal 63–4
Barygaza 104, 118, 119, 125
Bauli 145
Beni Salama 247
Berenike 115, 117
Berytus (Beirut) 136
Bet Eli'ezer 247
Beth She'arim 247
Biddenham 39
Bir Hooker 247
Bononia (Boulogne) 198
Brundisium 147
Bruttium 135
Byzantium 93, 121, 122, 128, 211

Caballio 135
Caesarea Maritima 59, 69, 70, 121, 124
Camarina 246
Camulodunum (Colchester) 111, 198
Carchedon 69
Carmania 240
Carthage 40, 46, 60, 86, 97, 103, 105, 120, 121, 124, 133, 187, 188, 189, 222, 227
Cassandreia 244
Catana 165
Caunus 148
Centumcellae (Civitavecchia) 70
Chaeronea 17, 212
Chalcis 24, 103, 165, 243
Chersonsesus 122
Cirencester 202
Citium 213
Clazomene 150, 208

Colchis 91, 108, 164
Colonia Claudia Ara Agrippinensium (Cologne) 59, 229
Colophon 213
Comum 212
Constantinople 133, 136
Corcyra 165, 237
Cordoba 213
Corinth 46, 69, 70–1, 77, 106, 121, 132, 135, 164, 166, 167, 176, 180, 182, 194, 200
Corycus 104
Cosa 70
Cotta 150
Cumae 92, 103, 122, 139, 147
Cyene 104
Cyme 238
Cyrenaica 171, 217, 240
Cyrene 122, 136, 171, 211, 233, 238
Cyrrha 97
Cythera 85
Cyzicus 69, 143, 165, 169, 173, 198

Damascus 74, 239
Delphi 33, 52, 167, 179, 237, 238
Dikaia 170
Diopolis 117
Dodona 237
Dolaucothi 65
Doriskos 241
Dover 78
Dyrrhachium 100

Eboracum (York) 59
Elea 169, 211, 213
Eleia 242
Elephantine 124
Eleusis 51
Elis 37, 166
Emporia 122
Ephesus 18, 19, 75, 76, 121, 136, 151, 182, 209, 210, 216, 218
Epidamnus 121
Epidaurus 40, 52
Epirus 93, 226
Eresus 212
Eretria 59, 139, 165, 170

Gades (Cadiz) 78, 111, 120, 122, 143, 149, 169
Gela 142, 167, 239
Genetiva Iulia 16
Gortyn 9

Hadrumentum 232
Halicarnassus 135, 176, 211, 237, 245
Herculaneum 48
Herpa 60
Hierapolis 53, 63, 221
Housesteads 33, 41, 58

Ilipoula 169
Irni 16

Jerash 63, 221
Jerusalem 136, 211

Kabeira 63
Kane 118, 120
Katakekaumene 119
Kato Zakro 40, 50
Kirrha 33
Knidos 69, 71, 79, 127
Knossos 40, 42, 50, 54, 57, 58, 158, 159, 166, 177
Koloe 118, 119
Korakesium 129, 130
Kyeneion 118

La Malga 60
Labraunda 144
Lade 239
Lampsacus 190, 212
Laodicea 117
Las Médulas 64–5
Laurentum 29, 221
Lechaion 69
Leontini 165
Leptis 150
Leuke Kome 116
Liguria 148, 215
Limyra 144
Limyrike 125
Lindum (Lincoln) 47
Lixus 150
Locris 19
Lugdunum (Lyon) 47, 105

Malaca 122
Malao 115, 118
Malia 42
Marea Philoxenite 247
Massilia (Marseille) 68, 69, 75, 92, 102, 121, 122, 165, 191, 212
Mecca 79, 80
Megalopolis 212

Mégara 42, 43, 50, 69, 99, 114, 149, 180
Meliboia 103
Messana (Zankle) 135, 167
Metapontum 165
Methyma 136
Miletus 71, 164, 165, 182, 208, 212, 241
Minnagara 116
Minturnae 20
Moscha 125
Mosyllon 117
Motya 32, 77, 171
Mundu 116, 118, 119
Muza 103, 117, 119
Mycenae 50, 163
Mylae 188, 197
Myonessos 244
Myos Hormos (Mussel anchorage) 118
Myra 124, 144
Myrtos-Pyrgos 40
Mytilene 69, 76, 179

Narbo 120
Naukratis 169
Naulochus 188, 193, 197, 247
Neapolis (Naples) 20, 41, 70, 76, 145, 245
Nelkynda 118
Nicea 29, 210
Nicomedia 77, 208
Nikopolis 195
Nîmes 30, 47
Nimrud 42
Novaesium (Neuss) 59

Okelis(Cella) 119, 126
Old Salpia 32
Olympia 52, 53, 229
Olynthos 44, 48, 57
Omana 118, 125
Opone 117, 229

Palaikastro 39
Palatia 112
Palmyra 13, 24
Panopeus 17
Paphos 76, 102
Patale 120
Patrae 90
Pella 182, 243
Pergamon 11, 42, 44, 45, 102, 210
Petra 116
Pharmakousa 130

Pharsalus 79
Phaselis 166, 171–2
Philae 135
Phocaea 169, 239
Phokis 169
Piazza Armerina 133
Pithekoussai 165
Plataea 32, 177
Pompeii 39, 48, 52, 145, 150, 248
Pons Aelius (Newcastle-upon-Tyne) 203
Porto Paulo 150
Pozzuoli 70, 120
Praeneste 208
Privernum 190, 246
Ptolemais 131, 136
Ptolemais Theron 117
Puteoli 70, 121, 147, 197
Pylos 42, 54, 92, 158

Ratae Corieltauvorum (Leicester) 55
Ravenna 32, 59
Reate 214
Rhapta 126
Rhegium 121, 135, 169
Rhypes 170
Rusellae 198

Salamis (Cyprus) 182
Salmydessus 68
Samosata 211
San Anastasia 112
Seleukia 212
Sestos 78
Side 166
Sidon 76, 122, 153, 182, 235
Sigerus 120
Sikyon 51
Sinope 71, 122, 141, 144
Siwah 243
Skione 93, 240
Sparta 2, 52, 90, 114, 162, 178, 179, 225
Spice Port 118
Stagira 208
Sullectum 78
Sulmo 105, 227
Susa 62
Sybaris 166
Syene 118
Sygaros 117
Syracuse 32, 70, 90, 97, 121, 122, 132, 142, 143, 144, 167, 170, 181, 183, 187, 208, 225

Tanagra 135
Taposiris Magna 229
Taras (Tarentum) 167–8, 235, 238, 244
Tarsus 184, 210, 243
Tartessus 169
Terina 171
Thebes 28, 32, 179
Thelepte 248
Thermopylae 175
Thessaloniki 221
Thina 118
Thurii 150
Timgad 59
Tingis 122
Tipha 89
Tiryns 50, 158, 236
Tivoli 53

Tourdetania 149
Trapezous 69
Troia 150
Troy 67, 71, 85, 90, 91, 92, 97, 100, 162, 163, 168, 192, 235
Tylissos 40, 42
Tyre 69, 75, 122, 151, 153, 181, 182

Utica 122

Veii 133
Venafrum 17
Vercovicium 33, 41
Vindobona (Vienna) 59
Viterbo 49

Zakik 247
Zama 189, 247

Territories, Provinces, and Countries

Achaea 170
Africa Proconsularis ("Libya"/North Africa) 13, 38, 41, 60, 66, 78, 85, 86, 97, 104, 111, 115, 121, 123, 124, 142, 149, 158, 164, 189, 215, 224, 228, 231, 233
Aithiopia 116–17, 149
Algeria 39, 60
Anatolia 53, 63, 164, 232
Apulia 32, 136, 163
Arabia Felix (Eudaimon) 116, 117, 118, 119, 120, 126, 149
Asia (Minor) 11, 24, 41, 42, 69, 71, 75, 97, 99, 113, 114, 129, 133, 136, 141, 144, 149, 166, 169, 178, 182
Assyria 65, 153, 235
Attica 131, 164, 178, 179, 236

Bactria 118
Boeotia 59, 78, 144, 162, 236
Britain 39, 46, 47, 56, 60, 65, 92, 111, 133, 134, 147, 197, 198, 199, 200, 201, 202, 203, 234
Brittany 104
Bulgaria 113
Bithynia 29, 77, 136, 186

Callaecia 65
Campania 46, 145
Cappadocia 60

Caria 144, 243
Cilicia (Rough Cilicia) 34, 60, 104, 105, 107, 126, 129, 190, 203, 211, 230, 243

Dalmatia 122, 150

Egypt 13, 29, 32, 68, 69, 80, 85, 88, 90, 97, 102, 107, 111, 113, 114, 115, 116, 117, 118, 124, 125, 126, 131, 137, 143, 144, 169, 170, 172, 186, 195, 223, 229, 244
Etruria 14, 45, 49, 52, 70, 105, 111, 114, 170, 198, 227

Galilee 234
Gaul 63, 68, 77, 122, 133, 152, 198, 201, 235
  Gallia Narbonensis 89, 121

Hispania (Spain) 34, 41, 47, 81, 106, 115, 132, 133, 164, 169
  Hispania Baetica 16, 149, 150, 211, 228
  Hispania Citerior 121
  Hispania Tarraconensis 65

Iberia 16, 149
India 7, 88, 96, 116, 117, 118, 125, 126, 151
Ionia 165

Italy 11, 19, 24, 33, 46, 59, 61, 65, 68, 70, 71, 97, 102, 108, 111, 117, 132, 133, 143, 147, 163, 164, 165, 167, 244

Judea 69

Lavinium 237
Levant 235
Libya 69, 85, 99, 120, 136, 149, 158, 186, 222
Lower Moesia 78
Lusitania 65, 106, 147, 149, 150, 248
Lycia 8, 113, 144, 148, 235, 243

Macedon 33, 44, 48, 166, 182, 186, 230, 235, 243, 244
Magnesia 103, 241
Mauretania 29, 153
Mesopotamia 24, 60, 69, 113, 114, 220
Mysia 91, 99

Numidia 56, 105, 227

Pamphylia 178
Pelasgia 103
Peloponnese 42, 51, 52, 54, 85, 170, 235
Persia 93, 178, 199, 241
Phoenicia 99, 151, 238
Phrygia 60, 69, 122

Scythia 107, 135
Somalia 126, 129, 230
Syria 41, 97, 172

Thrace 99, 100, 133, 144, 178, 244
Thessaly, 78, 163, 225
Transpadana 49
Tunisia 235

Umbria 106

Mountains and Ranges

Aetna 192, 246
Alps 216
Athos 76, 103
Aurès 60

Cameroon 86

Caucasus 149

Olympus 99

Taurus 141

Capes and Peninsulas

Pelorus 108

Sigeum 97

Trafalgar 197, 247

Islands

Aeaea 67
Amorgos 57
Andros 244
Ares 100–1

Ceos 11, 137
Cheldoniae 141, 148
Chios 147, 178, 211

Corfu 165
Corsica 169
Cos 171
Crete 9, 39, 40, 42, 50, 71, 85, 100, 114, 124, 127, 144, 146, 151, 157, 158, 159, 177, 236, 237
Cyclades 11, 57, 60, 94, 127, 170, 236
Cyprus 102, 113, 114, 115, 121, 141, 144, 146, 149, 178, 182, 232, 243

Cyzicus 148

Delos, 40, 41, 69, 129, 137, 152, 178
Dodecanese 185–6

Euboea 24, 59, 170, 178, 239, 241

Gyaros 137
Gymnesian 120

Herakles 1, 2, 11, 37, 85, 86, 90, 135, 143, 150, 171, 182, 188, 194, 231, 235

Ikaria 115
Ithaca 92, 101

Kasiterides 122

Lemnos 103
Lesbos 69, 147, 178, 179
Leucas 108
Leukate 92

Majorca 120
Malta 124
Melos 179
Menuthias 115, 141
Minorca 120
Myrína 103

Naxos 94, 165, 242

Ogygia 106
Ortygia 144, 157, 167

Parius 148
Paros 11
Pharos 68, 75, 78–81, 107
"Purple" 153

Rhodes 2, 87, 113, 121, 122, 124, 127, 128, 129, 130, 132, 136, 147, 152, 182, 185, 186, 222, 226, 235, 244

Samos 42, 59, 69, 115, 137, 209, 212
Samothrace 52, 183, 184, 185, 186, 244
Santorini 41, 158
Sardinia 120, 188
Scheria 49
Sicily 32, 67, 70, 100, 103, 108, 111, 113, 114, 124, 132, 143, 150, 166, 167, 169, 171, 180, 183, 187, 188, 193, 209
Skyros 178
Spice Islands 118, 119
Syros 137

Thasos 11, 17
Thera 122
Thrinacia 142

Ultima Thule 202

Harbors and Ports

Barbarikon 118, 125

Kenchreae 106

Munychia 71

Ostia/Portus 67, 68, 69, 70, 73, 74, 78, 79, 87, 97, 112, 120, 121, 122, 125, 133, 134, 145, 228, 231

Piraeus 67, 69, 71, 72, 90, 131, 184
Portus Vinarius 112, 228

Rivers

Almo 187, 244
Alpheius 37
Antioch-on-the-Orontes 46
Asopus 32, 135
Avon 57

Cayster 75
Clitumnus 106
Cothi 65

Danube 105
Durias 216

Ebro 20
Eleutherus 170
Eridanos 58
Euphrates 60, 75, 118, 187
Eurymedon 178, 242

Galaesus 187, 244
Ganges 106, 118

Hiberus 20

Indus 75, 106, 118
Issjel 77
Ister 26, 105, 106, 227
Istria 26

Jerash (Jordan) 63

Kaikos 99

Lixus 86

Mallos 60
Melas 60
Moselle 77

Nile 27, 28, 76, 85, 105, 118, 124, 135, 136, 140, 187, 223, 226

Orontes 46

Phaselis 70, 166, 171, 172
Phasis 124
Po 102, 144

Rhine 77, 105, 134, 187, 201, 228, 248
Rhone 105, 132, 134, 135, 147, 148, 201, 229

Saône (Arar) 77, 134
Severn 60
Strymon 144
Styx 135

Terracina 77
Tiber 45, 59, 60, 68, 73, 77, 135, 148, 187, 199, 217, 244
Tigris 75, 212
Tomerus 93

Zab 42

Lakes

Avernus 77

Como 140
Copais 144

Fucine 59–60, 200

Lucrine 147

Maiotis (sea of Azov) 141

Ptechae, 59

Gulfs/Bays

Ambracia 142
Arabian 116

Caietas 92
Cymean 239

Kopais 59

Lokrinos 147

Maeotic 104, 141
Malic 98

Pierian 98

Sachalitas 117
Salamis 86, 93, 149, 175, 176, 177, 181, 197,
    199, 200, 222, 237, 239, 241, 243
Saronic 71, 170, 175
Syrtis, 27, 68, 78, 107

Straits

Artemision 175, 241

Cimmerian 135

Hellespont (Bosporus) 102, 107, 114, 128,
    135, 143, 165, 190, 197

Pillars of Herakles (Hercules/Strait
    of Gibraltar) 1, 2, 85, 86, 121, 130,
    200

Messina, 120, 121, 143, 226

Seas and Oceans

Adriatic 68, 105, 123, 137
Aegean 11, 17, 37, 40, 68, 85, 99, 103, 107,
    113, 115, 136, 157, 163, 164, 166, 176, 177,
    178, 181, 199, 237
Atlantic 2, 102, 143, 150
Azanian 170

Baltic 113

Caspian 114, 187, 220
Cimmerian 135

Erythraean (Red Sea, Gulf of Aden, Arabian)
    68, 76–7, 85, 96, 103, 104, 111, 115, 116,
    117, 118, 119, 120, 129

Marmara 78, 165
Mediterranean ("our" sea) 1, 2, 102, 120, 123,
    130, 138, 157, 195, 229 *et passim*

North Sea 78

Phoenician 170
Pontus (Black/Euxine Sea) 68, 101, 102,
    105, 107, 125, 128, 136, 152, 226, 235,
    246

Southern 85

Index of Authors and Sources

Achilles Tatius
   Leukippe and Klitophon 80, 121, 136–7
Aelian of Praeneste (Claudius Aelianus)
   Nature of Animals (NA) 135, 140, 141, 143, 144, 170, 227, 233, 239, 240
   Various Histories 181
Aelius Aristides
   Oration Regarding Rome 125, 230
Aeschylus of Athens 197
   Eumenides 179
   Persians 177, 241, 243, 247
   Prometheus Bound 215
   Seven Against Thebes 179
   Suppliant Maidens 224
Aëtius 216
Alcaeus 179
Ammianus Marcellinus 53
Andocides
   On the Mysteries 96
Andreas, 26
Apicius
   On Cooking 239
Apollonius of Rhodes
   Argonautica 91–2, 100, 101, 102, 103, 108, 109, 226, 227
Appian of Alexandria 109
   Civil Wars 121, 130, 194, 195, 246, 247
   Mithridates 230
   Punika 222
   Roman History 106
   Spanish War 225
Apuleius
   Metamorphoses 106
Archestratus of Gela 142–3, 233
Archimedes 61, 132, 148, 208, 217
Aristophanes of Athens
   Acharnians 233
   Clouds 242
   Knights 87, 242
   Wasps 180
Aristotle of Stagira 139, 216
   Athenian Constitution 10, 58, 241
   Generation of Animals 147, 208

   History of Animals 141, 143, 144, 146, 208, 232, 233, 240
   Metaphysics 209, 244
   Meteorology 29–30, 96, 97–8, 99, 218, 225, 226
   Politics 31, 218, 239
   Posterior Analytics 106
Arrian of Nicomedia
   Anabasis 92, 227
   Indika 86, 93, 106, 224, 232
   Periplus of the Black Sea 68, 69, 91, 101, 102, 222, 225
Artemidorus of Ephesus
   On the Interpretation of Dreams 151, 221
Asconius 246
Athenaeus of Naucratis
   Table Talk 77, 103, 132, 142, 143, 183, 184, 209, 221, 231, 232, 233, 234, 239, 243, 244
Augustus
   Res Gestae 199, 245, 246
Aulus Gellius
   Attic Nights 99, 233
Aulus Hirtius [?]
   Alexandrian War 28
Ausonius, *Mosella* 63

Callimachus
   Against Apollonius 237
Cassiodorus
   Letters 38
   Varia 148
Cato the Elder
   On Agriculture 13, 104, 217
Catullus 248
Celsus, Aulus Cornelius 25, 34, 209, 218
Cicero, Marcus Tullius
   2 Verres 240
   On Divination 237
   On Oratory 245
   To Atticus 232
   To Quintus 13, 48–9, 61, 215, 217, 220
Claudian
   Gothic War 227

Columella 13, 32, 144, 145, 146, 217, 218, 233, 234
Cornelius Nepos
Iphicrates 33
Themistocles 222
Ctesibius of Alexandria
Pneumatics 62, 221

Demosthenes of Athens 242
Against Aristokrates 239
Against Callicles 9
Against Kallipos 136
Against Leptines 114
Dio Cassius 16, 60, 77, 93, 182, 197, 198, 199, 200, 201, 218, 223, 230, 233, 245, 246, 247, 248
Dio Chrysostom
Oration 225
Diodorus Siculus 71, 76, 121, 128, 144, 181, 182, 222, 230, 233, 239, 240, 241, 242, 243, 244
Dionysius of Halicarnassus
Roman Antiquities 135, 237, 245
Diophanes of Nicea
Geoponika 29
Dioscorides of Anazarbus 4, 24, 34, 149, 151, 171, 217, 218, 234, 240

Eratosthenes 109, 120
Euripides
Helen 90
Eusebius
Constantine 203

Firmicus Maternus
Error of Profane Religions 203
Florus 201, 246
Frontinus, Sextus Julius
On the Aqueducts of the City of Rome 2, 12, 13, 14, 16, 17, 18, 19, 24, 45–6, 48, 52, 216, 218, 220, 247
Stratagems 33, 34, 218
Fronto
On Wild Salt-Animals 140

Galen of Pergamon
Antidotes 218
Good and Bad Juices 217
On Preserving Health 217
On the Therapeutic Method 217
On Theriac for Piso 217
Greek Anthology 135, 220, 221

Hanno of Carthage 237
Herodian
Histories 217, 227
Herodotus of Halicarnassus
Histories 20–1, 32, 42–3, 52, 69, 76–7, 85, 86, 93, 104, 111, 135, 164, 167, 169, 170, 175, 176, 179, 188, 197, 216, 217, 224, 226, 227, 237, 238, 239, 240, 241, 242, 243, 247
Heron of Alexandria 95, 210, 221
Hesiod of Acragas
Theogony 210, 239, 248
Works and Days 96, 225, 229
Hipparchus 109, 210
Hippocratic Corpus
Airs Waters Places (*AWP*) 23, 24, 25, 27, 31, 210, 216, 217, 218
Aphorisms 216
Diseases 216
Epidemics 216
Humors 216
Regimen 217
Homer 98, 186, 206, 222
Iliad 78, 132, 139, 142, 162, 163, 211, 215, 224, 232, 236, 237, 242
Odyssey 67, 88, 91, 92, 94, 100, 101, 108, 113, 126, 127, 142, 164, 211, 220, 224, 227, 235, 236
Homeric Hymns 245
Horace (Horatius Flaccus)
Epodes 191, 193
Epistles 235
Odes 101, 201, 247
Satires 77, 233, 234
Hyginus
Fabulae 224, 240

Ibn Jubayr 79, 81

Josephus, Flavius, of Jerusalem
Jewish Wars 68, 79, 136
Life 137, 232
Juba II of Mauretania 29
Julius Caesar, Gaius
Civil Wars (*BC*) 101
Gallic Wars (*BG*) 92, 104, 134, 201, 218, 229
Justinian
Digest of Roman Law 12, 13, 14, 15, 19, 20, 136, 137, 214, 216, 231, 245
Juvenal 78, 87, 88, 147, 148, 233

Latin Anthology 201
Libanius
　Oration 136, 203
Livy
　From the Foundation of the City of Rome
　　93, 104, 183, 188, 189, 190, 216, 225,
　　237, 245, 246
Longus
　Daphnis and Chloe 90
Lucan
　Pharsalia 89, 92, 94, 106, 107, 193,
　　223
Lucian of Samosata
　Charon 135
　Dialogues of the Sea Gods 135–6, 238
　Hippias, Or the Bath 56
　How to write History 80
　Jupiter Rants 137
　Navigium 96, 121, 131, 225
Lucilius 135
Lucretius
　On the Nature of Things 226
Lycophron
　Alexandra 104, 163, 237

Manilius
　Astronomica 107, 150, 171, 247
Marcian of Heraclea
　Epitome of Menippus' Periplus of the Inner Sea 226
Martial 125, 147–8, 233, 247
Mela, Pomponius
　Chorography 26, 68, 86, 103, 105, 224,
　　227
Menippus of Pergamon
　Periplus of the Inner Sea 102, 226

Olympiodoros
　ad Aristotle *Meteorology* 226
Oppian of Anazarbus
　On Fishing, 139, 141, 225, 232, 233,
　　240
Oreibasios 217, 218
Orpheus
　Argonautika 225
Orphic Hymns 207
Ovid (Publius Ovidius Naso)
　Halieutica (*On Fishing*) 248
　Heroides 91, 94, 100
　Metamorphoses 8, 78, 88–9, 100, 101, 106,
　　224, 225, 227, 237, 239
　Trista ex Ponto 246

Palladius 218
Paul, Acts of the Apostles 103, 109, 124, 130,
　137, 230
Pausanias
　Description of Greece 17, 33, 42, 71, 89,
　　167, 216, 217, 220, 225, 235, 238, 241,
　　242
Periplus of the Erythraean Sea 68, 104,
　115–19
Periplus of the Inner (Mediterranean) *Sea*
　102
Petronius
　Satyricon 146
Peutinger Tablet 227
Philippus
　Palatine Anthology 246
Philo of Alexandria
　On the Confusion of Tongues 61
Philo of Byzantium
　On Harbor Construction 68
Philostratus the Athenian
　Apollonius of Tyana 88, 121, 136, 224,
　　227
　Lives of the Sophists 181
Philostratus the Younger
　Images 232, 233, 238
Pindar of Thebes
　Isthmian Odes 124, 176, 224, 241
　Nemean Odes 25–6
　Olympian Odes 26
　Pythian Odes 220
Plato of Athens
　Alcibiades 224
　Apology 242
　Critias 25, 217
　Euthydemus 240
　Gorgias 31, 88, 218
　Laws 9, 10, 11–12, 31, 60, 88, 140, 177,
　　215, 217, 218, 224, 244
　Phaedo 207
　Politics 143, 144, 242
　Republic 88, 142, 180, 224, 242
　Statesman 242
　Symposium 213, 242, 244
Plautus
　Truculentus 233
Pliny the Elder 8, 14, 23, 24, 25, 26, 27, 28, 29,
　30, 34, 38, 39, 41, 45, 59, 60, 64–5, 68,
　73, 79, 86, 98, 99, 100, 111, 120, 121,
　125, 126, 133, 139, 141, 144, 145, 147,
　148, 149, 150, 151, 152, 153, 166, 170,
　181, 191, 212, 215, 217, 218, 219, 224,

Index of Authors and Sources 285

225, 226, 228, 229, 230, 232, 233, 234, 235, 237, 240, 245, 247
Pliny the Younger
   Epistles 60, 70, 78, 106, 136, 140, 215, 221
Plutarch of Chaeronea
   Alexander 54, 153, 235
   Antony 140, 153, 182, 218, 246
   Caesar 247
   Cato the Elder 18, 26, 216
   Cimon 242
   Cleverness of Animals 62, 140, 212, 232, 233
   Demetrios 133, 182, 230
   Greek Questions 216, 232, 237
   Isis and Osiris 217
   Malice of Herodotus 176
   Marius 121
   Pericles 88, 114, 241
   Pompey 104, 129–30, 225, 230, 231, 246
   Precepts of Statecraft 224
   Roman Questions 105
   Romulus 247
   Solon 9, 179
   Table Talk 149, 217, 224
   Themistocles 10, 170, 241
   Theseus 178
   Timolean 135
   Whether Fire of Water is more Useful 2, 206
   Younger Cato 246
Polyaenus
   Strategems, 32–3, 244
Polybius of Megalopolis
   Histories 90, 104, 109, 127, 129, 181–2, 187, 188, 189, 224, 227, 229, 230, 243, 244
Procopius
   Buildings 105–6
   War Against the Goths 63, 121
   War Against the Vandals 121
Propertius 92, 95, 224, 247
pseudo-Aristotle
   Economics 230
pseudo-Aristotle
   Mechanical Problems 61, 94, 95, 225
pseudo-Aristotle
   On Marvelous Things Heard 86, 105, 224, 227, 233
pseudo-Skylax
   Periplus 68, 239
Pytheas of Massilia 2, 10, 102

Quran 81

Seleukos of Seleukia 2, 226
Seneca the Younger (Lucius Annaeus Seneca)
   Epistles 55, 89, 151
   Natural Questions 4, 23, 99, 216, 225, 226, 227
   Octavia 201
   On Clemency 233
Servius
   ad Aeneid 237
Silius Italicus
   Punica 103, 202
Sophocles of Athens
   Assembly of the Greeks 224
   Oedipus at Colonus 179
   Oedipus Rex 180
   Women of Trachis 231
Soranus 218
Strabo 2, 4, 16, 45, 59, 60, 63, 68, 69, 70, 71, 75, 77, 78, 86, 102, 107, 109, 116, 118, 120, 122, 128, 129, 135, 141, 144, 147, 149, 150, 152, 153, 170, 172, 201, 212, 216, 217, 218, 219, 222, 223, 224, 227, 229, 231, 232, 233, 235, 237, 238, 240, 246
Suetonius (Gaius Suetonius Tranquillus)
   Augustus 153, 188, 231, 245, 246, 247
   Claudius 60, 73, 77, 124–5, 200, 201, 223, 230
   Gaius (Caligula) 197–8, 198
   Julius Caesar 60, 130, 247
   Nero 55, 77, 145, 235, 247
   Titus 247
   Vespasian 248
Synesius of Cyrene
   Epistles 104, 136, 137, 233

Tacitus (Publius/Gaius Cornelius Tacitus)
   Agricola 202, 248
   Annals 73, 75, 77, 92, 145, 199, 200, 201, 230, 245, 247, 248
   Histories 102
Thales of Abdera 1, 39, 101
Theodosian Codex 229
Theophrastus of Eresus
   Causes of Plants 212, 217
   History of Plants 139, 213, 217, 223
   On the Winds 98
Theopompus of Chios 242
Thucydides of Athens
   History of the Peloponnesian War 34, 51, 71–2, 90, 93, 114, 123, 127, 157, 164–5,

167, 170, 176, 178–9, 180, 200, 222, 225, 230, 237, 238, 240, 241, 242, 246
Tibullus 201
Timosthenes of Rhodes
  On Harbors 68, 87, 99

Valerius Flaccus
  Argonautica 202
Varro, Marcus Terrentius
  Naval Books 87, 99, 226
  On Farming 31, 97, 105, 145, 233
Vegetius 26, 33, 87, 96, 97, 105, 107, 123, 224, 225, 226, 227, 229
Velleius Paterculus 103, 218, 247

Vergil (Publius Vergilius Maro)
  Aeneid 89, 92, 94–5, 100, 107, 108, 109, 187, 196, 203, 222, 244, 245, 246
  Georgics 61
Vitruvius
  On Architecture 23, 25, 31, 34, 38, 39, 45, 48, 54, 62–3, 68–9, 70, 97, 148, 152, 216, 217, 218, 235

Xenophon of Athens
  Anabasis 68
  Economics 90
  Hellenika 90, 127, 239
  Symposium 242

General Index

Achilles 78, 121, 136, 137, 224
Actaeon 218
Actian games 195
Aeëtes 108
Aeneas 89, 92, 94, 100, 107, 108, 187, 192, 196, 222, 237
aeölipile 63
Aeolus 88
Agamemnon 162, 163
agate 113, 117
Agatharchades 232
Agricola 202, 248
Agrippa, Marcus Vipsanius 15, 16, 52, 59, 101, 193, 195, 197, 199, 203, 227
Ahenobarbus, Lucius Domitius 145, 190, 245
Ajax 162, 163, 176
Albacore 143
Alban 53, 244
Albinovanus Pedo 201
Alcibiades 224
Alcinous 49
Alcmaeon 24
alder 38
Alexander of Macedon 2, 3, 33, 59, 92, 93, 116, 120, 127, 153, 175, 181, 182, 235, 243
allec 150
Allectus 203, 248
almonds 28, 113
amber 113
Amenhotep III 85
Amenophis 28
amethyst 153
Ammon 149, 237, 243
ammonia 146
Amphictyonic League 33
Amphitrite 171, 194, 195
amphorae 112, 114, 115, 122, 131, 132, 134, 151, 228, 231
amulets 147, 151, 159
Anchises 94, 195
anchovies 142
Antigonos II Gonatas 244
Antigonos Monophthalmos 33, 182, 183

Antiochus III 244
Antoninus Pius 53, 140, 243
Antony (Marcus Antonius) 33, 130, 140, 153, 182, 194, 195, 218, 243, 245, 246
Aphrodite 49, 132, 163, 183, 184, 232, 244
   Aphrodite Ourania 183
apoikiai 163, 164, 165, 169, 170, 172, 238
   see also settlement
Apollo 8, 11, 33, 52, 92, 144, 166, 167, 169, 171, 182, 195, 196, 237
   Apollo Actius 195
aquaphobia (rabies) 171
aqueducts 2, 12, 13, 14, 16, 17, 18, 19, 24, 25, 30, 34, 39, 41, 42, 43, 44, 45, 46, 47, 48, 49, 52, 53, 55, 59, 63, 64, 65, 66, 137, 210, 216, 220
   Aqua Alsietina 199
   Aqua Anio Vetus 46
   Aqua Appia 46
   Aqua Claudia 46
   Aqua Marcia 46
   Athens 48
   Lindum (Lincoln) 47
   Lyon 47
   Madra Dağ 44
   Olynthos 48
   Pont du Gard 47
   Segovia 47
   Serino 41, 46
arbutus 141
Arcturus 89, 96, 97
Arethusa 144, 157, 167
Ares 163, 176
Argo 1, 89, 91, 100, 102, 103, 104, 158, 180
Ariadne 94, 225
Ariarathes V (?) 60
Arion 167
Aristagoras of Miletus 164
arsenic 65
Arsinoë II 53, 183, 244
Artemisia of Halicarnassus 176
Artemis 8, 144, 218
Asklepius 52, 171
asphalt 50, 114, 220

Assurnarsipal II 42
Athena 163, 165, 166, 178, 180, 181, 182, 183, 238, 239, 240, 242, 243
   Athena Polias 178
athletes 26, 54, 140, 217
Atreus 90
Attalus II Philadelphos 75, 76, 186
Attis 169
Augeas 37
Augustus 14, 16, 17, 46, 52, 60, 70, 111, 130, 153, 188, 194, 195, 196, 197, 199, 200, 201, 203, 215, 218, 220, 231, 235, 243, 245, 246, 247, 248
   see also Octavian
Aulus Plautius 198, 202
Aurelian 229
Auriga 106
Avillius Caimus Patavinus, Gaius 49, 220

Bacchus 106
ballast, ballasted 92, 111, 113, 114, 132, 133, 225
barges 73, 76, 111, 133, 134, 135, 153
barley 29, 165
Barrea Soranus 75
baths/bathing 2, 15, 16, 23, 24, 25, 28, 33, 48, 53–7, 66, 106, 132, 201, 218
bechion 38
beer 24
bilge 62, 131
biremes 199
bitumen 25
boathouses 69
bogs 31, 33
Bomilcar 225
bonabos 86
Boötes 106, 107
Boreas (wind) 100
Boudicca 201
breakwater 69, 70, 73, 76, 77, 79, 222
Britannicus 198
Brutus, Sextus Iunius 106, 191, 245
buoyancy 217

cabotage 119, 120
Cadmus 180
Caesar, Gaius Julius 2, 26, 28, 33, 34, 46, 60, 69, 79, 89, 92, 93, 94, 100, 101, 104, 130, 134, 195, 198, 199, 201, 202, 220, 229, 247
caldarium 56
Caligula (Gaius) 73, 111, 197, 198, 247

Callias 178, 242
Callisto 218
Calypso 101, 106
camels 118, 119, 232
canal 2, 14, 28, 42, 58, 60, 67, 76-8, 82, 85, 200, 223
   Panama 78
canoes 85, 141
Canopus 107
Capella 106
Capitoline 46, 198
capotage 119
Caracalla 149
Caratacus 201
Carausius 202, 203, 243, 248
caravans 13, 232
cardium 158
Cassander 243
Cassandra 163
cassia 117
Castor 168
Cataenei brothers 192
Cato the Elder 13, 17, 18, 26, 105, 216
cedar 113
cement 41, 150
centaur 24, 94, 135, 231
Ceyx 88, 109, 225
Chaerephanes 59
chalcedony 141
Charon 135
charts 69, 107, 109, 108, 227
Charybdis 108
chimpanzees 86
China 231
cholera 26
Cicero, Marcus Tullius 12, 48–9, 61, 99, 136, 137, 215, 217, 220, 242, 246
Cicones 101
Cimon 178, 242
Cincinnatus 189
Circe 67, 94, 108
Claudian 198, 202, 227
Claudius 60, 70, 73, 124, 125, 198, 200, 201, 202, 208, 212, 223, 230, 232, 247
Clemens, Samuel (Mark Twain) 227
Cleopatra 140, 153, 195, 243, 246
Clodius Pulcher, Publius 246
Clytemnestra 163
coastlines/coastal waters 1, 28, 45, 67, 68, 70, 76, 79, 85, 86, 92, 97, 102, 103, 104, 105, 107, 111, 113, 114, 115, 119–20, 122, 123, 125, 138, 170–1

cockles 152
Commodus 29, 80, 235
compass 98, 99, 101
conch 171, 201
concrete 45, 47, 55, 70, 73, 150
Constans 203
Constantine 203
cookbooks 142
copper 113, 117, 118, 133, 243
coral 29, 151, 158
coriander 113
Corinthian War 90, 127
cork 140, 141
corvus (grappling gangway) 188, 245
cotton 117
crabs 147, 171, 172, 202, 240
Crates of Boeotia 59
crayfish 171
crocodiles 86, 115
Croesus 176
Cronus 86
cumin 113
curator aquarum 16, 45, 46
currents (surface currents) 28, 69, 70, 76, 102, 107, 118, 120, 123, 138, 143, 147, 186, 227
cuttlefish 141
Cyclops 67, 126
Cyex 225
Cynosura 106, 107
cypress 140
Cyrus II 68

dams 2, 60
Danaus 90
Darius I 21, 76, 175, 223
deforestation 25, 205
Deianeira 231
Delian League 127, 172, 175, 177–8
Demeter 11, 51
Demetrios Poliorcetes 77, 127–8, 175, 181, 182, 183, 184, 191, 243, 244
Diana 8, 218
diarrhea 33
didachoi 127, 182
Diocletian 153, 235, 243
diolkos 77
Diomedes 163, 237
Dionysius of Syracuse 32, 181
Dionysus 49, 194
Dioscuri 168, 183, 238

divers/diving 92, 93, 96, 113, 115, 122, 139, 140, 148, 171, 180
docks/dockyard 69, 71, 74, 75, 87, 132, 222
dogs 147, 151, 192, 235
dogs/dogfish (sharks) 142, 159, 191–2
dolphins 73, 79, 136, 139, 141, 158–9, 166–8, 173, 183, 189, 191, 197, 203
Domitian 53, 80, 125, 148, 202
Draco 224
drains/drainage 10, 15, 20, 21, 32, 39, 57, 58–9, 60, 69, 76
dredging 67, 74–6, 82
dropshafts 47
Drusus 77, 247
Duilius, Gaius 188, 245
dykes 32
dysentery 26, 150

eagle 103, 166, 169, 171, 182, 183, 193
ebony 113
ecliptic 97, 99
eels 141, 143, 144, 171
Elagabalus 248
elephants 86, 117, 118 133, 198
emerald 141
Epicurus/Epicureanism 4, 209, 211, 226
Erichthonius 180
erosion 25, 189
esparto 132, 140
estuaries 60, 73, 141, 144
Eupalinos of Mégara 43

faience 113, 158
fennel 29
fens/fenland 32, 60, 205
fermentation 150
ferries 135, 231
figs 38, 113, 146
filters/filtration 28, 29, 30, 35, 41, 48, 149, 218
fish tanks 132, 143, 144, 147
fisheries 144
fishponds 20, 144, 145, 146
flounder 145
frankincense 118
frogs 8, 38, 207
Fulvia 243

galingale 140
gangways 132, 134, 188
gardens 18, 41, 49, 132, 148, 234
garlic 29

garum 112, 139, 148, 150, 151, 154, 234
gazelle 133
Gelon of Syracuse 167
Gemini 168
Germanicus 201, 247
Gibraltar 85, 121
Gilgamesh 226
gilt-heads 145
glaciers 205
glass 111, 113, 117, 151, 171, 229, 233
Gorgon 201, 202
gorillae 86
gorse 65
grain 7, 61, 63, 71, 72, 73, 80, 112, 114, 122, 123, 124, 125, 127, 131, 132, 133, 135, 137, 138, 178, 180, 190, 234
granite 69
grapes 112, 113, 178
grapnels 193
Gratian 122, 123

Hadrian 33, 41, 53, 56, 60, 223
Hannibal 104, 189, 247
Hanno 2, 86, 164, 224, 237
harbors 2, 3, 12, 45, 67, 68, 69, 70, 71, 72, 75, 76, 78, 82, 85, 104, 112, 115, 118, 119, 128, 132, 162, 167, 170, 172, 181, 205, 222, 237, 239
Hasdrubal 188
Hector 162, 163, 237
Helen of Troy 89, 90, 91, 100, 162, 168, 235
Helenus 108–9
hellebore 33
hemp 40, 132, 141
Hephaestus 206
Hera 242
Herakles 1, 2, 11, 37, 85, 86, 90, 135, 143, 150, 171, 182, 188, 194, 231, 235
Hercules 63, 81, 130, 148, 200, 235
Hermes 245
Hermione 153
Herod 70
Herodes Atticus 53, 181
Hesiod 96
hexereis 181
Hieron II of Syracuse 132, 143, 144, 183, 187
Hippalos (wind) 120, 126, 229
hippocamps (seahorses) 159, 167, 194, 236
hippopotami 86
Horatius Cocles 189
Hortensius, Quintus 145
hurricanes 23, 24, 216

hushing 63, 65
Hyades 106
hydra 171
hydromel 29, 34, 147
hypocaust 54, 55, 221

ibex 133
ichthyphagoi 140, 147
Idomeneus 163
impluvia 41, 48, 65
Ino 167
irrigation 7, 16, 10, 42, 44, 60–2, 76, 206, 216
Isis (goddess) 80, 106, 217
Isis (ship) 96, 131, 225
ivory 113, 118, 133, 234
ivy 38

Janus 188, 190, 245
Jason 1, 89, 91, 108
Juba II 153
Julia Mammaea 243
juniper 29
Juno 24, 196, 224
Jupiter 101, 137, 149, 193, 202, 203, 224, 237
Justinian 3, 12, 14

keel 89, 101, 113, 131, 159
Kore 51

Laestrygones 67
lagoons 31, 141, 148, 222
lampreys 143
laser 26, 29, 239
latitude 102
Latona 8
latrines 15, 16, 28, 33, 53, 56, 57–8, 59, 66
laurel 29, 189
Lawrence of Arabia 4
lemboi 182
lemon 235
lentils 111
lettuce 29
lighthouses 67, 73, 78–81, 87, 133, 193
lilies 169
limestone 25, 39, 57, 58, 59, 71, 79, 228
liquamen 150, 154
lodestone 101
longitude 102, 226
Lucius Verus 20
Lucullus, Lucius 145, 146
Lysimachus 244

mackerel 141, 150
mahi mahi 139
mallow 34
Marcus Aurelius 20, 25, 53
Mardonius 32
Marius 121
Mars 199, 247
marshes/marshlands 12, 23, 24, 25, 26, 31, 32, 35, 60, 77, 99, 158, 207
Medusa 169, 239
Meleager 145
Melicertes 167, 238
Melqart 151
Menelaus 89, 90, 162
mercury 65
meridian 226
mills 60–3
Minerva 56, 201
mines/mining 16, 25, 33, 37, 61, 63, 64–5, 149, 175, 232
Minos 127, 157, 159
Minotaur 71, 178, 225
Minyan gray-ware 114
Mithridates VI 63, 129
moles 69, 70, 72, 73, 75, 78
mollusks 139, 145, 146, 147, 150, 151, 152, 235
monk seal 169, 209
monsoon 96, 125, 126
moon 80, 89, 98, 147
mosquitos 31, 32
mosses 38
Mt. Testaccio 73, 228
mulberry 215
mules 77
mullet 144
murena eel 141
murex (Tyrian) purple 129, 139–41, 145, 149, 151–4, 235
mussels 139, 141, 146, 147, 148

Narcissus, Tiberius Claudius 60
naumachia 199, 200
naval crown 201
navigators 79, 86, 101, 103–4, 106, 107, 180, 184
Nearchus 93
Necho II 69, 76, 85, 86, 223, 224
Nefertiti 113
Nelson, Horatio 231, 247
Neoplatonism 209
Neoptolemus 78, 163

Neptune 73, 134, 144, 148, 189, 190, 191, 194, 196, 197, 198, 248
Nereid 167
Nero 27, 53, 55, 73, 77, 145, 153, 199, 200, 217, 223, 235, 247
Nerva 45
Nessos 135, 231
Nestor 54
nets 140, 141, 142
Nike 52, 53, 167, 172, 183, 184, 185, 186, 243, 244
Numa Pompilius 191
nymphaea 53
nymphs 49, 167, 207, 238

oak 113
oases 66
obsidian 85, 117
Octavia 194, 195
Octavian 14, 15, 33, 182, 192, 193, 194, 195, 196, 218, 246
  see also Augustus
octopods 139, 142, 158, 168, 170, 171, 239
Odysseus 49, 67, 88, 91, 92, 94, 100, 101, 106, 108, 113, 126, 142, 224
Oedipus 179, 180
oenanthe 34
oikistes 164, 167, 169
Olenian Goat 106
olives 10, 111, 112, 113, 117, 124, 133, 138, 178, 228
onyx 118
orca 232
Orion 89
Orpheus 56, 225
Osiris 217
ostrich 113, 133
overshot wheel 63, 64
owls 165, 166, 207
oysters 139, 146, 147, 233

Palinurus 89, 107, 108
Palladium 163
Pallas 107
palm 196, 227
Panathenaia (Panathenaic Festival) 180–1, 242
parrot wrasse 142, 145
parsley 29
Parthenon 181, 242
Patroclus 215
pear 40

pearls 118, 139, 151, 234
Pegasus 166
Peisistratids 43
Peisistratos of Athens 51, 180
Peleus 171, 172
Pelion 103
Peloponnesian War 90, 92, 114, 177–9, 242
pennyroyal 29
peplis 29
perfumes 54, 55, 133
Pericles 88, 111, 114, 177, 178, 179, 241
Persephone 166, 169
Persian Wars 21, 93, 164, 172, 175–7, 199, 200, 225, 237, 238–39, 240, 241
Phaethon 225, 227
Pheidippides 225
Philip II of Macedon 181, 230, 243
Philoctetes 163
Phoceus 92
Phrixus 101, 226
piddick 147
piers 68, 70, 73, 222
pilgrims/pilgrimage 80, 81, 182
pilot fish 103
pilots 80, 88, 89, 103, 104, 105, 115, 119, 120, 129, 227
Pima 7, 8, 21
pirates/piracy 2, 3, 34, 112, 117, 119, 126–30, 157, 170, 172, 178, 190, 191, 193, 203, 230, 244, 245
Pleiades 89, 96, 106
Polaris 224
polenta 29
pollution 11, 19, 20, 23, 25, 27, 28, 47, 50, 65, 206, 214, 217
Pollux 168, 235
Polycrates of Samos 42
Polynesians 107
Polyphemus 126, 127
pomegranates 113, 166
Pompeius, Sextus 191–4
Pompey (Gaius Pompeius Magnus) 2, 33, 58, 80, 89, 93, 99, 100, 104, 106, 107, 126, 129, 130, 190, 191, 192, 203, 225, 230, 231, 245, 246
Poppaea 124, 230
poppy seed 92
ports 70, 71, 72, 73, 75, 78, 92, 94, 96, 105, 107, 112, 117, 118, 119, 120, 124, 125, 126, 132, 133, 137, 162, 187, 205

Poseidon 80, 104, 136, 167, 169, 171, 178, 183, 184, 195, 238, 239
  Poseidon Pelagaios 183
pozzolana 70, 73
Priam 163
Prometheus 1, 206, 214
Protesilaus 100, 162, 163
Ptolemy I Soter 128, 243
Ptolemy II Philadelphos 68, 76, 78–9, 244
Ptolemy III 132
Ptolemy IV Philopater 26, 77, 132–3, 182
Ptolemy Keraunos 244
pulleys 40, 80
pumps 47, 62, 221
Punic (Carthaginian) Wars 103, 104, 183, 187, 188–90, 225, 245

Qaitbay citadel 81
qanats 42
quadriremes 182, 185, 186, 200
quays 69, 73–4, 134
quinqueremes 188, 199

rafts 85, 106, 113, 118, 135, 137
ravens 103
Rea Silvia 231
reefs 93, 94, 102, 104, 122, 159, 179, 222, 239
rhinoceroses 118
rock-fish 145
Roman Civil Wars 33, 79–80, 89, 93, 100, 106, 130, 190, 243, 246, 247
Romulus 187, 245
rosemary 38
roses 125
rostral column 188, 195, 196, 197, 245

safflower 113
salinity 29, 32, 123, 146
salt 27, 29, 30, 77, 127, 140, 147, 148, 149, 150, 152, 215
salt-fish/salted fish 138, 139, 141, 146, 148–51, 154
sandstone 38
sardines 141, 149
Saturn 86
saunas 55
sawmills 63
Scaevola 189
scallops 139, 167
scarab 113
Scipio Africanus 55, 103, 247
scomber 150

scorpions 240
Scylla 91, 108, 159, 161, 191, 193
sea silk 151
sea urchin 92, 158
seals (animal) 92, 169
sealstones 159–62
seashells 113, 158
seasickness 86
seaweed 139, 140, 146
Septimius Severus 148
servitudes 13, 14, 20, 48
Servius Lupus, Gaius 81
Servius Tullius 153
settlement/"colonization" 2, 7, 46, 86, 103, 120, 122, 127, 157, 162–7, 169, 171, 189, 237
   see also apoikia
Severus Alexander 243
sewers/sewage 28, 39, 45, 57–9, 66, 217
shaduf 61
sharks 142, 159, 192
shearwater 180
shellfish 145, 146, 151, 152, 154
ship of state 179–80
ship sheds 69, 72, 195
shipwrecks 97, 107, 112, 117, 122, 228
   Cape Gelidonya 113
   Fourni 115
   Herodotus Abyssal Plain 132
   Madrague de Giens 112, 131
   Point Iria 113, 114
   Uluburun 113
shipyards 68
shrimp 171
silex 41
silicon 25
silk 118, 133
silphium 26, 217
siltation 68, 69, 74, 75, 76, 103, 105, 223, 228
sinkholes 59
sinter/sintering 25, 34, 46, 50
siphons 44, 47
Sirens 100
Sirius 152
Skiron 99
Skyllias 93, 225
Skyllis 93
slaves 11, 46, 59, 74, 117, 118, 129, 143, 145, 170, 192, 217
sluices 56, 59, 63, 65, 76, 144, 148
snails 147, 152, 154

snakes/serpents 119, 141, 171
snow 24, 27, 98
Socrates 4, 212, 242
sole 145
Solon 8, 9, 10, 179
solstice 96, 98
Sophists 181, 209
Sostratos of Knidos 79
soundings 103
Spartacus 246
sponges 30, 57, 113, 122, 139, 151
springhouses/fountain springs 11, 13, 16, 17, 39, 42, 43, 44, 48, 49–53, 55, 66
stopcock 34, 48
Sulis Minerva 56
sulfur 25, 33, 39, 55
swimming/swimmers 92, 93, 170
Syrakousia 132, 143

tamarisk 141
Tanit 86
Telemachus 92
Telephus 88, 90, 91, 96, 224
terebinth 113
Tethys, 207
Theagenes of Mégara 42, 50
Themistocles 10, 71, 162, 170, 175, 176, 179, 186, 222, 241
Theodosius 123
Theseus 71, 94, 178, 180, 225
Thetis 171, 172
thunnoskopos (tuna lookout) 143
Tiberius 145, 215
tides 77, 92, 99, 102, 104, 145, 146, 149, 201, 209
tin 113, 122, 133
Tiphys 91, 104
Titus 68, 136, 200, 247
tortoises 117, 170, 229
Tower of the Winds 97
Trajan 12, 73, 74, 76, 77, 81, 106
Trimalchio 146
triremes 69, 72, 93, 136, 164, 176, 181, 186, 188, 199, 200, 241
Triton 97, 104
tritons 53, 80, 169, 201
Trogodytes 8, 147
truffles 38
tufa 30
tuna 2, 85, 138, 139, 140, 141, 143, 144, 150, 153, 168, 169, 171, 172, 233

tunnels 42–7, 50, 59–60
turbot 145
turtles 141, 170, 240
Tyros 151

undershot wheel 62, 63
Ursa Major 106
Ursa Minor 86, 106, 107
usufruct rights 13, 21

Valentinian 123
Vandals 121
Vedius Pollo 143, 144
Venus 207, 226
Vespasian 136, 200, 202, 248
Vikings 101, 226
Vitellius 102

walnut 34
water divining 38
waterproofing 41, 45
waterwheels 28, 62, 63, 64
wetlands 59-60, 78, 200
whaling 141

wharf 73
wheat 103, 133, 187
wheels 105, 180, 181
whelk 147, 152
whipstaff 95
willow 38
winches 69
windlasses 131, 132
winds 24, 68, 69, 70, 71, 86, 87, 88–9, 92, 94, 95, 97–101, 102, 104, 107, 113, 114, 118, 120, 123, 124, 125, 136, 138, 142, 147, 181, 185–6, 196
windstorms 115
wine 24, 25, 26, 29, 34, 39, 111, 112, 113, 114, 115, 117, 127, 131, 132, 133, 134, 147, 152, 229, 231
wolffish (pike) 141, 144

Xerxes 77, 175, 176, 188, 197, 198, 200, 239, 240, 247

Zeus 80, 101, 144, 163, 166, 169, 171, 176, 180, 182, 183, 218
Zheng He 85, 231

www.ingramcontent.com/pod-product-compliance
Lightning Source LLC
Chambersburg PA
CBHW070751020526
44115CB00032B/1616